IOFFE INSTITUTE

1918–1998

Development and Research Activities

5/29/00
In St Petersburg

IOFFE INSTITUTE
1918 – 1998

Development and Research Activities

St Petersburg, 1998

Ioffe Institute: 1918–1998. Development and Research Activities.

The book is devoted to the 80th anniversary of the Ioffe Physico-Technical Institute—one of the oldest and most famous institutions of the Russian Academy of Sciences. The volume includes review articles on the main Institute's research areas carried out in the past and present and also some helpful data concerning the Institute and its staff.

This book was designed and composed by the Information Services and Publishing Department of the Ioffe Institute under the financial support of the Russian Foundation for Basic Research (grant No. 98-02-30075).

Managing editor: V. Grigor'yants
Compilers: B. D'yakov, V. Grigor'yants
Translation and language editing: M. Tagirdzhanov
Design and layout: N. Vsesvetskii
Desk editor: L. Solovyova

The following Institute's members also took part in preparation of the book as compilers, editors, and translators: N. Konstantinova, V. Kozub, G. Kulikov, G. Mikhailov, N. Nazina, G. Skrebtsov, T. Tisnek, A. Tsatsul'nikov, N. Vasil'eva, and A. Zabrodskii.

Label on board: The main building of the Institute.
Front endpaper: A group of the Institute's members near the front door of the Institute in the early 1930s.
Back endpaper: Celebration of the Institute's 70th anniversary. The first "parade" of the staff. The director Zh. Alferov and his assistant S. Konnikov are on horseback.

Information Services and Publishing Department
Ioffe Physico-Technical Institute
26 Polytechnicheskaya, St Petersburg 194021, Russia
Phones: (812) 247 6805, 247 9932
Fax: (812) 247 2135, 247 1017
E-mail: vgrig@eo.ioffe.rssi.ru
Web: http://www.ioffe.rssi.ru

ISBN 5-86763-125-7

Printed in Russian Federation

Contents

Part I

Introduction

The Physico-Technical Institute has now completed its first 80 years of service in the interests of science and technology. It traces back to a department of the same name of the State Institute for X-Rays and Radiology, founded in September 1918 by the initiative of Professors of M. I. Nemenov and A. F. Ioffe. A prominent physicist and research organizer, A. F. Ioffe run the department and then the institute for several decades laying the foundation for research in many areas of physics and technology in this country. So it is quite natural that the Institute bears his name since 1960.

The Institute, with initially only several tens of staff members, developed intensively in the following years and, owing to the managerial capacities of the Institute's first director and to the enrollment of young talented scientists, became one of the world's largest physical institutes conducting investigations in many fields of physics.

The Institute is rightfully considered cradle of the contemporary domestic physics, where, in Ioffe's "school," started their scientific careers the future Nobel Prize Winners P. L. Kapitsa, N. N. Semenov, and L. D. Landau, and worked such worldwide-known scientists as A. P. Alexandrov, E. F. Gross, Ya. I. Frenkel, G. A. Gamov, E. F. Gross, I. V. Kurchatov, I. E. Tamm (Nobel Prize Winner), Ya. B. Zel'dovich, and many others.

Fifteen research institutes have been created on the basis of the laboratories and branches of the Institute. Among them are the Institute of Atomic Energy, Institute of Chemical Physics, Urals, Siberian, and Ukrainian Physicotechnical Institutes, and also the Leningrad Institute for Nuclear Physics.

The Institute carried on various activities with brilliant success. Many important discoveries and inventions were first accomplished here. Early in the Institute's history, attention was attracted to the X-rays, solid state, cosmic rays, physical electronics, and chemical kinetics. Later, new fields of research opened up: semiconductors, nuclear physics, acoustics, and electrical and radio engineering. Creation of high-strength armor, development of first radars for aircraft location, successful protection of battle ships from magnetic mines by degaussing, and solution of the problem of isotope separation are examples of this contribution.

At present, investigations of the Institute's scientists cover nearly the entire field of the modern physics. The Institute gained worldwide recognition through experimental and theoretical work in the field of solid state physics, semiconductors, quantum electronics, astrophysics, plasma physics, and physical gas dynamics. The scientific community is well aware of the Institute's works on the physics of interplanetary space and cosmology, controlled fusion, and motion of bodies in atmospheres of the Earth and other planets.

The recognition of the high quality of works performed at the Institute is confirmed by the fact that more than a hundred of the Institute's members were decorated with top national and international scientific awards.

Among more than one thousand researchers there are four full members (academicians) and ten corresponding members of the Russian Academy of Sciences, more than two hundred persons having degrees of doctor of sciences, and about seven hundred—degrees of candidate of sciences[1]. Since 1987, the Institute has been headed by Aca-

[1] Since 1937, in the USSR and later in Russia, there have been two scientific degrees for scientists: candidate (lower) and doctor (higher) of sciences (in physics and mathematics, chemistry, technology, etc.) conferred by special boards of universities and other research institutions. A unique feature (distinctive of other countries' practice) is the obligatory approval of boards' decision by the Supreme Attestation Commission.

demician Zh. I. Alferov who is the fifth Institute's director in its history.

The Institute's work is mainly financed through the Russian Academy of Sciences (Department of General Physics and Astronomy), with which the Institute is affiliated. A wide variety of researches conducted at the Institute are also financed in the frame of national programs through the Ministry of Science and Technology of the Russian Federation, Russian Space Agency, and Ministry of Atomic Industry of the Russian Federation. Some of these programs are purely physical:

- Controlled fusion and plasma processes
- Topical directions of research in solid state physics
- High energy physics
- Physics of quantum and wave processes
- Physics of solid state nanostructures

and others are cross-disciplinary:

- Ecologically safe energetics
- Promising information technologies
- Global changes in the environment and climate
- Fundamental metrology
- Astronomy and fundamental space research
- Modern components based on acoustooptics and acoustoelectronics
- Future-technology components, new technologies of micro- and nanoelectronics

Many scientific works are financed on competitive basis by the Russian Foundation for Basic Research (RFBR). During the current year more than 180 investigations performed at the Institute are financed from this source. From the beginning of RFBR's activities, the Institute has always been among institutions leading in the number of competitions won.

In recent years, the participation in international projects is provided for by additional financing from various foreign national and multinational foundations and other institutions. In the first place, mention should be made of the International Science Foundation which financed about 100 scientific projects of the Institute's members in 1993–1994 and provided about 700 individual grants.

Some projects are financed by the organizations established to facilitate scientific and technological cooperation with the countries of the Former Soviet Union (FSU): CRDF[2], INTAS[3], INCO-COPERNICUS, and International Science and Technology Center in Moscow. The Volkswagen Foundation and NATO Science Program also provided a valuable financial support to the Institute's research. About 50 research projects are being conducted at the Institute with the financial support of these foundations and institutions in 1998.

The Institute's tradition is to pay particular attention to the development of a unique system of scientific-personnel training. Since 1919, the Institute, guided by A. F. Ioffe, has been the country's first institution to organize scientific education in cooperation with

[2]US Civilian Research and Development Foundation—a private, charitable organization established by the National Science Foundation in 1995.

[3]INTernational ASsociation for the promotion of cooperation with scientists from the Former Soviet Union, established by the European Community and its Member States.

the Leningrad Polytechnic Institute (now St Petersburg Technical University). Although not being a teaching body, the Institute promotes education of students in fundamental disciplines and allows them to practice scientific research, holding examinations, providing student's grants, and offering students an opportunity to hear competent lectures on many problems.

The system of personnel training have become particularly streamlined in recent decades, after creation of the Ioffe Institute-based Lyceum "Physico-Technical School," chair of optoelectronics at the St Petersburg Electrotechnical University, and four chairs at the Faculty of Physics and Technology of the St Petersburg Technical University. Students of the Lyceum familiarize themselves with the laboratories of the Ioffe Institute. Having entered the Universities, they practice at the Ioffe Institute, taking part in scientific research beginning with the 2nd–3rd year of education, and, after graduating, become postgraduates or research staff members of the Institute. The advantage of such an education system consists, in the first place, in purposeful scientific training and adaptation of young scientists both to research activities and to a specific scientific community. Every young scientist educated at these Institute-based chairs in fact becomes an Institute's researcher 3 years earlier than his equals in age coming from other high school institutions. First graduates who accomplished all the steps of the Institute-based education became postgraduates and staff members of the Institute in 1996.

In compliance with the common practice of conferring scientific degrees, special bodies—five Dissertation Councils dealing with candidate and doctoral dissertations in six fields of science—are founded at the Institute. These fields are: astrophysics and radioastronomy, physics and chemistry of plasma, physical electronics, solid state physics, physics of semiconductors and dielectrics, and theoretical physics. Leading scientists from the Institute and some other institutions, experts in the fields, are members of the Councils. A total of 24 doctoral and 107 candidate dissertations have been successfully backed at sessions of these Councils during the last four years.

The Institute constantly holds national and international scientific conferences, symposia, workshops, etc. and also has initiated several international meetings, some of which became widely known and traditional, namely, annual symposium "Nanostructures: Physics and Technology," biannual workshop "Fullerenes and Atomic Clusters," annual workshop on spin waves. After a certain interruption was resumed the annual "Winter School on Physics of Semiconductors," where leading scientists in the field deliver lectures to young staff members.

The Institute supervises and provides means for publication of four leading physical journals of Russia. The journals are translated into English and published in the USA by the American Institute of Physics under the titles of *Semiconductors*, *Physics of the Solid State*, *Technical Physics*, and *Technical Physics Letters*. Electronic versions of these journals are presented on WWW pages (http://www.ioffe.rssi.ru/). The Institute publishes scientific monographs in the framework of RFBR projects and maintains close relations with major foreign publishing houses specializing in physics.

Annually, a tree-stage competition of scientific works performed by the Institute's scientists is held. Ten best works are awarded the Institute's Prize and granted a privilege to be presented orally at a special session of the Institute's Scientific Council. One of the works presented at the session is honored with the highest Institute's award—Prize of the Scientific Council. This prize has been awarded annually since 1988, and all the awardees are listed on page 42.

The Ioffe Institute has high rating among physical institutes in the international scientific community. This enabled the Scientific Council to establish an honorary degree—Honorary Member of the Ioffe Institute. Annually, in September, when celebrating the Institute's birthday, new Honorary Members are elected from among prominent Russian and foreign scientists.

The present collected book is aimed at familiarizing the reader with the evolution and current state of the researches carried out at the Institute. Though not intended to be a guide to the works of the Institute, the book gives insight into the general aims of the studies, methods employed, successes achieved, and difficulties encountered. Obviously, it can only cover a small part of this wide field of knowledge. Even enumerating the works of outstanding merit would be rather tedious, and the book mainly appeals to the spirit of science conquering new fields and frontiers.

In the second part of the book the infancy and development of the Institute during the first 70 years is given in sufficient detail—with a brief analysis of the research conducted and the contribution made by the Institute to different fields of physics. The recent decade of the Institute's life is described only briefly since the third part of the volume includes review articles written by the Institute's leading scientists. These articles embrace a wide variety of topics and cover the main areas of research and results of a vast amount of arduous work by the Institute's members. Pitifully, this book's volume gives no way of including a greater number of review articles, and a certain amount of the Institute's investigations is only mentioned in the fourth part where the departments' fields of research are listed.

The fifth part—Personalia—presents a short account of the Institute's administrative board, and information about over a hundred physicists of the academician rank who ever worked at the Institute, and also lists the persons who have been awarded various prizes and medals for the research carried out at the Institute. The last part of the book is devoted to the honorary membership at the Institute.

Part II

Development of the Institute

The first 70 years

B. B. D'yakov

The scope of this chapter is defined by the desire to give a concise and comprehensive outline of the most important scientific events of the Institute's history. There are a large number of sources of information about its history, including the literature, archives, and personal evidence. Some paragraphs and stories are taken therefrom, others are proposed by the leading Institute's specialists, with their own materials and drafts of the text. All these sources have been extremely helpful, but a large number of topics were left out of consideration. (The contributors will be mentioned at the end of the chapter.)

The items are presented mostly in chronological order. The general themes are overviewed as a whole over the main periods of the Institute's activities: in the traditional mainstream areas of research and in new fields.

The historical overview of the Institute's organization is given in brief. The names of the numerous Soviet institutions and government bodies are referred to in a transliteration used in English sources with minor exceptions; those, we hope, do not obscure their content.

There are a number of historical studies and some helpful reviews of the Institute history (Frenkel 1996, Frenkel and Tuchkevich 1978, Alferov 1998). Most of them are indebted to the late V. Ya. Frenkel (1930–1997), an outstanding historian of science, a staff member of the Institute since 1959. His thoroughly documented works also contributed largely to the present review.

Early years

Foundation and birth of traditions

The Ioffe Physico-Technical Institute of the Russian Academy of Sciences has experienced a rich 80-year history. It was founded in 1918 as the State Institute of X-Rays and Radiology under the auspices of the People's Commissariat[1] for Education. Initially there were three departments: Medico-Biological, headed by M. I. Nemenov, a prominent science manager and X-ray physicist; Physico-Technical, headed by A. F. Ioffe, then a professor of the Petrograd Polytechnic Institute; and Radiological, headed first by P. S. Kolovrat-Chervinskii, a physicist experimenting in radioactivity, and later (after his death) by V. I. Vernadsky, a member of the Russian Academy and a prominent public figure.

Initially, the Physico-Technical Department and later the Institute had been nestled in several rooms in the Petrograd Polytechnic Institute's building. In November 1921, the three departments became independent institutes existing till now (the other two are the

[1] Russian title for ministries in 1917–1946.

Institute of X-Ray Radiology and Cancer, and the Radium Institute), and the Physico-Technical Department became the State Physico-Technical X-ray Institute. When moving into the nowadays main building (this occurred in 1923), its staff numbered in tens.

The next step consisted in the foundation of the Leningrad Physico-Technical Laboratory (LPhTL) of the Research and Technology Department of the Supreme Council for National Economy (SCNE) affiliated with the Institute and also headed by Ioffe himself. This body was intended to link the scientific research at the Institute with industry. This principle, established by the founder of the Institute, became *its first tradition.* The Laboratory itself existed for not too long, but the relation was prolonged and maintained.

Under its present name Physico-Technical (*FizTech* in popular Russian abbreviation) the Institute was baptized in 1930 when the LPhTL was associated with the Institute within the SCNE system. In the late 1920s–early 1930s, new important changes in the Institute's structure occurred: in 1927, the Heat Technology Department, headed by M. V. Kirpichev, became the Heat Technology Institute (now the Central Boiler and Turbine Institute); in 1928, came into being the Ukrainian Physico-Technical Institute in Kharkov (then Ukrainian capital), the Siberian Physico-Technical Institute in Tomsk, the Ural Physico-Technical Institute in Sverdlovsk (now Ekaterinburg), and the Dnepropetrovsk Physico-Technical Institute, with a number of scientists from *FizTech* forming the nuclei of these institutions.

In 1930, the Acoustic Department headed by N. N. Andreev was separated, the next year came the turn of the Electrophysical Department headed by A. A. Chernyshev and N. N. Semenov's department which became the Institute of Chemical Physics. From January 1931 till October 1933 the last two with their home Institute existed as a Combine of Institutes under the People Commissariat for Heavy Industry.

These transformations and renamings are history now. Nearly to-day's appearance was acquired by *FizTech* after the final separation of the affiliated institutions and its transfer to the Academy of Sciences of the USSR in 1939. The name of its founder and the first director—Abram Fyodorovich Ioffe—crowned the Institute on December 3, 1960.

The fundamental and applied research, which was the very beginning of the Institute, was associated with Ioffe himself. This was favored by the activities of Ioffe's seminar at the Polytechnic Institute the names of whose participants went down in the history of physics in a wide sense and of the Ioffe Institute, in particular: P. L. Kapitsa, N. N. Semenov, Ya. I. Frenkel, P. I. Lukirskii, I. V. Obreimov, Ya. G. Dorfman, and others. In 1919, almost simultaneously with the organization of the Institute Ioffe initiated creation and became Dean of the Physico-Mechanical Faculty of the Polytechnic Institute, contemplated already before 1917 by A. F. Ioffe and S. P. Timoshenko, a professor of the Petrograd Polytechnic Institute and a prominent Russian specialist in mechanics, Ioffe's compatriot and friend (in this way the physics and mechanics were united in the name of the faculty). The first students and graduates of this faculty were incorporated in the research staff of the new Institute. Among them were Yu. B. Khariton, V. N. Kondrat'ev, A. F. Valter, G. V. Kurdyumov, A. I. Shal'nikov, G. A. Grinberg. Thence *the second tradition* of the Ioffe Institute—well aimed training of its own future staff members, was established in the very first year of its existence.

Abram F. Ioffe in 1932.

First works

The initial fields of research—mechanical and electrical strength of solids and electrical properties of dielectrics (A. F. Ioffe), chemical kinetics (N. N. Semenov), quantum physics (Ya. I. Frenkel), X-rays (P. I. Lukirskii)—were the areas of interest of the Institute's founder and its first staff members. At that time Ioffe reviewed his accumulated experimental data on dielectrics in a fundamental paper *Zum Teil in Gemeinschaft mit W. C. Röntgen. Elektrizitatsdurchgang durch Kristalle* (Ioffe 1923), and his ideas on the ionic conductivity of crystals were developed and theoretically substantiated by Frenkel (1926). These works assumed the existence and the influence of a vacancy in crystal lattice on the conductivity (the term "hole" was introduced for such a vacancy and the ensemble of an interstitial atom and a hole was named Frenkel defect). This gave rise to *the third tradition* of the Ioffe Institute—unity of theoretical and experimental methods of investigation.

At that same time a series of books was published under a general title "The Structure of Matter" by Frenkel (1922, 1924) which, together with his "Theory of Relativity" (Frenkel, 1923), not only reviewed the latest advances in physics, but even served as a higher school manual. This series was a forerunner of later Frenkel's monographs translated into many languages and used to educate several generations of physicists of many countries.

Another aspect of Ioffe's personal scientific activities were his investigations of mechanical properties of crystals, providing a tool for understanding the nature of elasticity, plasticity, and strength of solids. Among the incentives to this work was the then contradiction between the theoretical (M. Born) and experimental strength limits of solids. Having applied an X-ray technique (Laue method) to rock salt, Ioffe not only resolved this controversy, but also opened up an essentially new avenue of research (Ioffe, *et al.* 1924). This work employed an original technique for improving the strength of solids (by removing imperfect surface layers) and approaching the theoretical strength limit. The result obtained was so impressive that it was named Ioffe effect. To expand the investigations of mechanical properties of materials A. F. Ioffe invited N. N. Davidenkov to create the Material Strength Department at the Institute.

These works gave impetus to the corresponding research activities at the Institute and mostly served as their basis to impart, from the very beginning, to these investigations a characteristic feature—they dealt with practical problems. In this way, the second component of the Institute's name (Technical) started to be filled with real substance. A. F. Ioffe involved more and more staff members in investigations of the electrical properties of solids. I. V. Kurchatov and P. P. Kobeko started their studies of the electrical properties of dielectrics as early as 1926. They also carried out classical works on ferroelectrics in 1929–1933. The interest in this kind of materials was aroused, in particular, by the development of the Soviet Acoustics School headed by another leading scientist of the Institute, N. N. Andreev, and having among its prominent representatives B. P. Konstantinov, and was mainly associated with piezoelectric crystals. One of such crystals, the Rochelle salt, with the highest permittivity, became the object of thorough investigations. It was of significance that Kurchatov and Kobeko (1930) enlarged this class of materials and developed the first theory explaining their properties, and also initiated further research and discoveries in this field[2]in 1944–1945.

Investigations of A. F. Valter, N. N. Semenov, and V. A. Fock aided in creating a theory of heat breakdown in dielectrics.

During the first ten years of the Institute in being, fundamental results were obtained in the field of *magnetic phenomena.* In 1920, Kapitsa and Semenov advanced the idea of measuring the atomic magnetic moment by deflecting a beam of neutral atoms in magnetic field. Pitifully, their experiment could not be accomplished in due time because of the scarcity of the Institute's experimental base at the moment and this idea was published with two year delay (Kapitsa and Semenov 1922, see, also, interesting features in Semenov 1924) after the well-known experiment of O. Stern and W. Gerlach. Ya. I. Dorfman's concept of paramagnetic and ferromagnetic resonances advanced in 1923 was supported by his experiments and Ya. I. Frenkel's theoretical interpretation of ferromagnetism as being due to electron magnetic moment orientation in ferromagnetic crystal (Frenkel 1928, Dorfman and Frenkel 1930). Studies on this subject were continued by Landau (1930) who developed a quantum theory of diamagnetism of free electrons in metals.

Mention should be made of M. P. Bronstein and Ya. I. Frenkel's study of the possibility to observe, in principle, the effect of resonance absorption of radiation whose frequency corresponds to the energy of electron transition in magnetic field (Bronstein and Frenkel

[2]Ferroelectric properties of barium titanate, discovered by B. M. Vul, a former Institute staff member, and theory of ferroelectricity, developed by V. L. Ginzburg, a physicist from Moscow.

1930). The photoelectromagnetic effect was discovered by Kikoin and Noskov (1934). It consists essentially in a build-up of electric field in illuminated dielectric samples placed in magnetic field at low temperature. Immediately after, Frenkel (1934) furnished its explanation. These works are worth mentioning, as the role of minority carriers in semiconductors was originally indicated in them.

The general problems of *physics of metals* were considered by Ya. I. Frenkel. Noteworthy are G. V. Kurdyumov's works on the phase crystalline-structural mechanism of the martensite transformation and on thermoelastic equilibrium in phase transformations in metallic alloys (the Kurdyumov effect) which became fundamental for heat treatment of steels. Also, investigations of X-rays themselves were conducted in: P. L. Kapitsa's work on the method of X-ray reflection in crystals, where it was proposed to use a bent crystal to produce a monochromatic beam of hard X-ray radiation, Ya. I. Frenkel's work on theory of X-ray dispersion, A. I. Alikhanov and L. A. Artsimovich's study on total internal reflection of X-rays from thin metallic films in 1933. As it was found much later, the phenomenon studied by them enables creation of devices of X-ray optics (X-ray telescopes).

In the field of *chemical kinetics*, the N. N. Semenov's works of 1925–1929, reviewed in his book on chain reactions (Semenov 1935) and crowned afterwards by the Nobel Prize in chemistry in 1956, were of fundamental value. These works were concerned with self-ignition of gases, making distinction between two types of reactions, namely, thermal and chain (divergent or branching) ones. Critical phenomena of mutual conversion of these types were first discovered in studying oxidation of phosphorous vapor by Yu. B. Khariton and Z. F. Val'ta—on minor changes of ambient conditions they observed that this reaction triggered an explosion. According to Semenov, the chain (branching) process takes place when in the course of this reaction the active species initiating this process is reproduced in sufficient amount. Otherwise, this chain terminates and the reaction decays. As it was the rule with the priority works at the Institute, these ideas were applied afterwards to other fields of physics, namely, for calculating chain nuclear reactions by Ya. B. Zel'dovich and Yu. B. Khariton in 1939–1941.

At the Institute, the *physical electronics* originated in P. I. Lukirskii and N. N. Semenov's early works on the photoeffect and electron scattering by the liquid Hg surface (Lukirskii and Semenov 1923, Lukirskii 1926) to become later one of the main traditional avenues of research. To analyze the electron velocity distribution, P. I. Lukirskii developed and put in practice an original technique of spherical capacitor (Lukirskii 1927). In the same years, investigations of ionization of gases and vapors were initiated (Semenov 1923, Lukirskii 1923). P. I. Lukirskii's works also gave rise to research in the field of *electron optics* at the Institute, continued by G. A. Grinberg and his followers beginning with the 1940s (theory of electron focusing in electric and magnetic fields).

Note also the work of V. N. Kondrat'ev in 1924, who was the first in our country (and the second in the world) to study ionization of salt vapors and to introduce a mass-spectrometer into investigations of chemical reactions. In the 1950s, *mass-spectroscopy* was developed further by N. I. Ionov and B. A. Mamyrin (see pp. 221–232).

In the 1920s, experimental studies of the adsorption kinetics, in which the critical temperature of adsorption was determined for a number of materials were carried out (N. N. Semenov, Yu. B. Khariton, and A. I. Shal'nikov).

The main achievements of the first ten years of the Institute's activities accomplished through creative effort of about two hundred scientific staff members of the Institute

(170 were at the age below 25–27 years) were summarized in a collection of abstracts on electron theory, X-ray spectroscopy, ionization of salt vapor, mechanical and electrical properties of dielectric crystals. Original papers of the Institute's members were published in the leading scientific journals. For example, in Germany alone and only in 1926, 80 papers were published. (By that time it was still possible to summarize the quantitative "harvest of works," cropped at *FizTech*, but in its later history this task became impracticable!)

Certainly, outstanding works of the Institute's scientists in different areas of *technology* are also worth mentioning. Many of these embodied previous basic research. Others were original applied studies, often performed in advance of their time. To the latter belonged L. S. Termen's inventions of radiotelegraph devices, alarm signal systems to detect an approaching human at a distance of 4–8 m, and especially his famous musical instrument, the TermenVox, the first in the history electronic musical instrument (1920). In addition, L. S. Termen developed a mechanical system of "far vision" (television) and demonstrated it to K. E. Voroshilov, one of the Soviet military leaders. In Termen's laboratory was working A. P. Konstantinov, inventor of the first electron-beam tube with charge accumulation and one of the founders of the State Precision Time Service.

In the early 1920s, A. A. Chernyshev (head of Technology Department) developed power-line carrier telephony and invented a discharge arrester for protecting weak-current lines from the hazard of high voltage lines, a power line fault detection and repairing method, later used at the Volkhov Hydro, was developed in his department. It was A. A. Chernyshev who invented the indirectly heated cathode. Works on SiC-based self-adjusting resistors (later called varistors) were successfully carried out by I. V. Kurchatov and his co-workers in the early 1930s to give rise to their industrial manufacturing. His investigations of ferroelectricity, mentioned above, gave rise to application of ferroelectrics in technology—in 1929, V. P. Vologdin, I. V. Kurchatov, and P. P. Kobeko proposed to use the Rochelle salt as an automated tuning regulator in radio receivers.

Radio engineering was an especially wide area of application of investigations conducted at the Institute including A. P. Konstantinov's works on TV system, A. A. Chernyshev's inventions in this field (his creative power was polyphonic indeed and he was named "the Russian Edison!"), D. A. Rozhanskii's "Fundamentals of Physics of Radiowave Propagation." The foundations of practical radar engineering were laid and designing of a radar was started (Yu. B. Kobzarev).

Noteworthy among other valuable advances in technology is the development by M. V. Kirpichev of a simulation method widely used in boiler industry.

Applications in technology also resulted from traditional studies of the *mechanical strength of solids*, conducted at the Institute. For this purpose, the Institute's staff headed by A. F. Ioffe developed the fundamentals of the theory of strength of solids. After famous Ioffe's achievements in reaching the NaCl strength limit, A. P. Alexandrov and S. N. Zhurkov closely approached the theoretical strength limits of glass and quartz filaments by etching them. A. P. Alexandrov and P. P. Kobeko studied polymers (influence of temperature on polymer deformation, concept of polymer relaxation states, describing abrupt changes in the structure of "large" molecules as a result of thermal energy fluctuations), N. N. Davidenkov studied the brittle failure of steels at low temperature and transition to plastic failure at higher temperatures.

UNION OF SOVIET SOCIALISTIC REPUBLICS

REPORTS

of

The Phys.-Techn. Roentgen Institute

and

The Leningrad Phys.-Techn. Laboratory

1918—1926

SCIENTIFIC CHEMICAL TECHNICAL PUBLISHING OFFICE
SCIENTIFIC-TECHNICAL DEPARTMENT OF THE SUPREME COUNCIL OF NATIONAL ECONOMY

LENINGRAD

1926

The title page of a collection of brief reports on main fields of Institute's research, published in 1926.

In 1931, A. V. Stepanov started investigations of defect formation and build-up of electric potential on faces of ionic crystals as a result of *plastic deformation of crystals*, which are considered direct continuation of Ioffe's works. Stepanov advanced a revolutionary fundamental idea of a double role of plastic deformation in crystalline materials. He was the first to declare that the plastic deformation, making crystals stronger, gives rise to their fracture. Later these points formed a basis for the present-day dislocation theory of strength of solids. It was mentioned more than once that the series of Ya. I. Frenkel and T. A. Kontorova's papers (e.g., Frenkel and Kontorova 1938) concerned with the theory of plastic deformation and twinning became a basis not only for the physics of dislocations, but for the physics of solitons as well.

At the new frontiers

Semiconductors

The beginning of the 1930s was marked not only by rearrangements of the Institute's structure, but also by an essential reshaping of "minds" and "crystallization" of the main fields of the Institute's research. Studies of *semiconductors* and *nuclear physics* were initiated at that time and essentially enlarged later. At the same time, other fields of research transformed into independent Institutes mentioned above, with the addition of the Institute of Agricultural Physics (headed by A. F. Ioffe himself).

Relating in his Jubilee review (Ioffe 1957) to the main scientific achievements of that period, A. F. Ioffe mentioned the following.

> In 1927, cuprous oxide, the first classical semiconductor material, and rectifiers of alternating current were produced. Simultaneously, it was found that cuprous oxide produces current under illumination. It is at that time that systematic studies of semiconductors started in the USSR. At the Physico-Technical Institute these investigations were initiated by I. V. Kurchatov and K. D. Sinel'nikov. Another subject for investigation of the semiconductor properties was copper sulfide, using which B. V. Kurchatov and Yu. A. Dunaev developed a novel type of rectifiers. Then, a turn of selenium rectifiers came. Concerning the photocells, Ya. P. Maslakovets and B. T. Kolomiets established that photocells based on thallium sulfide and silver sulfide were the most effective ones. Under sunlight they exhibited an efficiency as high as 1.1% instead of the common 0.05%. In this way the problem of solar cells arose for the first time.

The main scientific problem of semiconductor physics "attacked" by *FizTech's* scientists at that time was to determine the properties of the barrier layer. This was enabled by the creation of V. P. Zhuze's artificial barrier layer of silicon dioxide on a cuprous oxide substrate, and development of B. I. Davydov's barrier layer theory based on the diffusion mechanism. A. V. Ioffe and A. F. Ioffe's experiments confirmed the diffusion theory of the barrier layer. At the same time, a quantitative agreement with the experiment was observed at the boundary of two semiconductors of different (p- and n-) types. In their pioneering work, V. P. Zhuze and B. V. Kurchatov in 1932 separated the semiconductor conduction into intrinsic and impurity components.

Beginning with 1931 was developed Ya. I. Frenkel's model of excited states in a crystal—the "excitons" (Frenkel 1931). This excitation is not related to any single atom in the crystal lattice but travels in the crystal bulk in the form of an excitation wave. Below, it will be shown that this state was experimentally discovered by E. F. Gross and

A. F. Ioffe (third row, center) with his co-workers of the 1930s. Between them are: Yu. A. Dunaev, B. V. Kurchatov, V. P. Zhuze (from Ioffe to the left), B. T. Kolomiets, and A. R. Regel (men in the first row, from left to right).

N. A. Karryev in 1951 when studying absorption spectra of cuprous oxide and other compounds by optical techniques.

Investigations in another "semiconductor" field, that of thermoelectricity, were also commenced. The efficiency of semiconductor thermoelectric cells was increased to 3% by 1940. The theory of these was developed by Ioffe himself in 1950. Later, a number of thermoelectric devices were designed, introduced in practice, and successfully developed. Also in his Jubilee article, Ioffe mentioned B. T. Kolomiets's successful works on thermistors and microthermistors (thermal resistors).

The Institute's activities manifested themselves in a permanent seminar on semiconductors and organization of six conferences on semiconductors during the period from 1931 till the beginning of the War.

"Nuclear revolution"

The birth of *nuclear physics* at the Institute can be dated with D. V. Skobel'tsyn's studies in the field of cosmic radiation (e.g., Skobel'tsyn 1929). He was the first to place a Wilson chamber in magnetic field (following an idea of P. L. Kapitsa) and revealed the existence of the high-energy component of cosmic rays (showers in the present-day terminology), judging from deflections of particle tracks.

A. I. Alikhanov's work on positron detection and annihilation of positron–electron pairs proved to be an important step, and the coincidence circuit designed in Alikhanov's laboratory was the first one in the USSR. With its use, positrons emitted by an active

F. Joliot-Curie, A. F. Ioffe, I. Joliot-Curie, D. V. Skobel'tsyn, and S. I. Vavilov in Leningrad, 1934.

residue of Th were detected, and the spectrum of positrons emitted by RaC as a result of internal conversion was studied.

The true golden age of the research in nuclear physics is related to 1932, *annus mirabilis* of the world's physics, when the neutron, positron, heavy hydrogen isotope, and artificial transmutation of nuclei were discovered and a cyclotron was created. The reaction of the Institute's scientific community followed immediately. On December 16, 1932, A. F. Ioffe created a group of researchers to deal with nuclear physics, with I. V. Kurchatov first his deputy and then head of the direction. In 1933, this group became a department (with I. V. Kurchatov as head and D. D. Ivanenko as seminar secretary). Conferences on nuclear physics were initiated in 1933, when the first one took place at the Institute headquarters (with I. V. Kurchatov as Chairman of the Organizing Committee) immediately acquiring an international status to prove the authority of physics in our country.

At the end of the 1920s, the success in determining the atomic structure with the help of the just born nonrelativistic quantum mechanics furnished an explanation for the atomic nucleus structure. All the new discoveries in nuclear physics, especially, that of the neutron (1932) made the concept of atomic structure highly dynamic, with essentially new approaches to the old problems required. This was very urgent so long as some scientists proposed to give up the conservation laws on the atom-nucleus scale in view of the arising problems (alpha-particle emission from nucleus, continuous spectrum of beta–decay, strong penetrating radiation of presumably gamma-rays, which in fact turned out to be neutrons).

At the Institute, primarily D. D. Ivanenko, I. V. Kurchatov, and A. I. Alikhanov were engaged in these problems (confirmation of energy conservation law was obtained in experiments with positron–electron annihilation). Although the alpha–decay problem was solved (by G. A. Gamov in 1927 during his traineeship in Gettingen) without any reference to the discovery of a new particle, the neutron, explanation of this cumbersome bundle of problems demanded a more revolutionary approach including this new particle. At this very moment the idea of the atomic nucleus as consisting of elementary heavy particles, i.e., protons and neutrons, was advanced by D. D. Ivanenko in a several-line Letter to Editor under the title "The Neutron Hypothesis" (Ivanenko 1932). This idea withstood the test of time but was not adopted immediately (in contrast to Gamov's theory of alpha–decay)—among the leading physicists W. Heisenberg was the only one to advocate and develop it. He proposed to consider the proton and the neutron as two states of a "nucleon." The second part of the proton–neutron model, i.e., the birth of an electron in beta–decay similarly to that of photon, rather than its escape from the nucleus, was proved experimentally by the discovery of cascade showers in cosmic rays (1933) when a chain of electron–positron pairs was found. Together with P. Dirac's theory requiring the occurrence of a positive electron, this model led to the present-day fundamental concepts of the structure of nuclear matter. Taking into account that the showers just mentioned were first observed by D. V. Skobel'tsyn in 1927 with his new technique, it should be taken as proved that the Institute's scientists made a major contribution to solving one of the greatest problems of the structure of matter. It is in 1933 at the First All-Union Nuclear Conference (and then, at the Solvay Congress the same year) that this model was accepted universally.

In addition to the proton–neutron model of the nucleus, as a response of the Institute to the challenge of the world's science, nuclear isomerism was discovered (Kurchatov *et al.* 1935) for Br isotopes (i.e., the existence of metastable states of nuclei and isomeric transfer as a novel kind of nuclear transformations), the field theory of nuclear forces and a theory of uranium fission (Frenkel 1939) were developed, and spontaneous fission of uranium was discovered by G. N. Flerov and K. A. Petrzhak, and reported in the May issue of Physical Review (Flerov and Petrzhak 1940) in the form of a few-word cablegram

Spontaneous Fission of Uranium

> With 15-plate ionization chambers adjusted for detection of uranium fission products we observed 6 pulses per hour, which we ascribe to spontaneous fission of uranium. A series of control experiments seem to exclude other possible explanations. Energy of pulses and absorption properties coincide with fission products of uranium bombarded by neutrons. No pulses were found with UX and Th. Mean lifetime of uranium follows ten to sixteen or seventeen years.
>
> Physico-Technical Institute (F),
> Radium Institute (P),
> Leningrad. U.S.S.R.,
> June 14, 1940 (by cable).
>
> Flerov
> Petrjak

In 1934, a "small" cyclotron, for the moment a replica of E. O. Lawrence's cyclotron, was put in action at the Institute by M. A. Eremeev. A high-voltage accelerator operating at 500 kV (I. V. Kurchatov, G. Ya. Shchepkin) and a quadratron (Ya. L. Khurgin, L. A. Artsimovich) were built and construction of a large cyclotron was commenced. This will be of great significance for atomic research in the War years. Later, first information was obtained on fission parameters (T. I. Nikitinskaya) and the mean number of neutrons emitted in fission of uranium nucleus (L. I. Rusinov, G. N. Flerov), with these data being close to the present-day results.

The abstract of Ya. I. Frenkel's paper "On the splitting of heavy nuclei by slow neutrons" was published in Physical Review in 1939 and a more comprehensive presentation followed in Russian journals. N. Bohr and J. Wheeler's fundamental paper was issued in the next volume of the Physical Review, with a reference to this paper (Frenkel 1939) as a well independent one. The mechanism of asymmetric fission of the nuclear drop[3] was described in terms of the electrocapillary model of the nucleus, when an additional neutron excites the fundamental nucleus oscillation mode which is unstable for a large ratio of the volume energy of the heavy nucleus to its surface energy. Disintegration is also possible in the unclassical case of an instability—in the presence of a barrier (activation energy)—owing to the tunneling effect. If the mean lifetime is large compared with that necessary for energy redistribution over the degrees of freedom of a nucleus, its surface energy may decrease as the excitation energy is lowered. At zero activation energy this may lead to an explosive disintegration of the nucleus.

Further, such topics as intrinsic fission neutrons, delayed neutrons, resonant (selective) neutron absorption, and decay of the fission products appeared. All nuclear scientists of that time were concerned with these problems.

Applications of physics in *technology* resulted in a unique way in an outstanding work of the leading scientists of the Institute on ship degaussing, initiated by B. A. Gaev. In 1936, studies of the principles of ship protection against magnetic contactless mines and torpedoes were commenced under leadership of A. P. Alexandrov. This work started with the development of devices to determine the magnetization of models and real objects. Ship degaussing was achieved by passing current through coils and using feedback control with account of field variations. This method was successfully tested in 1937–1939, which was of great importance for battleship protection during the forthcoming War.

Wartime years

With the outbreak of the War the work on *the ship degaussing* became actually a heroic labor, often performed in battle actions, of the whole Institute's staff in all the fleets and divisions of the Soviet Navy (A. P. Alexandrov, I. V. Kurchatov, B. A. Gaev, V. M. Tuchkevich, A. R. and V. R. Regel, P. G. Stepanov, V. A. Ioffe, M. M. Bredov, and others). The degaussing techniques were also modified with account taken of the ship class, its battle orders, and operating conditions. In 1941, Tuchkevich gained inventor's certificate on his original degaussing method using no coils. The recognized result of this work was that no loss of battleships through contactless magnetic mines was suffered with the use of the Institute's degaussing system[4].

The studies in *nuclear physics* already performed at the Institute, their timely organization and enhanced effort had the greatest impact on the history of the USSR when

[3]The nuclear drop model was advanced by G. A. Gamov in his book on nuclear structure and radioactivity published in Oxford in 1931 (well before the discovery of the neutron).

[4]For comparison, it is worth mentioning as an example that only during September, 1939—December, 1940 the British Navy lost no less than 5 destroyers hit by contactless magnetic mines from the total irreversible loss in action of 39 destroyers.

in 1943 a decision was made to launch a nuclear project with I. V. Kurchatov (supported by Ioffe) as scientific director. I. V. Kurchatov and his first ten co-workers from *FizTech* (A. I. Alikhanov, M. I. Kornfeld, L. M. Nemenov, P. Ya. Glazunov, S. Ya. Nikitin, G. Ya. Shchepkin, G. N. Flerov, P. E. Spivak, M. S. Kozodaev, V. N. Dzhelepov) initiated the Laboratory Number Two. This event coincided surprisingly in time with the beginning of the USA effort in Los Alamos whereto R. Oppenheimer called his first twelve collaborators.

The real success in *radar engineering* (as a continuation of studies initiated in 1934 and creation of the first field radars) under direction of Yu. B. Kobzarev, in the field of *armor protection* (interaction between armor and projectile, investigation of the projectile rebound and a protection grid with the purpose to obtain an optimal interlayer gap) with the participation of V. L. Kuprienko and F. F. Vitman, and in creation of *thermoelectric generators* (Yu. P. Maslakovets, A. F. Ioffe) also had a considerable effect.

As for that part of the Institute's staff who remained in besieged Leningrad, they also continued research under the guidance of P. P. Kobeko, and the results obtained were of significance in the War time: production of food surrogates for the hungered city, studies of the strength of ice on the Lake Ladoga's "Way to Live" and the determination of the optimal speed of traffic on ice in relation to ice properties (P. P. Kobeko, V. A. Ioffe, and N. M. Reinov); scanning of mine fields with compressed air flow, and manufacture of a new antibiotic to cure anaerobic infection (M. V. Glikina).

This antigangrene medicine (named *Preparation P*) was grown from appropriate sorts of soil bacteria. The first positive results were obtained by the late 1941. The effect was more pronounced as compared with the newly discovered penicillin. Much attention was given to the yet unknown biochemical processes occurring during the convalescence period. *Preparation P* was recommended for use in field hospitals. The mortality was lowered from 20% to 10–12%.

The fundamental physics was represented by a classical book "Kinetic theory of liquids" (Frenkel 1946). Note also his important work on formation and dynamics of crystal surface (Frenkel 1945), important for the present views on semiconductor nanostructures, together with experiments on crystal face formation on the surface of a rock salt ball (Lukirskii 1945).

Postwar development

Traditional fields

The previous period of the development of the Institute in particular and the domestic physics as a whole inseparably connected the Institute with the primary tasks of the economy and defense and country's scientific and technology culture. Therefore, the research was continued and extended in those areas of physics in which the Institute had played the main and significant role, and new investigations were initiated and urgent special problems were solved as a response to the imperatives of the postwar time.

Nuclear physics

Even having lost from its staff a large number of scientists and engineers who were transferred to the atomic project, the Institute continued investigations in the field of *nuclear physics*. First, the new Institute's cyclotron was put in action and is functioning to the present day, and, first of all, microgram samples of Pu and Np were obtained with its help in 1946 (D. G. Alkhazov) to be used for the needs of the atomic project. Work on the theory of accelerators led to the creation in 1948 of a phasotron, the first in the USSR (D. G. Alkhazov, D. M. Kaminker, and others), with the proton and deuteron beams having energy of 22 MeV and intensity of 0.2 μA, and a synchrotron with electron energy of 100 MeV, on which a gamma-ray beam was obtained from the inner target in 1953 (A. P. Komar) and studies of photonuclear reactions were started. Further effort in nuclear and high energy physics was undertaken when A. P. Komar became director instead of A. F. Ioffe in 1950. In 1954, development of a research nuclear reactor of *VVR* type[5], which became the most elaborate research pile in the USSR on being put into operation in 1960, was started under guidance of L. I. Rusinov. In 1959, work was initiated on a synchrocyclotron with proton energy of 1 GeV, the largest in the world (D. G. Alkhazov, N. K. Abrosimov, A. P. Grinberg, K. V. Donskoi, D. M. Kaminker).

The diversity and scale of problems of the neutron physics, accelerators, and nuclear engineering and the scope of research into Coulomb excitation, nuclear fission, and weak interaction were so wide at the laboratories of the Institute's branch in Gatchina, and the number of novel research techniques used was so great that this led, in 1972, to the formation of a new "daughter" institute—Institute for Nuclear Physics—again following the true Institute's tradition.

The Institute's cyclotron was renewed to become a modern device with unique experimental capabilities, with beams of ions of different energies in the range from protons to Ar. With these heavy ion beams, fundamental studies of the Coulomb excitation of nuclei (D. G. Alkhazov, D. S. Andreev, A. P. Grinberg, Yu. P. Gangrskii, K. I. Erokhina, I. Kh. Lemberg, K. A. Ter-Martirosian) were accomplished. Along with these studies, works on application of techniques commonly used in nuclear physics, in other fields of technology were undertaken in cooperation with a number of the Institute's laboratories.

Emission electronics

Problems in another traditional field of the Institute's research, emission electronics, were investigated actively in the postwar years, mainly through the efforts of L. N. Dobretsov since 1951. He made an outstanding contribution to the understanding of thermoemission and secondary electron emission (Dobretsov and Gomoyunova 1966).

As in his principal area of investigation, nuclear physics, A. P. Komar deserves mentioning here owing to his contributions to the *electron microscopy*. Under his guidance a field ion microscope (first in the USSR and the second in the world) was created in 1952, attaining atomic resolution. In 1961, A. P. Komar solved the enigmatic problem of a magnificient uniformity of images of different organic molecules. He proposed a waveguide origin of these images, presenting a new nontrivial confirmation of the electron's wave nature.

At the Dukel'skii's laboratory were discovered the previously incredible negative ions of alkali metals (V. M. Dukel'skii, N. I. Ionov, E. Ya. Zandberg). Then, at the Ionov's laboratory, investigations of the surface ionization were commenced. A theory was developed of surface ionization of any particles on various surfaces (Ionov 1972). A 10^6-fold increase of the degree of ionization by the electrical field effect was discovered,

[5] *VVR* is Russian abbreviation for water-cooled and water-moderated nuclear reactor.

ionization of organic molecules of various sorts was investigated, and high-sensitivity devices were created to detect these molecules (E. Ya. Zandberg, V. I. Paleev, U. Kh. Rasulev).

Also, the nonequilibrium surface ionization was studied (N. I. Ionov, B. N. Shustrov). To investigate various processes on the surfaces, a fundamentally new method of thermodesorption spectroscopy was developed on the basis of a time-of-flight mass-spectrometer (N. I. Ionov, V. N. Ageev, Yu. K. Ustinov) and used to study the interphase boundaries (N. I. Ionov, M. A. Mittsev). Important progress was made in the field of formation and processing of effective field point electron and ion emitters (V. N. Shrednik).

Solid state physics

In the postwar years, investigations in the field of material strength were significantly expanded. The method of profiled-crystal growth from melted alloys (A. V. Stepanov's method) became widely used. A theory of capillary formation of materials in molten state was developed to create a technique for controlling crystal growth. The influence of the symmetry of a heat field on the symmetry of a growing crystal was taken into consideration and the heat fields and thermoelastic strain fields in drawn crystals of different shapes were studied experimentally and theoretically. This gave an insight into the mechanisms of dislocation formation and provided a means for controlling the perfection of a profiled-crystal structure (Nikanorov and Regel 1981).

Stepanov's technique was applied industrially for obtaining profiled germanium and sapphire crystals for semiconductor technology, optics, and lighting engineering and silicon crystals for solar cells. The possibility was demonstrated of using this technique to produce articles of complicated cross section from Al- and Mg-based alloys for heat exchangers and other purposes, articles reinforced with high-strength fibers, and those made of eutectic alloys. The elasticity of crystals with interatomic bonds of various types was investigated in a wide range of temperatures up to their melting points (S. P. Nikanorov).

Data on mechanical properties and crystal bulk deformation at liquid helium temperatures were obtained by O. V. Klyavin. These investigations turned out to be highly important for clarifying the origin of interatomic binding forces, constructing a theory of strength, and meeting the technological demands of that time. In the 1950s, the kinetic theory of strength was developed by S. N. Zhurkov's school (see pp. 160–169).

"Semiconductor revolution"

Fundamental works and achievements in another important home field of research at the Institute—*physics of semiconductors*—deserve particular treatment, as their history cannot be reviewed in any concise form. Here, only the most characteristic outstanding studies, maintaining the Institute's priority in this field, will be considered.

These again were connected with Ioffe's name and works, but he was forced to leave the Institute for reasons independent of him and dismissed his directorship in 1950 to create a separate laboratory which became the Institute of Semiconductors of the Academy of Sciences (*IPAN*, in popular Russian abbreviation). Some of Ioffe's co-workers continued their work at this new institute, so their results obtained there will be briefly outlined in what follows.

Semiconductor materials and devices

In the War time the demand for thermistors gave rise to B. T. Kolomiets's research on materials suitable for these purposes and the urgently needed result was achieved with compound semiconductors. This was one of the reasons why new success was achieved in semiconductor research at the Institute.

In the early postwar years, studies of the properties of a number of intermetallic compounds were initiated by A. F. Ioffe. The most significant was the discovery of several valuable classes of semiconductors, primarily the $A_{III}B_{V}$ compounds. Investigations of their electrical properties were initiated by N. A. Goryunova. In 1950, at a conference on semiconductors held in Kiev A. R. Regel reported that InSb binary compound possesses semiconductor properties; in the discussion, B. I. Boltaks and V. P. Zhuze suggested that other compounds of this group are also good semiconductors. An experimental proof of the semiconductor properties of the $A_{III}B_{V}$ compounds was obtained independently and somewhat later by Heinrich Welker of the Institute of Physical Chemistry in Munich.

Another important task was to develop methods for purification, growth, doping, and synthesis of other new semiconductor compounds with 3, 4 and more components. The work done on diamond-like semiconductors allowed systematization of manifestations of semiconductor behavior in various types of compounds of this kind (Goryunova 1963). The development of methods for single crystal purification and growth is to be considered a great success of D. N. Nasledov and his co-workers.

The significance of I. Bardeen and W. Brattain's discovery of the transistor effect in 1947 was correctly evaluated at the Institute, and, in 1950, works on fabrication and study of a p-n junction in Ge were initiated. After point-contact diodes and transistors were fabricated in 1950–1951, the first Soviet single crystals of Ge and transistors with p-n junctions, high-voltage Ge diodes, first electronic circuits with transistors, and transistor radio receivers were developed by the Institute's scientists under the guidance of V. M. Tuchkevich in 1952–1953.

In 1954, first Soviet planar photodiodes of original design, superior to those fabricated abroad in sensitivity and active area, were produced. S. M. Ryvkin advanced a phenomenological theory of photocell rectifiers with p-n junctions.

In the same year, the first Soviet power semiconductor devices—Ge semiconductor rectifiers for currents of up to 500 A and a reverse breakdown voltage of 150–200 V—were developed (Zh. I. Alferov and V. M. Tuchkevich).

After this success with Ge devices, studies of single crystals of the other leading material, silicon, were undertaken under the guidance of V. M. Tuchkevich in 1954. In 1955–1956, V. E. Chelnokov developed the diffusion technique and Zh. I. Alferov—the alloying method to obtain a p-n junction in silicon. The former was applied to fabrication of Si cells intended for solar energy conversion, and the latter used to obtain high-voltage silicon diodes. The practice led to a conclusion that materials possessing high parameters at higher temperature were needed to ensure effective conversion. Gallium arsenide was found to be a material of this kind, and photocells and converters based on it were produced at the Institute in 1957 (D. N. Nasledov and B. V. Tsarenkov).

Here, an important, but little-known example of the Institute's practical achievements in this field is worth mentioning. In 1958, some new semiconductor devices, rectifiers for 1000 A and 300–400 V, were installed at the first Soviet nuclear-powered submarines.

In the 1960s, Ge and Si-based semiconductor devices were developed and successfully put into practice (V. M. Tuchkevich, I. V. Grekhov, V. E. Chelnokov, V. B. Shuman)[6].

New discoveries

Studies of nonequilibrium electronic processes in GaAs culminated in the discovery of effective band-to-band radiative recombination with a possibility of stimulated emission by D. N. Nasledov, A. A. Rogachev, S. M. Ryvkin, and B. V. Tsarenkov (Nasledov *et al.* 1962). In investigations of recombination radiation spectra, a dependence of gap width on injected carrier density was revealed (A. A. Rogachev, S. M. Ryvkin).

Noteworthy are works on electronic phenomena in semiconductors with narrow forbidden gap and semimetals (B. T. Kolomiets), which resulted in the development of a new field of semiconductor physics—the physics of gapless state of semiconductors (V. I. Ivanov-Omskii, B. L. Gel'mont).

Ioffe himself summed up his studies of semiconductors, performed in cooperation with his pupils and collaborators, in the well-known monographs on electronic semiconductors (Ioffe 1960) and thermoelements (Ioffe 1957a). A. F. Ioffe also called attention to a new class of amorphous and liquid semiconductors which were studied by A. R. Regel (in 1946 was discovered the fundamental role of the short-range order in charge carrier transport, the Ioffe–Regel rule). It is these works that initiated research in the field of physics of disordered systems to bring outstanding results. First came a discovery of a vast class of vitreous chalcogenide semiconductors by Goryunova and Kolomiets (1955), which initiated their wide investigations in a number of laboratories all over the world. These semiconductors found application in technology and attracted widespread attention to the fundamental problem of compatibility of the powerful apparatus of the band theory with the lack of long-range order in glasses and other disordered systems. The answer was given by A. N. Gubanov's theoretical works, according to which the band pattern is mainly determined by the short-range order or the first coordination spheres, with the lack of long-range order leading to the so-called tails of the density of states, extending into the forbidden gap.

World's interest in the physics of disordered systems quickened in the 1960–1970s. By that time, the Ioffe Institute became an internationally known center of research into disordered systems. The strongest school formed in the field of theoretical and experimental studies of doped and compensated semiconductors. This school's concept of a Coulomb gap near the Fermi level, predicted by Efros and Shklovskii (1975) and discovered and studied experimentally by Zabrodskii (1977), had a famous and dramatic history of acceptance upon the initial negation by Sir N. F. Mott, a Nobel Prize winner, and others. This gap arises owing to the attraction of a hopping electron to a hole at its site and governs the transport and dynamic properties inherent in the insulating state at low temperature.

Important results related to construction of physical models describing the behavior of shallow impurities in semiconductors were obtained by Yu. V. Shmartsev and coworkers in 1973. These results are, firstly, construction of a model of impurity centers related to nonequivalent band extrema and, secondly, creation of a model explaining the specific features of the influence exerted by isovalent impurities of III–V semiconductors. The latter model enabled fine control over electrical, optical, and mechanical properties of semiconductors without introduction of electrically active impurities.

[6]For more detail, see pp. 119–131.

Ioffe's new institute

Investigations closely related to research in the field of semiconductors at the prewar Physico-Technical Institute were continued at *IPAN* under the guidance of A. F. Ioffe. Here, studies of *IPAN's* scientists are mentioned, since *IPAN* merged with the Ioffe Institute in 1972. In 1954–1956, the phenomenon of deep penetration of keV-range ions into semiconductors was discovered (M. M. Bredov and A. R. Regel with coworkers). This phenomenon stimulated the development of an implantation technique which became an effective instrument for fabrication of microelectronics components.

In the late 1950s, works on neutron scattering in solids were initiated by I. V. Kurchatov and A. F. Ioffe. With the use of the time-of-flight spectrometer created at the Institute, a method of direct determination of phonon spectra for coherently scattering media was developed, spectra of semiconductors, metals, and solid solutions were obtained, and the primary role of the short-range order in phonon spectrum formation was established (B. A. Kotov, N. M. Okuneva, V. Oskotskii). The method found wide use in the USSR and abroad.

In the late 1950s—early 1960s, G. A. Smolenskii initiated works on magnetism in nonmetals. Single crystals of garnet-type ferrites with a narrow FMR line were obtained and garnet ferrites with Va and Bi, exhibiting a huge Faraday effect, were synthesized. Magnetic ordering was revealed in a number of ferroelectrics. The nature of weak ferromagnetism was investigated and a transparent ferromagnetic $RbNiF_3$ was synthesized. Optical spectra of fluorides with magnetic ordering were investigated and a significant decrease in acoustic wave damping in crystals with impurity addition was revealed.

In the early 1960s, A. G. Gurevich initiated studies of ferro- and antiferromagnetic resonances and spin waves. Oscillations of spin density were investigated by the NMR method in a number of fluorides, and the phenomenon of ferrimagnetism induced by magnetic field well above the Curie point was revealed as well as scattering of parametric spin waves by nuclear magnetic momenta (M. P. Petrov, R. V. Pisarev, V. V. Lemanov).

In 1961, theorists V. L. Gurevich and Ya. A. Firsov predicted a new phenomenon—oscillations of magnetoresistance in semiconductors in strong magnetic fields, related to inelastic scattering of charge carriers by optical phonons, i.e., the magnetophonon resonance. This prediction initiated experimental work at *IPAN*, which confirmed the new phenomenon in semiconductor kinetics (S. S. Shalyt, R. V. Parfeniev, V. M. Mudjaba). A thorough experimental investigation of the magnetophonon resonance in various semiconductors and kinetic effects, combined with the development of the theory of this phenomenon, laid the foundations of the method of magnetophonon spectroscopy of semiconductors.

In the mid-1960s, investigations of tellurium—an elementary semiconductor with pronounced anisotropy of physical properties—was initiated by S. S. Shalyt and I. I. Farbshtein at the laboratory of low temperatures. A technique for growth of perfect Te crystals (of record-breaking purity) was developed along with doping methods. Experiments on galvanomagnetic properties of the crystals obtained and the developed theory of anisotropic semiconductors (G. E. Pikus) established the electron spectrum of Te and revealed a topological phase transition of the Fermi surface for holes on changing the doping level, magnetic field, and uniform compression. Later, the unknown electro-optical effects (circular photo-galvanic effect and induced optical activity) were predicted (N. S. Averkiev and G. E. Pikus) within the frame of this theory and discovered experimentally as well.

A novel phenomenon, electroacoustic echo, was discovered (S. N. Popov, N. N. Krainik) at Smolenskii's laboratory in 1970. Fundamentals of acoustoelectronics and acoustics were developed by Yu. V. Ilisavskii and V. V. Lemanov.

In 1972, after merging of *IPAN* with the Physico-Technical Institute, investigations of rare-earth semiconductors commenced and a know-how of their synthesis was developed. The semiconductor–metal phase transition discovered in thin films of SmS gave rise to the development of a new class of materials with alternating valence of rare-earth ions, and highly sensitive heat probes were invented. In 1981, V. P. Zhuze, V. M. Sergeeva, and A. I. Shelykh registered, together with scientists from several Kharkov Institutes, the discovery of chemical inertness of metal impurities in semiconductors with stoichiometric vacancies. A research school of rare-earth materials, having a worldwide reputation, was founded and is successfully functioning now in our country.

New wave: heterostructures

The development of semiconductor materials allowed investigations of single crystals with heterojunctions, i.e., two semiconductors with different chemical composition inside the same single crystal. The present achievements in the field of semiconductor heterostructures were preceded by the research of the 1930s, beginning with above mentioned work by Zhuze and Kurchatov (1932) on the intrinsic and impurity conduction in semiconductor heterostructures and the first theory of current rectification at a metal–semiconductor interface, based on the tunneling effect (Frenkel and Ioffe 1932). This mechanism was realized in heavily doped p-n structures. The first diffusion theory of a rectifying p-n heterojunction, laying a foundation for W. Shockley's theory of the p-n junction, was published by B. I. Davydov in 1939. In the late 1940s, intensive studies on intermetallic compounds were started at the Institute. Later, phenomena occurring in p-n structures on injecting carriers to large distances were understood, which allowed control over the most important parameters of semiconducting crystals. Thus, the idea of using heterostructures in semiconductor electronics emerged.

Various concepts concerning the application of heterostructures in solar cells evolved in the 1960s. Then, the concept of the double-heterostructure laser was formulated by Alferov and Kazarinov (1963), with the possibility of attaining a high density of injected carriers and a population inversion by "double" injection. It remained to find a heterostructure in which the predicted effects (superinjection of carriers, optical and electron confinement) could be observed. By the end of 1966 was found an "ideal" heterostructure—a lattice-matched system for GaAs—in the form of AlGaAs solid solutions.

The unique injection properties of wide-gap emitters and the phenomenon of superinjection were confirmed experimentally, and stimulated emission in AlGaAs double-heterostructures was observed along with the luminescence properties and carrier diffusion in a smooth heterojunction. Diagonal tunneling-recombination transitions directly between holes from the narrow-gap component and electrons from the wide-gap component of a heterojunction were discovered. As a result, many research teams all over the world have conducted investigations of heterostructures and devices on the basis of III–V compounds since 1971.

A search for new lattice-matched heterostructures (Alferov *et al.* 1971) revealed the possibility of obtaining various lattice-matched heterojunctions by using quaternary III–V solid solutions, which permit the lattice constant and the width of the bandgap to be varied independently. Various InGaAsP compositions were investigated and found to be among the most important materials for a number of applications: photocathodes and especially lasers operating in the infrared for fiber optic communications and in the visible range.

Further progress was also due to the Institute's scientists. The basic concepts of the distributed-feedback (DFB) laser were first advanced (Alferov *et al.* 1971a), supported by the idea of a diffraction grating formed on the surface of the waveguide layer to produce not only distributed feedback, but also well-collimated radiation at the output. Kazarinov and Suris (1972) confirmed the feasibility of single-mode lasing in DFB lasers by developing a detailed theory. Then, the first semiconductor lasers with narrow beam grating coupled output and DFB lasers were demonstrated (Alferov *et al.* 1974 and 1975). The interest in heterostructures with a stepped band structure (type-II heterojunction) also increased, resulting in the creation of an injection laser of this type, opening up excellent opportunities for construction of efficient coherent light sources in the infrared. Now, it is a well-established fact that the introduction of the heterostructure concept resulted in the discovery of fundamental physical phenomena, dramatic improvement of characteristics of essentially all the known semiconductor devices, and creation of new types of these.

Special tasks

Here, other studies performed at the Institute should be mentioned, which do not fall into the previously described pattern of its traditional fields of research, but, to say nothing of their practical value and urgency, must be highlighted with regard to the originality of ideas and the audacity of researchers.

A direct consequence of the success achieved in these works was the foundation of two novel areas of research at the Ioffe Institute—astrophysics and gas dynamics—whose organization and progress were much promoted by the efforts of, respectively, B. P. Konstantinov and Yu. A. Dunaev.

Isotope separation

Since 1947 the Physico-Technical Institute had undertaken a strong attack on the problem whose solution gave rise to the present-day industry of light isotopes, and manufacture of thermonuclear fuel, beginning with a wide search for methods of fabrication and analysis of this important product, to the point of organizing the industry itself. An original approach was laid in the foundation of this work, since the already existing techniques employed in the USA were energy consuming and had a low production rate. In the USSR these works were initiated by I. V. Kurchatov; several Institute's laboratories were involved together with the chair of isotopes at the Leningrad Polytechnic Institute and some other institutions. In an incredibly short time the problem of lithium-6 as the primary material (lithium hydride) for deuterium and tritium components of the thermonuclear fuel was solved on the industrial scale. The development of nuclear warfare provided a parity of weapons between the USSR and the USA and prevented a world war. The role of the team headed by B. P. Konstantinov was so great, but these works were so little known for the reason of secrecy that, here (maybe for the first time) mention should be made of their main participants from the Ioffe Institute (Konstantinov's, Ionov's, and Dunaev's laboratories) and from other institutions: B. A. Gaev, V. N. Dyn'kov, K. V. Donskoi, G. Ya. Ryskin, A. P. Votinov.

Space technology research

In another field of research of the day, the space research, one of the most important was the technology of high temperature coatings for landing space vehicles and missiles, developed by Yu. A. Dunaev, a founder of gas dynamics research at the Institute, with the participation of T. V. Sokolova and A. P. Obukhov in 1953–1958. His solution of this problem was in the first place favored by investigations of changes in the aggregative state of substance, aimed at distributing the enormous heat flows and mechanical stresses (arising in motion at cosmic velocities in the atmosphere) over all degrees of freedom of phase transitions inherent in the applied organic materials and compounds, rather than allowing them to cause rapid disintegration and sputtering of the coating. The energy of the incoming gas flow should be expended to evaporation in a boundary layer and ablation of the "sacrificed" surface layers of the coating. The higher viscosity of these coatings protected them from being carried away rapidly when entering the dense atmospheric layers of the Earth and planets. Apart from a study of their composition, structure formation, and fabrication methods, it was of importance to develop a reliable technique for laboratory and field tests of the coating samples—any failure had to be ruled out in manned flights! These works have laid a foundation for the Institute's successful gas dynamics research, including hypersonic aerodynamics, shock waves, ballistic and shock tube experiments to model space flights and landing.

In the 1960s, Dunaev also headed studies of thermionic conversion aimed to obtain energy sources for space missions. He organized this work on the national scale, including a most fruitful experimental and theoretical support from V. G. Yuriev's laboratory at *IPAN* and a nuclear pile experiment.

New areas of activity

Spectroscopy of semiconductors

The development of the theory of condensed state led to new discoveries. Shortly after the War, E. F. Gross, a prominent optician, organized the optical laboratory at the Institute. In 1951, Gross and Karryev (1952) discovered the optical spectrum of excitons in a semiconductor—cuprous oxide. This sensational discovery of the time was preceded by experiments of V. P. Zhuze and S. M. Ryvkin who supposed the exciton mechanism of Cu_2O conduction. (Ya. I. Frenkel himself was aware of this event but was already not alive to see its triumph.) The discovery of the linear spectrum of exciton together with the first experiments on magneto-optics, electro-optics, piezospectroscopy, and exciton dynamics (B. P. Zakharchenya, A. A. Kaplyanskii, and S. A. Permogorov) gave rise to a new field of science—optics and spectroscopy of semiconductors and dielectric crystals. It may be stated without exaggeration that the experimental studies carried out at Gross' laboratory and, after his death, at the Zakharchenya's and Kaplyanskii's laboratories, combined with theoretical works of V. I. Perel and G. E. Pikus and their coworkers, exerted a decisive influence on the worldwide development of the spectroscopy of crystals.

In the early 1970s, the Ioffe institute was engaged in extensive investigations of the optical orientation and alignment of electrons, nuclei, and excitons in semiconductors.

These works were first conducted for bulk crystals and then extended to quantum-size structures. New phenomena—optical detection of NMR, cooling of nuclear spin system by light to μK, spin-dependent scattering of carriers, manifestation of the anti-crossing of levels under orientation and alignment, and many others—were discovered at the Institute's laboratories headed by V. I. Perel, G. E. Pikus, B. P. Zakharchenya, A. A. Kaplyanskii, V. I. Safarov, A. N. Titkov, and P. G. Baranov. Beginning with the mid-1970s, this prominent school of optics, established at the Ioffe Institute, also included R. P. Seisyan's laboratory, engaged in a number of exquisite investigations in spectroscopy of magneto-excitons. In very beginning of the 1970s, a new field of research came into existence owing to the discovery of the alignment of angular momenta of hot electrons caused by polarized light in bulk and quantum-size semiconductors (Zakharchenya *et al.* 1982). This phenomenon made it possible to measure ultrashort times (down to femtoseconds) of carrier relaxation under steady state excitation of a semiconductor by light.

Nonequilibrium processes in semiconductors

The semiconductor physics and technology have been influenced essentially by the Institute's investigations on nonequilibrium processes in semiconductors guided by S. M. Ryvkin (since the 1960s) and V. M. Tuchkevich (since 1981). A convective current instability of plasma was discovered in semiconductors (Ivanov and Ryvkin 1958), giving rise to a huge number of publications and revealing other types of instabilities in this matter. When studying the spectra of recombination radiation, A. A. Rogachev discovered the dependence of the bandgap on the density of injected carriers. Basically new non-cooled detectors of far-infrared radiation were created on the basis of the effect of photon drag of electrons discovered with the participation of A. A. Grinberg, A. A. Kastal'skii, A. M. Danishevskii, and I. D. Yaroshetskii. L. G. Paritskii's group developed unique devices for infrared imaging—semiconductor photographic devices and image converters. The semiconductor radiation physics was stimulated and its high level was maintained by investigations of nuclear exposure of semiconductors, carried out by T. V. Mashovets and the S. V. Novikov's section from the Gatchina branch. N. B. Strokan's and O. A. Matveev's groups accomplished important priority works on nuclear radiation detectors based on Si and CdTe. In the 1980s, A. G. Zabrodskii developed a method of neutron transmutation doping of Ge and Si-Ge solid solutions, with the peculiarities of decay of radioactive isotopes and induced transmutation impurities' kinetics resulting in the motion of the Fermi level through the forbidden gap states, which was used in precision experiments concerned with semiconductor and nuclear properties of materials under investigation and their isotopes. Here mention should be made of V. M. Tuchkevich and I. S. Shlimak's project on neutron doping of Si crystals on industrial scale in a high-power nuclear pile which was, however, rejected after the Chernobyl accident in 1986. But Yu. L. Ivanov's many years' search for stimulated infrared and submillimeter radiation via intraband transitions in p-Ge in crossed electric and magnetic fields succeeded to produce tunable lasers for this frequency band.

Astrophysics

Many concepts underlying modern cosmology, nuclear astrophysics, and superdense star physics were put forward by the outstanding scientists Ya. I. Frenkel and L. D. Landau in the 1920–1930s. Radiative equilibrium of stars and application of the second law of thermodynamics to the Universe as a whole was studied by Bronstein and Landau (1933). In the 1940s, L. E. Gurevich and A. I. Lebedinskii developed a theory of planet formation and studied the problem of nova and supernova star explosions. After a prolonged interruption, the theoretical and experimental astrophysics again became Institute's active area of research in the early 1960s. B. P. Konstantinov and M. M. Bredov formulated a wide program of astrophysical investigations based on new experimental possibilities and concepts of modern astrophysics.

Investigations of micrometeorites in the near-Earth space (1962–1972) rejected the widely accepted concept that a dense dust cloud exists around the Earth (B. P. Konstantinov, M. M. Bredov, E. P. Mazets). These new data indicated that in most cases there is no need for any special anti-dust protection of space vehicles. In 1986, two *Vega* missions studied in detail the dust coma of Halley's comet in a dust particle mass range of ten orders of magnitude (R. L. Aptekar, E. P. Mazets, G. G. Petrov, D. D. Frederiks, and others). In 1971–1976, the diffuse background of metagalactic gamma-rays in the low and medium energy range was investigated in several space experiments. Reliable data on the intensity, energy spectrum, and isotropy of diffuse radiation were obtained for the first time and the anisotropic component of the diffuse background, related to the emission of our Galaxy, was revealed (S. V. Golenetsky, V. N. Ilyinsky, E. P. Mazets, and others). Since 1974, intensive exploration of gamma-ray bursts has been conducted (see pp. 211–220). A number of balloon experiments with magnetic spectrometers were performed (1974–1978) and antiproton component of cosmic rays was detected (V. A. Romanov and others).

In 1962–1972, A. M. Romanov, E. V. Myakinin, and others studied the problem of generation of high energy neutrons and gamma-rays in the Earth's atmosphere by means of balloon experiments. Using the satellites *Kosmos-561* and *Kosmos-731*, hard gamma-radiation from the Galaxy core and diffuse radiation were studied by E. I. Chuikin, G. M. Gorodinsky, E. M. Kruglov, and others. In 1990–1991, in the framework of international collaboration using *Gamma-1* observatory, high energy gamma-rays of solar origin and those coming from Vela and Geminga pulsars were detected (E. I. Chuikin and others).

A series of experiments performed aboard the space stations *Zond*, *Soyuz*, and *Kosmos* (in the late sixties) revealed sporadic fluxes of tritium and helium-3 of solar origin (G. E. Kocharov, V. O. Naidenov). As a result, a new class of solar flares based on plasma effects was revealed. In the early 1970s, priority experiments on the lunar surface were performed using *Lunokhod-1* and *Lunokhod-2* vehicles (B. P. Konstantinov, M. M. Bredov, G. E. Kocharov, V. O. Naidenov, S. V. Viktorov, and others). The chemical composition of the Mare Imbrium and Crater Le Monnier was determined for the first time. In experiments aboard the satellites *Prognoz* and *Coronas*, X-ray precursors of solar flares in the soft energy range and fine burst structure in the hard range were detected. The occurrence of a series of elementary bursts of small duration of 100 ms in the hard range was demonstrated (Yu. N. Charikov, G. E. Kocharov, G. A. Matveev, Yu. N. Starbunov, and others).

In 1963–1969, studies of interactions between the relativistic nuclei of Fe group in cosmic rays and Ag and Br nuclei were undertaken using the photoemulsion chambers. Evidence of pronounced collective effects in these interactions of heavy nuclei at small collision parameters were obtained. With solid track detectors the energy spectra of Fe group in the galactic cosmic rays of low energies and the Fe ions in the solar cosmic rays in a range of 7–200 MeV per nucleon were measured for the first time aboard the space stations *Salyut* and *Mir* (Yu. F. Gagarin and N. S. Ivanova).

Based on the idea of Konstantinov and Kocharov, a new field of astrophysics—paleoastrophysics—has been developed at the Ioffe Institute since 1965, using natural archives of cosmic radiation: tree rings, corals, ice cores, etc. At present this avenue of research is well known and explored in many countries. High-precision measurements of radiocarbon contents in annual tree rings yielded pioneering results on the temporal variations of cosmic ray fluxes for the last 400 years. These data revealed a theoretically unpredicted effect of cosmic ray modulation during the period of the deep minimum of solar activity (1645–1715). Annual tree rings gave a timing of high energy gamma-rays generation during the 1006 Supernova explosion. An increase of the cosmic ray intensity in the near-terrestrial region due to supernova explosion was discovered. Both findings are of fundamental importance for the theory of cosmic ray origin. These results have been highly appreciated by the scientific community. In studying the temporal variations of the cosmogenic isotope generation, a 2400-yr cycle of climatic and solar activity changes was revealed (V. A. Dergachev and G. E. Kocharov).

In 1962, the theoretical astrophysics department was organized at the Ioffe Institute by A. Z. Dolginov (headed by D. A. Varshalovich since 1986).

Among the works done in the Department one should first mention those performed by E. B. Gliner. He was the first who understood the role of the physical vacuum in cosmology and used this concept in studying the earliest stages of evolution of the Universe. Later, these ideas were employed in constructing the most popular inflation models of the Universe.

A new astrophysical phenomenon, which appeared to be widespread in Nature, spin alignment of atoms and molecules in interstellar gas clouds, stellar envelopes, comets, and upper layers of the atmosphere was discovered and examined by D. A. Varshalovich. It affects the relative intensities of spectral lines, being thus important for evaluating the chemical composition, ionization, temperature, and density of a gas.

Great attention is being given to studies of neutron stars—unique objects composed of superdense matter with superstrong magnetic fields (G. G. Pavlov, D. G. Yakovlev, A. I. Tsygan, Yu. A. Shibanov, V. A. Urpin, and others). The models of neutron-star atmospheres, cooling of neutron stars, electrodynamics of their magnetospheres, and evolution of their magnetic fields have been studied in detail. For instance, a number of neutron-star atmosphere models has been created, widely used by astrophysical community through Internet; a realistic theory of ohmic decay of crustal magnetic fields in neutron stars has been developed. Comparison of the theory with observational data provides important information on fundamental properties of superdense matter in stellar interior.

Another direction is the investigation of distant quasars, the most energetic emitters whose spectra were formed about 10 billion years ago (D. A. Varshalovich, S. A. Levshakov, A. Yu. Potekhin, and A. V. Ivanchik). An analysis of these spectra enabled reconstruction of the chemical composition of matter and physical conditions in the early Universe. A new class of astrophysical objects was discovered, molecular (H_2) clouds with high redshifts. It was proved that (contrary to the predictions of some contemporary theories) the fundamental physical constants have not changed during the cosmological evolution and have the same values in different casually disconnected regions of the Universe.

Much work has been done in the physics of cosmic rays. The foundation of the transport theory of cosmic rays in interplanetary and interstellar medium was laid by classical articles by A. Z. Dolginov and I. N. Toptygin. This work was continued and extended by A. M. Bykov. In particular, he predicted gamma-ray nuclear emission lines from star-formation regions, which have been confirmed by subsequent observations of the Orion nebula by the Compton orbital observatory.

Holography

In the late 1960s, holography experienced its new birth, when lasers appeared, which simplified essentially the problem of formation and registration of patterns and radiation interference. Initiated by B. P. Konstantinov, who was the first to envisage success in this field, works on holography were started at the several laboratories headed by A. N. Zaidel, A. P. Komar, S. B. Gurevich, and Konstantinov himself. The Scientific Council on holography was created at the Institute under the auspices of the Division of General Physics and Astronomy of the Academy of Sciences of the USSR, with the aim to enlarge and coordinate the efforts of Soviet scientists. Then, beginning with 1973, studies on optical information processing were undertaken by M. P. Petrov and at Dunaev's, Zakharchenya's, and Kolomiets's laboratories. The coordination of these efforts on the Institute level was enhanced upon invitation of Yu. N. Denisyuk, a well-known discoverer of holography, and foundation of the laboratory of optoelectronics and holography in 1988.

In the first stage of development of laser holography, the laws and applications of static holograms were under investigation. Yu. I. Ostrovskii was the pioneer of holographic interferometry. As a practical application of this technique, he recorded processes of plasma formation. By that time the scope of holographic applications had formed: optical elements, 3D images, perfection of radiation sources (first semiconductor lasers), use of semiconductor recording media (glassy chalcogenides) for recording holograms, receivers of incoherent-to-coherent light converters (controllable space modulators of light), photothermoplastic recording devices, creation of photorefractive crystals to analyze vibration, holographic information processing (S. B. Gurevich). The first holographic devices (laser interferometric resolvometers) were created at the Institute, and much effort was undertaken to develop holographic cinema and TV by the initiative of Konstantinov. Halftone information transfer via binary channel was achieved with holograms.

Atomic radiospectroscopy

The atomic radiospectroscopy group was organized in 1960 with the aim to conduct radiospectroscopic and optical studies of gases and solids. Since its founding, the group has been led by R. A. Zhitnikov.

During the first decade the work was aimed at creation of systems with stabilized atoms (in crystalline and amorphous molecular matrices) and the study of their properties by EPR method and by optical spectroscopy. The research on the production of new dye impurity centers and defects in crystals were done by the same techniques. Since 1962, studies have been made on the optical orientation and magnetic resonance of atoms in gases, vapors, and plasma. New quantum electronic devices are being constructed, in particular, quantum magnetometers based on the optical orientation of atoms. A specific feature of these researches is the combination of radiospectroscopic and optical methods which provide highly informative and reliable results.

Successfully developed were investigations by optical orientation and magnetic resonance of atoms on the spin-dependent elastic and nonelastic interactions in gas discharge plasma containing metastable atoms of helium or other rare gases and also in alkali-helium plasma and in alkali metal vapors. Thus spin exchange and penning ionization were studied in various atomic and atomic-molecular collisions. Presently in the investigations optical orientation of atoms by laser radiation is actively employed (see Dmitriev *et al.* 1984 and references therein).

The matrix-isolated H, D, and N atoms were obtained by condensation method in H_2D_2, N_2, and rare gas cryogenic crystals and studied by EPR. The matrix shifts of g-factors and hyperfine structure constants of the atoms were measured and explained theoretically (Zhitnikov and Dmitriev 1987).

Unstable local paramagnetic centers with lifetimes $\tau = 10^{-2}$–10^{-4} s are discovered by EPR in the cryocrystals of rare gases, Ne, Ar, Kr, and Xe, condensing from the gas phase and trapping products of the gas discharge running in the same gas or in He during condensation. The unstable paramagnetic centers are interpreted as being local metastable excited $np^5(n+1)$s atomic-type states in rare gas cryocrystals, which form by trapping the gas discharge products as the cryocrystal grows, and are subject to the action of the anisotropic electric field resulted from the crystal surroundings distorted by the center.

Plasma physics and atomic collisions

The development of works on plasma physics, electronics, and atomic collisions was much promoted by the solution of the most important problem of the wartime and postwar years—isotope separation by the electromagnetic method. A systematic investigation of collisions of atomic particles (atoms, molecules, and their ions) with one another and with electrons and protons was started in the 1950s by N. V. Fedorenko at the laboratory headed by V. M. Dukelsky. These studies were stimulated by needs of nuclear fusion, magnetic separation of isotopes, and investigations of the upper atmosphere and space. The physical significance of data on processes occurring in atomic collisions lies in the fact that they are the main source of information about the atomic structure and the mechanism of interaction between particles. Furthermore, these processes determine to a great extent the parameters of all low-density media, such as gases and laboratory and astrophysical plasmas.

In this connection, unique techniques and facilities were developed by N. V. Fedorenko, V. V. Afrosimov, Yu. S. Gordeev, and M. N. Panov at the Ioffe Institute to study elementary atomic collision processes involving transitions from both the outer and inner electronic shells of interacting particles. Many-electron transitions, correlation effects, and quantum interference between different inelastic interaction channels have been observed.

These pioneering studies played a decisive role in the development of the present concept that the mechanism of inelastic interaction between atomic particles consists in the formation of a quasi-molecule by colliding particles. The quasi-molecular concept of inelastic interaction of atomic particles became a widely used way to consider a variety of elementary processes in atomic collisions. A wealth of data on the cross sections of different elementary processes in atomic collisions, including excitation, ionization, and electron capture, has been obtained for the first time at the Institute. These data formed a database of information about the most important types of atomic collision processes.

The experimental techniques and atomic collision data were also used at the Ioffe Institute by V. V. Afrosimov, A. I. Kislyakov, M. P. Petrov to develop and improve new methods of hot plasma diagnostics in controlled fusion research. These methods, the so-called neutral particle or corpuscular diagnostics and charge-exchange spectroscopy, became an indispensable part of plasma diagnostics at the majority of the largest fusion installations all over the world.

The rapidly progressing application of atomic collision techniques and information have resulted in the development of a new diagnostics of the composition and electronic and crystal structure of solids at the laboratories headed by Afrosimov and Gordeev. Different versions of these methods frequently provide information that cannot be obtained by other means and require no calibration or use of reference samples.

Nuclear fusion

In 1958 I. V. Kurchatov initiated studies of thermonuclear fusion at the Institute. This problem was under the overall supervision of B. P. Konstantinov. First, original effective methods for hot plasma diagnostics were developed (Afrosimov *et al.* 1961). This aim was especially urgent since data on plasma parameters were lacking at all or were controversial for different installations. The diagnostic techniques were developed in three directions simultaneously: corpuscular, microwave, and optical methods.

The corpuscular methods intended to investigate the ion component of the fusion plasma (Afrosimov and Petrov 1967) are based on analysis of atomic fluxes emerging from the plasma and on injection of atoms into the plasma. Microwave methods are based on the interferometric analysis of microwave emission and scattering in the plasma (Golant 1968). Particularly noteworthy is the original multiple frequency reflectometric method for determining the plasma density profile.

A number of fine spectroscopic methods (A. N. Zaidel, Yu. I. Ostrovskii, E. Ya. Shreider) and holographic interferometry have been employed to investigate with high spatial and temporal resolution the electron density distribution in dense plasma. These methods were successfully applied to study the neutral current layer dynamics (Zaidel and Ostrovskaya 1977). The electron component diagnostics based on the Thomson scattering of laser radiation and the local method for determining the neutral component of plasma, using the resonant fluorescence effect, have been developed and widely used (G. T. Razdobarin, V. K. Gusev).

Golant's laboratory was established in 1968 to study the additional nonohmic plasma heating, EM waves–plasma interactions, and energy and particle transport in tokamaks. In these fields, most interesting results have been obtained (Golant and Fedorov 1989) Small tokamaks: TUMAN-2, TUMAN-2A, FT-1, FT-2 have been constructed and put into operation. Top-priority experimental and theoretical results were obtained at the highest level of the world's controlled fusion research. A good number of these results are of value to the present day. Experiments on plasma compression on fast magnetic yield build-up have been carried out on the TUMAN-2 and TUMAN-2A tokamaks (Kaganskii 1976). Adiabatic plasma compression and the possibility of strong electron and ion heating were revealed and an essential enhancement of plasma thermal isolation under compression was achieved. In 1980, the larger TUMAN-3 tokamak was launched and experiments were performed on compression along the minor and major radii and combined compression. (For more detail, see pp. 190–199.)

Microwave heating experiments have been accomplished on the FT-1 tokamak under conditions of the electron cyclotron resonance. The possibility was demonstrated of central plasma heating both by an ordinary EM wave supplied from the source side of low magnetic field and by an extraordinary wave arising on the source side of high magnetic field. Criteria of accessibility of the central plasma regions to waves of the low-hybrid (decimetric) band were established. An approach to analyzing the propagation of these waves in the tokamak plasma was developed on the basis of the ray trajectory method. The interaction of low-hybrid waves with both ion and electron components was demonstrated experimentally on the FT-1 and FT-2 tokamaks (V. N. Budnikov, V. E. Golant, M. M. Larionov). Techniques involving interaction of high-frequency waves with the plasma became widely used (A. D. Piliya, V. I. Fedorov, E. Z. Gusakov).

Small-scale fluctuation diagnostics based on the increase of the EM wave scattering cross section in the vicinity of a hybrid resonance, the so-called diagnostics of enhanced microwave scattering, was proposed and tested in model experiments. Enhanced scattering by low-frequency plasma turbulence and low-hybrid waves was investigated on the FT-2 tokamaks. The effect of linear transformation of EM waves into plasma waves was investigated in nonuniform plasma. A peculiarity of this process, namely, formation of local singularities of the EM field in whose vicinity a linear transformation occurs in two-dimensional magnetoactive plasma, was demonstrated for the first time both experimentally and theoretically. A theory of parametric decaying instabilities of nonuniform plasma was developed. The stabilization effect of plasma nonuniformity, furnishing saturation within linear approximation, was observed. The possibility was predicted that absolute parametric instabilities can be formed when presenting the cut-off points of the interacting waves, together with low-frequency fluctuations. The developed theoretical features have been confirmed by special model experiments.

Acknowledgments

The author is grateful to Zh. I. Alferov, V. E. Golant, A. G. Zabrodskii, and B. P. Zakharchenya for their continued support, valuable comments, and very useful contributions. Also included are historical fragments written and supervised by the leading Institute's scientists: V. A. Bokov, G. E. Kocharov, V. B. Konstantinov, V. V. Lemanov, V. O. Naidenov, S. P. Nikanorov, R. V. Parfeniev, A. I. Slutsker, D. A. Varshalovich, A. Ya. Vul', and R. A. Zhitnikov. Especially precious and helpful were advice and comments of V. I. Ivanov-Omskii and R. A. Suris. Valuable corrections were made by S. A. Gurevich and A. G. Gurevich.

A special debt of gratitude is to L. F. Gavrikova and A. M. Studenkov for a search for innumerable documents, sources, and facts.

It is clear that, having supported with so much help, author alone is responsible for the possible errors and shortcomings.

References[7]

Afrosimov V. V., V. A. Glukhikh, V. E. Golant, A. M. Zaidel, E. G. Komar, B. P. Konstantinov, G. M. Malyshev, N. F. Malyshev, N. A. Monoszon, A. M. Stolov, and N. V. Fedorenko, 1961,

[7] Authors' names in non-Russian entries are given in the transliteration of the day, otherwise, the present transliteration is used (e.g., Frenkel J., Frenkel J. I., and Frenkel Ya. I. is the same person).

Sov. Phys.-Tech. Phys., **5**, 1311.

Afrosimov V. V. and M. I. Petrov, 1967, *Zh. Tekh. Fiz.*, **37**, 1995 (in Russian).

Alferov Zh. I. and R. F. Kazarinov, 1963, *Inventor's Certificate* No. 181737, dated as 30.03.1963 (in Russian).

Alferov Zh. I., V. M. Andreev, S. G. Konnikov, V. G. Nikitin, and D. N. Tret'yakov, 1971, *Proc. Int. Conf. on the Physics and Chemistry of Semiconductor Heterojunctions and Layer Structures*, Budapest, October, 1970, ed. G. Szigeti, (Budapest: Academiai Kiado) vol. 1, p. 93.

Alferov Zh. I., V. M. Andreev, R. F. Kazarinov, E. L. Portnoi, and R. A. Suris, 1971, *Inventor's Certificate* No. 392875, dated as 19.07.1971 (in Russian).

Alferov Zh. I., S. A. Gurevich, R. F. Kazarinov, M. N. Mizerov, E. L. Portnoi, R. P. Seysian, and R. A. Suris, 1974, *Sov. Phys. Semicond.*, **8**, 541.

Alferov Zh. I., S. A. Gurevich, N. V. Klepikova, V. I. Kuchinskii, M. N. Mizerov, and E. L. Portnoi, 1975, *Sov. Phys. Tech. Lett.*, **1**, 286.

Alferov Zh. I., 1998, *Semiconductors*, **1** 1–14.

Bronstein M. P. and Ya. I. Frenkel, 1930, *ZhRFKHO. Part Phys.*, **62**, 485 (in Russian).

Bronstein M. and L. Landau, 1933, *Phys. Z. d. SU*, **4**, 114.

Dmitriev S. P., N. A. Dovator, and A. I. Okunevich, 1984 *Zh. Tekh. Fiz.*, **54**, 578–582.

Dobretsov L. N. and M. V. Gomoyunova, 1966, *Emission Electronics*, Moscow: Nauka, (in Russian).

Dorfman J. and J. Frenkel, 1930, *Nature*, **126**, 274.

Efros A. L. and B. I. Shklovskii, 1975, *J. Phys.* C, **8**, L49.

Flerov G. N. and K. A. Petrzhak, 1940, *Phys. Rev.* **58** 89; reprinted in *40 Jahre Kernspaltung*, ed. H. Wohlfarth, (Darmstadt: Wissenschaftliche Buchgesellschaft), 1979 p. 195.

Frenkel Ya. I., 1922, *The Structure of Matter.*, Part I, Petrograd: Seyatel', (in Russian).

Frenkel Ya. I., 1923, *Theory of Relativity*, Petrograd: Mysl', (in Russian).

Frenkel Ya. I., 1924, *The Structure of Matter.*, Part II, issues 1 and 2, Petrograd: Seyatel', (in Russian).

Frenkel J., 1926, *Z. Phys.*, **35**, 652.

Frenkel J., 1928, *Z. Phys.*, **49**, 31.

Frenkel J., 1931, *Phys. Rev.*, **37**, 17.

Frenkel J. I., 1934, *Phys. Z. d. SU*, **5**, 597.

Frenkel J., 1939, *Phys. Rev.*, **55**, 987.

Frenkel J. I., 1945 *J. Phys.*, **9**, 392.

Frenkel J. I., 1946, *Kinetic Theory of Liquids*, Oxford: Clarendon Press.

Frenkel J. and A. Ioffe, 1932, *Phys. Rev.*, **39**, 530.

Frenkel Ya. I. and T. A. Kontorova, 1938, *Zh. Eksp. Teor. Fiz.*, **8**, 89 (in Russian).

Frenkel V. Ya. (compiler), V. M. Tuchkevich (ed.), 1978, *A. F. Ioffe Physico-Technical Institute. 1918–1978*, Leningrad: Nauka.

Frenkel V. Ya., 1996, *Yakov Il'ich Frenkel*, Basel: Birkhauser Verlag.

Golant V. E., 1968, *Microwave methods of plasma research*, Moscow: Nauka, (in Russian).

Golant V. E. and V. I. Fedorov, 1989, *RF Plasma Heating in Toroidal Fusion Devices*, New York: Plenum Press.

Goryunova N. A. and B. T. Kolomiets, 1955, *Zh. Tekh. Fiz.*, **25**, 984 (in Russian).

Goryunova N. A., 1963, *Chemistry of Diamond-Like Semiconductors*, Leningrad: LGU, (in Russian).

Gross E. F. and N. A. Karryev, 1952, *Dok. Akad. Nauk SSSR*, **84**, 261 (in Russian).

Ioffe A., 1923, *Annalen d. Physik*, **72**, 461.

Ioffe A. F., 1957, *Usp. Fiz. Nauk*, **63**, 527 (in Russian).

Ioffe A., 1957a, *Semiconductors Elements and Thermoelectric Cooling*, London: Infosearch Ltd.

Ioffe A., 1960, *Physics of Semiconductors*, London: Infosearch Ltd.
Ioffe A. F., M. V. Kirpicheva, and M. A. Levitskaya, 1924, *ZhRFKhO. Part Phys.*, **56**, 489 (in Russian).
Ionov N. I., 1972, *Surface Ionization and its Applications*, Oxford: Pergamon Press.
Ivanov Yu. L. and S. M. Ryvkin, 1958, *Zh. Tekh. Fiz.*, **28**, 775 (in Russian).
Ivanenko D., 1932, *Nature*, **129**, 798.
Kapitsa P. L. and N. N. Semenov, 1922, *ZhRFKhO. Part Phys.*, **50**, 149.
Kaganskii M. G., 1976, *Adiabatic Plasma Compression in Tokamaks*, Leningrad: Nauka (In Russian).
Kazarinov and R. A. Suris, 1972, *Sov. Phys. Semicond.*, **6**, 1184.
Kikoin I. and M. Noskov, 1934, *Phys. Z. d. SU*, **5**, 586.
Kurchatov B., I. Kurchatov, L. Mysovskii, and L. Rusinov, 1935, *C. R. Acad. Sci., Paris*, **200**, 1201.
Kurchatov I. and P. Kobeko, 1930, *Z. Phys.*, **66**, 192.
Landau L., 1930, *Z. Phys.*, **64**, 629.
Landau L., 1932, *Phys. Z. d. SU*, **1**, 285.
Lukirskii P. I., 1923, *ZhRFKhO, Part Phys.*, **55**, 93 (in Russian).
Lukirskii P., 1926, *Z. Phys.*, **37**, 308.
Lukirskii P. I., 1927 *ZhRFKhO, Part Phys.*, **57**, 458 (in Russian).
Lukirskii P. I., 1945, *C. R. Acad. Sci. URSS*, **46**, 274.
Lukirskii P. I. and N. N. Semenov, 1923, *ZhRFKhO, Part Phys.*, **55**, 107 (in Russian).
Nasledov D. A., A. A. Rogachev, S. M. Ryvkin, and B. V. Tsarenkov, 1962, *Fiz. Tverd. Tela*, **4**, 1062 (in Russian).
Nikanorov S. P. and V. R. Regel (eds), 1981, *Producing of the Profile Single Crystals and Articles by Stepanov's Method*, Leningrad: Nauka (in Russian).
Semenov N. N., 1923, *Usp. Fiz. Nauk*, **3**, 449 (in Russian).
Semenov N., 1924, *Z. Phys.*, **30**, 151.
Semenov N., 1935, *Chemical Kinetics and Chain Reactions*, N.Y.: Oxford University Press.
Skobel'tsyn D. V., 1929, *C. R. Acad. Sci., Paris*, **189**, 55.
Zabrodskii A. G., 1977, *Sov. Phys. Semicond.*, **11**, 595.
Zaidel A. N. and G. V. Ostrovskaya, 1977, *Laser Methods of Plasma Diagnostics*, Leningrad: Nauka, (in Russian).
Zakharchenya B. P., D. N. Mirlin, V. I. Perel, I. I. Reshina, 1982, *Usp. Fiz. Nauk*, **136**, 459 (in Russian).
Zhitnikov R. A and Yu. A. Dmitriev, 1987, *Zh. Eksp. Teor. Fiz.*, **92**, 1913 (in Russian).
Zhuze V. P. and B. V. Kurchatov, 1932, *Phys. Z. d. SU*, **2**, 463.

Recent decade: 1988–1998

I. A. Merkulov

The recent history always looks ambiguous. Consequences of decisions which were and were not made are not clear yet, the participants of the events are still alive, have not lost the fighting spirit, and often have dramatically differing vision of the past. For these reasons, we are going to dwell here only on basic trends in the life of the Ioffe Institute during the recent decade.

Though it is a fairly trite remark, but this period of time has demonstrated once more the key role of the "filthy lucre" in the scientific progress. The chart on this page, which shows the dynamics of financing of the Institute, also reflects the state of minds of many of the Ioffe's scientists. The late 1980s were the time of hopes and rapid progress in basic and applied research. The Physico-Technical Institute celebrated its 70th anniversary. The scientific achievements of the Institute attracted great attention both in the USSR and abroad. Laboratories of the Institute were grouped together with regard to direction of their research in four divisions headed by Zh. I. Alferov, V. E. Golant, V. V. Lemanov, and B. P. Zakharchenya (the today's structure of the Institute is presented in Part IV). The system of continuous education in physics was being formed under the auspices of the Institute. Small business came into being at the Institute. All these activities resulted in an almost twofold increase in the budget of the Institute.

The leading scientists of the Institute were invited to work at scientific centers of the USA and Europe, and they did not say politely: "No." It seemed that a spurt to new horizons of science was being prepared, but, as the ensuing years demonstrated,

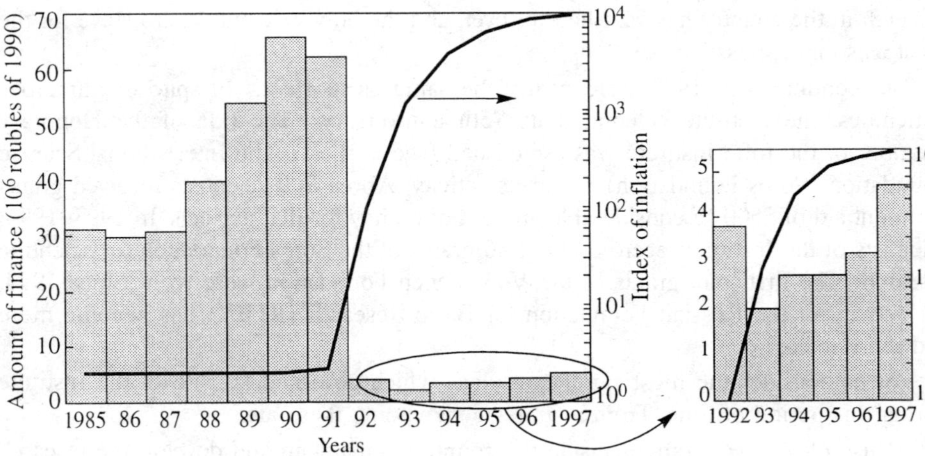

Financing of the Institute in 1987–1997.

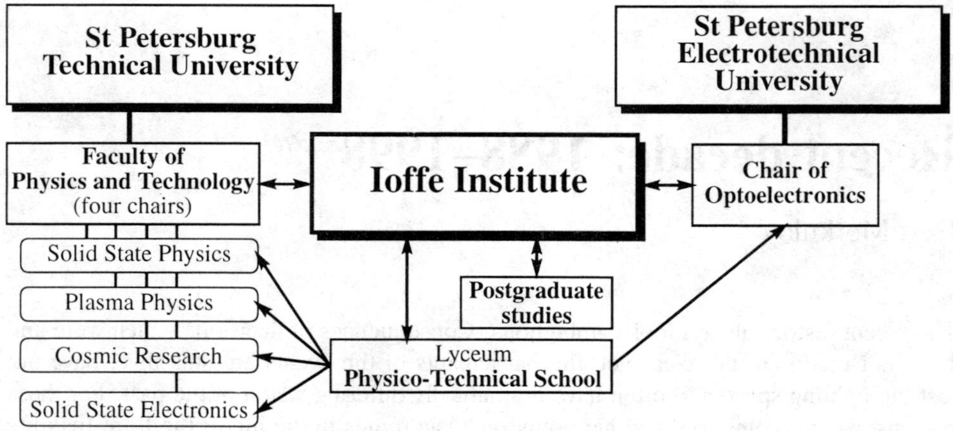

The structure for education in physics formed at the Ioffe Institute by the beginning of the 1990s.

this was a delusion.

The peak of funding was in 1990–1991. At briefings held twice a month, the Institute's authorities informed scientists of various projects to organize applied and basic research at the Institute. Feelings about the future became more and more gloomy. Presentiment of the approaching catastrophe forced many scientists to look for nonscientific activities. The board of directors underwent changes.

Then the fussy days of the coup of August 19, 1991, came. In autumn, the Academy of Sciences of the USSR became the Russian Academy of Sciences. The Institute's authorities and Scientific Council met regularly, and the information leaflets describing the main events of the preceding month were issued before every meeting. The "portfolio" of orders for applied research for the next five-year period was formed. However, these plans were doomed to failure. The financial crisis of 1992 was approaching.

The salary was miserable and, in addition, was not paid regularly. There were problems with payments for electricity, heating, and water. The equipment became out of date and needed repair. Many people believed that the social demand for scientific research in the country has vanished for ever, and the only way out was to leave for the West as soon as possible.

The conditions in 1993 were nearly the same as in 1992. In spite of numerous difficulties, the Institute celebrated its 75th anniversary. The title of the Honorary Member of the Ioffe Institute was established (see Part VI). The International Science Foundation (Soros Foundation) began its activity. About 700 scientists received grants of urgent aid of \$500 (a considerable amount of money for that period). In 1993–1994, scientists of the Institute received about 80 grants of the Soros Foundation for scientific research. The first joint grants of the Volkswagen Foundation were won. Since 1994, the activity of the Russian Foundation for Basic Research (RFBR) has become more and more intensive.

Let us consider the most significant ideas which governed the life of the Institute during this period of time. From a strategic viewpoint, these are:

(i) use of all the resources inside the country to maintain and develop the research activities at the Institute,

Presentation of the *Soros Professor* diploma to the Institute's scientists. V. Chelnokov, A. Gurevich, V. Andreev, B. Mamyrin, A. Rogachev, and I. Merkulov at the St Petersburg Scientific Center of the Russian Academy of Sciences.

(ii) contacts with international foundations, most intensive development of joint investigations with foreign scientific centers and industry,

(iii) continuous efforts aimed at bringing "fresh blood" to the Institute and ensuring the succession of generations of scientists and scientific schools.

The predominant idea among the totality of the hourly problems which arose and were solved during those hard times was that "the show must go on." More than once it seemed that only the vacations of the whole staff of the Institute could solve the financial problems stifling the Institute. Calculations unambiguously demonstrated that this decision was inevitable. However, the opinion of the Director was indisputable: "The Institute must continue its work. On no account must the research be stopped. The meetings of the Institute Scientific Council are to begin with a scientific report of one of the departments." Since then such a report has been presented at every meeting.

What were the solutions of the above-described strategic problems in the process of tactical struggle? The time when sputniks were flying and the armed forces were enjoying substantial budgets had gone with the wind, and the resources available were shrinking dramatically. What remained was associated with minor activities of the civilian and local industry. The scientists of the Institute actively participated in 13 federal and interdepartmental scientific programs. The number of grants of the RFBR received by the Institute's scientists in 1998 exceeded 170. In spite of the crisis in the industry, several other Russian foundations supported the applied researches conducted at the Institute. Among these were pressing of shavings by means of hydromagnetic forces, production of fullerenes, and development of LED-based traffic lights.

There was a moment when it seemed that the number of scientific papers published by the Institute decreased. However, it turned out that this decrease was of just a formal character, because in many cases the scientists had foreign co-authors. These papers were not included in the lists of publications of the Institute and thus were overlooked by the official statistics. On the verge of the staff reduction of 1996 the major part of the scientists presented full lists of their publications, and their number turned out to be twice as high.

Be as hard as it may, every year the Ioffe Institute arranges a competition for the best scientific work. Here we restrict ourselves to listing the works which have won the most prestige Institute's award—the Prize of the Scientific Council.

The Prize was established in 1988. Its first winners were A. M. Vasil'ev, S. V. Ivanov, P. S. Kop'ev, N. N. Ledentsov, B. Ya. Mel'tser, and V. M. Ustinov who presented the work "Heterolasers based on short-period superlattices."

In 1989, the Scientific Council awarded S. D. Baranovskii, B. L. Gel'mont, V. G. Golubev, V. I. Ivanov-Omskii, and D. V. Osutin for the work "Investigations of electron correlations in the impurity band of a semiconductor by far infrared spectroscopy."

In 1990, the Prize winners were V. V. Afrosimov, A. N. Zinov'ev, and A. A. Korotkov with their work "Plasma spectroscopy using injection of atom beams."

I. A. Merkulov, D. N. Mirlin, M. E. Portnoi, I. I. Reshina, V. F. Sapega, and A. A. Sirenko won the Prize of 1991 for the series of works "Photoluminescence of hot electrons and electron–hole pairs in the structures with quantum wells."

In 1992, the best work was "Microscopy of instabilities of a crystal surface with competing electrostatic and 'capillary' forces" by V. G. Butenko, Yu. S. Vlasov, O. L. Golubev, E. L. Kontorovich, and V. N. Shrednik.

In 1993, the Prize was given to astrophysicists D. A. Varshalovich, S. A. Levshakov, and A. Yu. Potekhin for the series of works "Basic physical constants: the upper limits of the probable variation rates of $\alpha = e^2/hc$ and $\mu = m_e/m_p$."

The series of works by B. B. Krichevtsov and R. V. Pisarev "Nonlinear optics of magnetics" was the best work of 1994.

In 1995, the Scientific Council accepted an unusual decision to divide the Prize between the authors of the book "Superlattices and Other Heterostructures. Symmetry, and Optical Phenomena," E. L. Ivchenko and G. E. Pikus, and the authors of a series of original investigations "Intercalation of two-dimensional graphite films on metals atoms and molecules" N. R. Gall, E. V. Rut'kov, A. Ya. Tontegode, and M. M. Yusufov.

In 1996, the Prize winners were A. V. Kavokin, K. V. Kavokin, I. A. Merkulov, and D. R. Yakovlev with their series of works "Magnetic polarons in semimagnetic-semiconductor heterostructures."

Finally, in 1997 the Prize was awarded to M. V. Gomoyunova, I. I. Pronin, and N. S. Faradzhev for the work "Imaging of the atomic structure of near-surface layers by electron focusing."

The international cooperation is one of the most important, complicated, and interesting part of the activities of a scientific institution. All the aspects of scientific work

At the Symposium on the Physics of Heterostructures, St Petersburg, 1990. From left to right: D. Mitchell (USA), J. Nishizawa (Japan), K. Tissen (GDR), R. Dupuis (USA), Yu. Gulyaev (USSR), Zh. Alferov (USSR), Chairman of the Leningrad Executive Committee A. Bolshakov, L. Keldysh (USSR), J. Ripper (Brazil), A. Gippius (USSR), M. Panish (USA), V. Afrosimov, S. Konnikov (USSR), Nobel Prize Winner K. von Klitzing (FRG), A. Fursenko (USSR), Nobel Prize Winner L. Esaki (Japan).

manifest themselves here most clearly. These are the necessity to keep abreast with the latest developments of colleagues, the possibility to work with the most modern equipment, the combination of fast and free exchange of results with the know-how, state and commercial secrets kept safe, friendly aid and hard competition. This form of scientific activities of the Institute is traditionally governed by the Director and the first Deputy Director. Yu. S. Gordeev who had been working in close cooperation with the Director during eight years tells about the development of this form of activities:

> Physico-Technical Institute has been developing international scientific contacts for a long time. However, before the "perestroyka" the main role was played by the scientific aspect of the cooperation, whereas the economic problems were not so significant. Radical changes occurred in 1992.
> The economic crisis broke out in our country. It became evident that nobody would finance the science in Russia. The Director became aware of the fact that the most appreciable and well-timed financial support could be obtained by using the international contacts. The aim was to have scientific products that could find a ready market outside the Russia.
> We went to the USA in February, 1992. We visited the scientific centers of the National Academy of Sciences, the American Physical Society, National Science Foundation, and Pentagon. In the White House we met President's Advisor on Science and Technology A. Bromley. The visit resulted in the first cooperation programs with the American laboratories and enterprises.
> Negotiations with people from the industry and government of the South

At the Workshop "Fullerenes and Atomic Clusters," St Petersburg, 1995.

Korea started in the late 1980s, and the first agreements were concluded in 1991. Simultaneously with this, business relations were established with Chinese scientific institutions.

Not every proposal was accepted. We had to refuse our consent to those suggestions which were beyond the scope of our main fields of interest and the contracts with insufficient funding.

The Institute's authorities and heads of laboratories encountered very complicated problems. In addition to the traditional participation in the international scientific investigations, applied work had to be carried out and technological procedures were to be elaborated and handed over to the customers. Owing to this, the Institute got the finances necessary for continuing its research activities. In 1992–1993, I. V. Grekhov visited the USA and China and concluded some agreements. The laboratories headed by I. S. Tarasov, P. S. Kop'ev, S. A. Gurevich, and others were engaged in investigations in the framework of the agreements with scientific centers, universities, and industrial enterprises.

Another important problem encountered by the Institute during that period was the problem of brain-drain. Before the beginning of 1998, more than a hundred of the Ioffe scientists had been abroad for long periods for scientific work. Typically they were experienced and active researchers having scientific degrees. Earlier, just these scientists had played the most active role in the education of future scientific generations, implementation of the ideas, and laying foundations for new scientific ideas.

The international cooperation, grants of international foundations, and contracts made it possible to carry out the investigations which have not been funded by the state for a long time. More than $1 500 000 have come from 101 grants of the Soros Foundation.

Director of the Institute, Academician Zh. I. Alferov delivers a lecture on the advances in physics in the 20th century.

A pupil of the Lyceum in a laboratory.

The building of the Educational Center of the Ioffe Institute is being constructed.

Twenty eight INTAS grants have provided nearly \$600 000 for equipment and salaries. The grants of the International Science and Technology Center obtained in the framework of the conversion program have allowed us to design and begin the construction of a tokamak of a new type. Various laboratories of the Institute carry out investigations in the framework of more than fifty contracts with foreign centers and enterprises. Every year the Institute arranges several international conferences in different areas of physics.

Under the conditions of the falling prestige of science in Russia, the problem of improvement of the Institute's system for continuous education in physics has become of major significance. Along with maintaining the traditional forms of probation, the Director has succeeded in maintaining and developing a new system which includes the Lyceum "Physico-Technical School", the Faculty of Physics and Technology at the St Petersburg Technical University, and the Chair of Optoelectronics at St Petersburg Electrotechnical University. Schoolchildren and students attend lectures of the leading scientists of the Institute and work actively at the Institute's laboratories. The situation when a young specialist joining the Institute after having graduated from a Institute-based chair has already published several papers in the leading scientific journals is quite typical (every year nearly 50 young specialists join the Institute). In 1998, the building for the Educational Center where the pupils of the Lyceum and students will study and begin their scientific activities will be ready. In the lecture halls and laboratories, they will learn the fundamentals of the experimental physics and get the initial knowledge necessary for more advanced studies in special subjects and for the scientific work at the Institute's laboratories. The halls of the new building will allow arrangement of large-scale scientific conferences in future.

To feel the Institute's life, one needs to walk along its corridors, work in its laboratories and library, attend seminars. No paper, chapter, or book can replace this "communion to the flesh and blood of science." You are welcome to come, watch, and work. And we shall believe that our task in writing this paper is fulfilled if the reader is eager to exclaim together with us (following Mark Twain): “Gentlemen! The reports on our death are exaggerated.” The 80-year-old Physico-Technical Institute is alive and continues its active work!

Part III

Scientific Reviews on Advanced Areas

Theory of semiconductors

E. L. Ivchenko and V. I. Perel

The development of the theory of semiconductors at Physico-Technical Institute was initiated by classical works by Ya. I. Frenkel', B. I. Davydov, and I. M. Shmushkevich. L. D. Landau started his scientific activities at this Institute as well. In the 1950s–60s the sectors involved in the solid state theory at Physico-Technical Institute and the Institute of Semiconductors were headed by A. I. Ansel'm, A. I. Gubanov, and L. E. Gurevich. Their contribution to the physics of semiconductors deserves a separate review which could be very instructive. In this review we make an attempt to represent the basic ideas and results belonging to their "children, grand-sons, and great grand-sons in science". The works related to the theory of metals are also included to such an extent to which they concern degenerate semiconductors as well. To diminish the reference list we tried to cite when possible books and review articles containing a detailed bibliography.

Symmetry. An effective knowledge of methods based on symmetry analysis implies not only professional abilities of the researcher, but to some extent is close to art. Indeed, a non-specialist is often surprised by the fact how many conclusions can be made based on minimal information about the symmetry of a system. A number of fundamental results concerning the method of invariants, the core of the group theory analysis, belong to G. L. Bir and G. E. Pikus [1]. Their formula for deformational contribution to the Luttinger Hamiltonian, corresponding to the Γ_8 band in semiconductors with diamond and zinc blende structure, is well known. The deformational potential constants a, b, and d are now commonly accepted and used in original papers, monographs, and handbooks.

Waves. Studies of waves of different types in semiconductors and metals constitute an important part of works by theorists of the Institute.

O. V. Konstantinov and V. I. Perel' (and independently of them Egren) were the first to demonstrate in 1960 the possibility of electromagnetic waves (helicons) propagating in metals in strong magnetic fields. A detailed investigation of the wave properties and electromagnetic excitations of other sorts in metals was carried out jointly by V. G. Skobov and E. A. Kaner [2]. Later a great number of experimental and theoretical papers were concentrated on this topic.

The giant quantum oscillations of sound absorption as a function of magnetic field, predicted by V. L. Gurevich, V. G. Skobov, and Yu. A. Firsov [3], became an important source of information about the Fermi surface structure and have been a subject of intensive experimental studies. The oscillations originate from the Landau quantization of the energy spectrum of electrons responsible for the absorption and also occur in the case of absorption of slow electromagnetic waves [4].

Of particular interest is a cycle of papers concerned with non-linear theory of absorption, amplification, and generation of ultrasonic waves in semiconductors and metals via drift of electrons in electric field (Yu. M. Galperin, S. V. Gantsevich, V. L. Gurevich, V. D. Kagan, V. I. Kozub, B. D. Laikhtman). The amplification takes place when the drift velocity of electrons in external electric field is larger than the sound velocity. Studies of electron coupling with sound waves were recently continued for semiconductor nanostructures. The oscillations of a dc. current originated due to electron drag by the ultrasonic wave as a function of the gate voltage were predicted [5].

The theorists of the Ioffe Institute predicted the existence of completely new waves in semiconductors: thermomagnetic [6], recombination [7], and trap recharge waves [8], and investigated conditions for their autogeneration. All these waves were later found experimentally. The recharge waves were observed in photorefractive crystals [9]. Stimulated light scattering by these waves was studied by A. S. Furman [10].

V. A. Kiselev obtained important results concerning the theory of additional light waves in thin semiconductor films [11]. The results allowed to demonstrate for the first time the existence of additional waves at the exciton resonance region in a direct rather than indirect way, based on optical reflection spectra for thin CdSe films. E. L. Ivchenko and V. A. Kosobukin generalized the theory of exciton polaritons to periodical structures of quantum wells and superlattices including resonant Bragg structures [12, 13].

In the recent years it has been shown [14] that the hydrodynamical approach can be under certain conditions applied to description of the electron liquid in a field effect transistor channel. It was found that the main equations of the theory are completely similar to those describing waves in shallow water. In this system certain boundary conditions allow excitation of two-dimensional plasmons in the terahertz region ("electron flute"). This behavior has not been observed yet.

Transport. One can characterize the magneto-phonon resonance predicted by V. L. Gurevich and Yu. A. Firsov (see [15]) as a well-known phenomenon. The resonance takes place every time when a distance within some pair of Landau levels becomes, in the course of variation of magnetic field, equal to the optical phonon energy. In this case the probability of electron–phonon scattering in the semiconductor has a maximum. The maxima of the scattering probability manifest themselves experimentally as maxima of transversal magnetoconductivity. Until now this phenomenon has been observed in tens of semiconductor compounds while the magnetophonon spectroscopy has become one of the most effective methods for studying the energy spectra of semiconductors. A contribution to the theory of conductivity in quantizing magnetic field was also made by M. I. Klinger [16].

There are crystalline materials where even at high temperatures the carrier mobility increases with temperature by an activation law. Thus, in a perfect (not disordered!) crystal there is a possibility of hopping transport of carriers between the lattice sites. This is related to a strong coupling of carriers with lattice vibrations (phonons): a carrier is "dressed" in a phonon "coat" and forms a small polaron. A need has arisen to give a perfectly new description of transport in the lattice (discrete) **r**-space rather than in **k**-space like in the Boltzmann transport equation. This program has been realized by Yu. A. Firsov, V. V. Bryksin, E. K. Kudinov, and I. G. Lang. The theory they have developed describes hopping transport, tunneling under barriers, and movement within a narrow polaron band. The Hall effect is in this case controlled by a magnetic flux passing through closed trajectories formed by forward and backward chains of polaron transitions between the sites. This result was obtained as early as 1963, much earlier than a similar mechanism was considered in the theory of weak localization. Optical absorption was studied as well, which allowed interpreting the data for TiO as a first experimental evidence to support the existence of small polarons. These papers were reviewed in [17, 18].

In materials with narrow allowed bands the electron spectrum in the presence of an external electric field has the form of a Stark staircase; the wavefunctions in the electric field direction are localized and the transport has a form of phonon-assisted hopping even without the polaron effect. The transitions acquire a resonant character if energy spacing between some of the Stark levels is a multiple of the optical phonon energy. Such a non-monotonous current dependence on the electric field was predicted in [19] and was called electrophonon resonance. The effect was observed in ZnS crystals. The giant electrophonon resonance was observed in some artificial superstructures based on quantum wells and zeolites.

In recent years the electrophonon resonance in semiconductor quasi-1-dimensional nanostructures has been considered. When the spacing within a pair of neighboring 1-dimensional subbands of lateral quantization coincides with the optical phonon energy, and the bias applied to the structure exceeds this energy, generation of optical phonons is possible, resulting features in the current–voltage curve [20].

O. V. Konstantinov *et al.* [21] showed that at high enough density of the surface states fixing the Fermi level in 2D system there is a step-like magnetic field dependence of the Hall voltage, typical of the quantum Hall effect. A similar factor can lead to step-like behavior of the capacitance of a metal-semiconductor superlattice structure as a function of devastating bias applied to the superlattice [22]. The conductivity of a 2D-system with electron–phonon scattering was studied for a situation when electron–phonon coupling cannot be considered as a perturbation [23].

One should also mention investigations of the ballistic charge transport through nanostructures [24].

Fluctuations. Semiconductors in equilibrium obey general Callen–Welton relations expressing the current fluctuations in terms of conductivity. In the absence of equilibrium, in particular, in the presence of an electric field or under illumination no such universal relation is available. This problem was studied by S. V. Gantsevich, V. L. Gurevich, and R. Katilius. In particular, it was established that there exists a specific correlation of fluctuations due to electron–electron interactions in the non-equilibrium state [25]. These works were complemented by a cycle of papers concerned with the theory of shot noise in semiconductor nanostructures [26]. Another cycle [27] suggested a new source of mesoscopic fluctuations in a 2D quasiballistic channel related to correlations between successive acts of boundary scattering. At high enough temperatures this mechanism dominates over the Universal Quantum Fluctuations. M. E. Raikh and I. M. Ruzin have studied fluctuations of transparency of non-homogeneous potential barriers [28].

Interference of electron waves in semiconductors and metals. This effect underlies the weak localization phenomenon which has been studied extensively in recent years. The first evidence of its role in semiconductors was obtained in a work concerned with resonant light scattering within the exciton resonance band in GaSe [29]. An exciton created resonantly by incident light can emit light in opposite direction only changing its momentum to the opposite one. Such a change can take place in the course of multiple elastic scattering of the exciton by several impurity centers. The two waves representing the same exciton and passing these centers in opposite sequences have the same phases for backscattering and amplify each other due to interference.

A. G. Aronov *et al.* [30] proposed the experiment directly illustrating the role of the interference in semiconductors. In a thin hollow conducting cylinder it is possible to have interference of waves representing the same electron rounding the cylinder in the opposite directions. This leads to oscillations of the cylinder's resistance as a function of external longitudinal magnetic field. The experiment carried out at the Institute for Physical Problems (Moscow) completely confirmed the theory. The interference effects are limited by inelastic processes breaking the wave function phase. A. G. Aronov *et al.* have shown that electron–electron interaction can be an effective mechanism for phase breaking [31].

The role of spin-orbital interaction and corresponding spin relaxation was noted already in the first papers concerned with weak localization. A detailed theory of these effects in reference to 2D-structures was developed by G. E. Pikus *et al.* [32]. It was shown in particular that contributions of different mechanisms of spin relaxation, e.g., caused by the absence of inversion center in the composite material and by the quantum well asymmetry, are not additive and can even compensate each other. N. S. Averkiev *et al.* [33] developed a theory of weak localization for a situation when the spin-orbital interaction cannot be considered weak perturbation, the scattering is anisotropic, and more than one subband of lateral quantization is occupied. The magnetic field behavior of magnetoresistance in the case when the backscattering probability cannot be considered in diffusive approximation was studied in [34]. V. V. Afonin *et al.* predicted weak localization in a random field of acoustic phonons [35]; the effect was soon observed in CdS crystals.

Photogalvanic and photoelectric effects. It is known that A. F. Ioffe was always interested in photoelectric effects and was himself involved in the 1930s in studies of photoelectromotive forces. The first theoretical papers concerned with the effect of electron drag by photons considered the photocurrent as a direct Hall current originating in alternating electric and magnetic fields. In the quantum theory developed by A. A. Grinberg the drag current is a result of momentum transfer from photon to electron in the course of light absorption. In addition to the drag effect, two more effects are possible in crystals without inversion center; these effects were called linear and circular photogalvanic effects. The photocurrent excited by circularly polarized light and changing its direction in the course of a change of the polarization sign was predicted by E. L. Ivchenko and G. E. Pikus and was soon observed in tellurium [36]. The second rank pseudotensor describing this effect is non-zero only in gyrotropic crystals where the components of polar and axial vectors transforming according to equivalent representations exist. The theory of natural gyrotropy, or optical activity of semiconductors was developed by E. L. Ivchenko and G. E. Pikus while the Coulomb effects were studied by A. G. Aronov and A. S. Ioselevich. The circular photogalvanic effect and the inverse effect—the current-induced emf—have expanded a circle of phenomena characteristic of the gyrotropic crystals. A review of papers concerned with the photogalvanic effects is given in [36].

A new effect—resonant photocurrent in semiconductors due to a quantum interference—was observed [37]. The theory developed by A. P. Dmitriev showed that the photocurrent is related to interference of direct and indirect optical transitions from localized states to continuous spectrum.

R. I. Lyagushchenko and I. N. Yassievich [38] showed that some experimental data on photoconductivity and photomagnetic effect can be explained taking into account that it is the majority carriers heated in course of their interactions with photoelectrons (or photoholes) which give the main contribution to a photocurrent in doped semiconductors rather than the photoexcited carriers themselves. As a result, the photoconductivity is caused by a change of mobility of the majority carriers due to their heating while the photomagnetic effects comes to Nernst effect related to a gradient of effective electron temperature. On this basis it appeared possible to explain the quantum Shubnikov–de Haas and Gurevich–Firsov oscillations observed for the photomagnetic effect.

Disordered systems. At low temperatures charge carriers in doped semiconductors are localized on impurities. The impurities form a disordered system with a random spatial distribution of localized states and with a significant scatter of binding energies. The calculation of the density of states and transport for such a system is a complicated, but very interesting problem. The contribution of theorists from the Ioffe Institute to the solution of this problem was generally recognized and served as a basis for setting and interpretation of a lot of experiments in our country and abroad. The results of these investigations are summarized in the monograph [39]. We will restrict ourselves to the results most interesting from our point of view.

A. L. Efros and B. I. Shklovskii applied the percolation theory to the problem of hopping conductivity which allowed them for the first time to find the correct dependence of the resistance on donor concentration and degree of compensation for temperatures where the conductivity is described by an activation law.

According to Mott, at low enough temperatures the activation law for the temperature behavior of resistance is replaced by the law $\rho \propto \exp(T_0/T)^{\alpha}$, where $\alpha = 1/4$. This happens because, on the one hand, the conductivity is due to tunnel hops of electrons from occupied donor states to empty ones in a narrow band close to the Fermi level and, on the other, the lower the energy of the impurity–impurity transition, the larger the characteristic hopping length. Mott has assumed that the density of states near the Fermi level is constant and does not depend on energy. A. L. Efros and B. I. Shklovskii showed that as a result of electron–electron interactions the density of states near the Fermi level is devastated and vanishes at the Fermi level. (A "soft" gap which was called the Coulomb gap.) This fact changes the exponent α in the Mott law and allows one to relate the temperature T_0 to the electron localization radius. Now the concept of the Coulomb gap is evidenced by a great number of experiments and is a conventional one. It has a general character and is valid for all disordered systems with localized electrons, e.g., for the situation of the quantum Hall effect [40].

The presence of randomly situated impurities in partly or strongly compensated semiconductors leads to "tails" of density of states in the forbidden gap which manifests itself as a smearing of the fundamental absorption edge. A. L. Efros and B. I. Shklovskii studied this effect in detail.

One of characteristic features of disordered materials is a strong frequency dispersion of conductivity at small frequencies. V. V. Bryksin studied this problem with the help of the effective medium approach and derived a universal law [41] which excellently describes the frequency dependence of conductivity for very different materials—from doped semiconductors to transition metals oxides. The origin of the frequency dispersion for the disordered materials is basically different from the one described by the Drude–Lorenz formula. Indeed, a direct current flows along optimal curved trajectory prescribed by the percolation theory. At non-zero frequencies a displacement current starts to be of importance in the trajectory segments with the lowest values of local conductivity. As a result the average conductivity increases with frequency and becomes complex. G. Boettger and V. V. Bryksin also studied nonlinear effects in the framework of the percolation theory. It was shown in particular that the average conductivity decreases with increasing voltage [42]. This effect was observed for doped semiconductors, amorphous oxides, and other materials. An impressive nonlinear effect in the hopping conductivity is the current instability stipulated by the declining part of current–voltage curve [43], predicted in the paper [44]. The effect originates due to pressing of electrons by electric field to dead ends of the percolation cluster where they take no part in the conductivity. The nonlinear high-frequency conductivity was studied in [45].

Glasses. The discovery of glassy semiconductors by B. E. Kolomiets and N. A. Goryunova at the Institute in the mid-1950s stimulated the development of a corresponding theory. The first monograph summarizing the theoretical studies of the Institute researchers up to 1963 was the book by A. I. Gubanov [46]. It introduced an important concept of "fluctuational local states" splitting from the allowed band due to the disorder and having sizes exceeding the interatomic distance.

Since P. Anderson showed that the effective interactions of electrons in glassy semiconductors can become attractive due to smallness of local stiffness coefficient, the theory of soft interatomic potentials has been developed by M. I. Klinger and V. G. Karpov [47] (see also a review article [48]). V. L. Gurevich *et al.* have shown that a coupling with soft vibrational modes causes some features of light scattering in dielectric glasses, and also features of thermal conductivity temperature behavior observed in some glasses (see review articles [49]).

The monograph [50] summarizes the results of theoretical studies carried out at the Institute, concerning impurity and defect states in glassy semiconductors and electrical instabilities in these materials. The analytical solution of the problem of a current filament formation for exponential temperature dependence of conductivity obtained by B. L. Gel'mont and K. D. Tsendin is given as well as the results by K. D. Tsendin evidencing a decisive role of hybridization of defect wave functions and wave functions of surrounding atoms in the appearance of a soft atomic potential and negative effective Hubbard energy. On the basis of the statistics developed by K. D. Tsendin it was shown that doped glassy semiconductors are self-compensated, i.e. the number of active impurity centers coincides with a high degree of occurancy with the number of intrinsic defects (dangling bonds).

Optical orientation of electrons and nuclei. In the course of interband absorption of circularly polarized light the angular momentum of photons is transferred to photoexcited electrons. This leads to optical orientation of electron spins observed by Lampel in France in 1968 and later extensively studied at the Ioffe Institute. As a result of energy relaxation the photoexcited electrons quickly relax to the bottom of the conduction band. The degree of its spin orientation under stationary illumination is rather high, because spin relaxation time is typically of the order of, or even larger than the lifetime with respect to recombination. Repeated application of the selection rules to interband radiative recombination leads to a conclusion that the photoluminescence also appears to be circularly polarized (at least partly).

A detailed theory of optical orientation of free carriers has been developed [51]. The interaction of spin-polarized electrons with the crystal lattice nuclei has also been investigated in a cycle of works carried out jointly with experimentalists of the Institute. With the help of hyperfine interaction the oriented electrons align the nuclear spins while the latter in their turn affect the electron spins. This inverse effect can be considered to be an effect on the electrons of an effective magnetic field proportional to the average nuclear spin. Thus, there is a possibility to make conclusions about the nuclear orientation from the polarization properties of photoluminescence, and in particular to register nuclear spin resonance (NSR) by optical methods. In contrast to standard methods the optical registration of electron spin resonance and NSR needs no strong magnetic fields and very low temperatures, because the magnetic field must only split the spin sublevels while their selective occupation under the optical orientation of electrons takes place with no effect of the external field. The optical orientation of electrons and nuclei was a subject of detailed experimental studies.

In other words the appearance of nuclear polarization in course of interactions with the oriented electrons in a magnetic field is a result of cooling of nuclear spin system. This system can be considered as being in thermodynamic quasiequilibrium and isolated from lattice vibrations because spin–spin interaction of nuclei is much stronger than spin–lattice coupling. The existence of strongly coupled electron and nuclear spin systems leads to many nonlinear effects like hysteresis, self-supported oscillations, etc. In this field the main contribution to the theory was by I. A. Merkulov (see a review article [52]).

Optical orientation and alignment of excitons. The first experiment on optical orientation of excitons was made by E. F. Gross *et al.* on CdSe in 1971. At the same time this was also the first observation of optical orientation in a semiconductor with non-cubic symmetry. G. L. Bir and G. E. Pikus noted in 1972 that along with the orientation of angular moments of excitons by circularly polarized light it is possible to align their oscillating dipolar moments in the course of resonant pumping by linearly polarized light. The alignment creates on exciton state in which the average spins of electron and hole vanish, but there exists spin correlation. Owing to this fact linear polarization exists only for geminate recombination. But if the exciton is formed by coupling of electrons and holes created at different acts of photon absorption, the memory concerning the linear polarization does not exist.

In the Faraday geometry, magnetic field leads to circulation of the exciton oscillating dipolar moment and suppresses the linear polarization (the Hanle effect). For bound excitons the depolarization is described by monotonous Lorenz dependence on the Larmour frequency. The picture becomes more complicated for resonant excitation of free excitons: since the directions of exciting beam and registered secondary radiation are different, the exciton created with a wave vector of initial photon suffers before the act of radiation one or more acts of momentum relaxation. A consistent analysis taking into account such a cascade process allowed explaining the change of sign of linear polarization related to exciton radiation in longitudinal magnetic field, observed for GaSe crystals [29]. E. L. Ivchenko, G. E. Pikus, and L. V. Takunov calculated the dependence of the degree of radiation polarization on the number of the LO-phonon replica for optical excitation of hot excitons. A review of works concerned with optical orientation of excitons in bulk semiconductors can be found in [53, 54].

In the last decade polarizational optical spectroscopy has confirmed its efficiency in investigations of the fine structure of exciton levels in low-dimensional semiconductor systems. A vast body of experimental data concerning the fine structure of quasi-zerodimensional excitons has been obtained, the excitons of this type include excitons localized in quantum wells or superstructures as well as electron-hole pairs excited in quantum dots. According to experiment there is, as a rule, rather small splitting of the radiation doublet corresponding to quasi-1D exciton into two linearly polarized sublevels. However, this splitting is large in comparison to the natural width of the sublevels and manifests itself in experiments on exciton optical orientation, quantum spin beats, optically detected magnetic resonance, and anti-crossing of exciton sublevels in magnetic field. In a series of joint papers by E. L. Ivchenko, I. L. Aleiner, S. V. Goupalov, A. V. Kavokin, A. Yu. Kaminskii, and U. Rössler (the results published in a book [12] and in review papers [55, 56]) the theory of electron-hole exchange interaction in a quasi-zero-dimensional exciton was developed based on the quarter-of-century old results by G. E. Pikus and G. L. Bir [1] concerning bulk excitons and explaining the splittings in question. In the course of development of the theory, generalized boundary conditions for hole envelope wave function were derived in reference to GaAs/AlAs superlattices and a relation between coefficients appearing in these boundary conditions and microscopic parameters of the tight-binding method was established for the first time.

Momentum alignment. B. P. Zakharchenya, V. I. Zemskii, and D. N. Mirlin found in 1977 that the hot photoluminescence in GaAs crystals excited by linearly polarized light appears to be partly polarized. The question arose in what way photoelectrons preserve memory concerning the linear polarization of light. The answer was given by V. D. Dymnikov, M. I. Dyakonov, and V. I. Perel: the interband excitation creates electrons with anisotropic momentum distribution. Experimental studies of the optical alignment of electron momenta, its destruction by magnetic field, its dependence on energies of exciting photons, as well as on doping level and temperature allowed for the first time a study of the energy of hot electrons and momentum relaxation with characteristic time scale in the subpicosecond range by direct methods. A significant contribution to these investigations was made by Institute's theorists D. N. Polyakov, Yu. T. Rebane, I. N. Yuassievich, and M. A. Alekseev. The main results are summarized in a review paper [57].

In recent years the methods of optical alignment were for the first time applied to studies of quantum wells and superlattices (see, e.g., [58]).

The anisotropy of carrier momentum distribution manifests itself in the polarization of radiative recombination and fundamental absorption accompanied by tunneling of carriers in semiconductors. In [59] the experimentally observed polarization of radiation in so-called diagonal tunneling in p-n junction [60] was explained for the first time. The effect is related to the fact that recombination tunneling processes are dominated by electrons and light holes with momenta normal to the barrier. The inverse effect, i.e., anisotropy of light absorption is observed in the situation of Keldysh–Franz effect; the corresponding theory was reported in [61].

Spin relaxation. For a long time it was believed that the main mechanism of free carrier spin relaxation in semiconductors without paramagnetic centers is the Elliot–

Yafet mechanism related to spin-orbital interaction in the course of electron scattering by phonons and crystal static defects. Institute's theorists have contributed to studies of this mechanism (S. T. Pavlov, Yu. A. Firsov, V. N. Abakumov, I. N. Yassievich).

Investigations of the optical orientation extended significantly the possibilities to study spin relaxation. In this connection two new mechanisms that can dominate in some circumstances were suggested at the Ioffe Institute. One of them—the mechanism proposed by Bir, Aronov, and Pikus—plays the main role in p-type semiconductors at high enough hole concentration and takes place if an exchange interaction in the course of nonequilibrium electrons collisions with holes is taken into account. The second mechanism, known as the Dyakonov–Perel mechanism, is related to features of electron band structure in semiconductors without inversion center. The effective Hamiltonian for such semiconductors contains terms which are odd with respect to the wave vector and spin-dependent; these terms can be interpreted as coupling of the electron spin with a momentum-dependent internal magnetic field. It is this field which is responsible for the spin relaxation. Fast chaotic changes of the field lead to its dynamical averaging. As a result, the spin relaxation time, appears to be inversely proportional to the momentum relaxation time rather than proportional to it, as is the case for the Elliot–Yafet or Bir–Aronov–Pikus mechanisms. As it was shown by E. L. Ivchenko, the Dyakonov–Perel mechanism of spin relaxation is effectively suppressed by external magnetic field due to cyclotron rotation of electron momenta. M. I. Dyakonov and V. Yu. Kachorovskii considered this mechanism for quantum well structures [62]. A review of works concerning spin relaxation is given in [63].

Light scattering. The development of experimental technique for optical studies of semiconductors opened up possibilities to study different processes of resonant light scattering for excitation into the fundamental absorption band. As an example one can mention the observation of higher order lines (up to 7th–9th order!) in the scattered light spectrum, the order corresponding to the number of emitted optical phonons with weakly diminishing cross-section. A significant contribution to experimental studies of multiple-phonon resonant Raman light scattering was made by researchers of the laboratory of E. F. Gross at the Institute. To describe these processes which—as it should be emphasized—obey neither the standard theory of non-resonant light scattering nor the photoluminescence theory, the general theory of scattered radiation in semiconductors and insulators has been developed [64, 65]. An important feature characterizing these processes is that the role of intermediate states is played by real states like electron-hole pairs or excitons (rather than by virtual states as in non-resonant processes). The general theory was successfully applied to theoretical description of resonant Raman scattering of different types. It allowed in particular predicting a sharp increase of the cross-section of multiphonon light scattering in strong magnetic field [66]. The reason is that the magnetic field reduces the volume occupied by an electron-hole pair and thus increases the scattering probability. The predicted effect was soon observed experimentally.

Experimental investigation of resonant light scattering with a flip of the hole spin $\pm 3/2 \to \mp 3/2$ at the acceptor in GaAs/AlGaAs quantum well structures of p-type posed to the theory a set of extremely difficult questions. From the point of view of angular momentum conservation the backscattering should be forbidden in the Faraday geometry because the projection of the photon moment to the direction of the principal axis of a structure is not changed or is changed by ± 2 while the projection of the hole momentum is changed by ± 3. E. L. Ivchenko proposed two original scattering mechanisms which lift the restriction mentioned above: (i) exchange interaction of the "flip-stop" type between a hole within localized exciton and a hole on acceptor nearby and (ii) scattering assisted by additional acoustic phonon which causes a spin-flip for an electron in the complex "exciton on neutral donor" excited by light [12]. The scattering behavior in inclined magnetic field, predicted by the theory, was confirmed by a specially performed experiment [67].

A sharp, almost threshold-like increase of intensity of resonant single-phonon light scattering in quantum well structures in magnetic field was observed in the work by D. N. Mirlin *et al.* A transparent quasi-classical description of this effect was proposed by I. A. Merkulov and V. I. Perel [68]. The event of scattering as a whole can be considered as a result of three virtual processes: creation of an electron-hole pair, emission of an acoustic phonon, and recombination of the pair. Since at the moment of creation the electron and the hole have oppositely directed momenta and their recombination is possible only at distances of the order of the de Broglie wavelength, the time for phonon emission without magnetic field is very small. In magnetic field they move after the creation along the same circumference in opposite directions and can recombine on meeting after the period $t = 2\pi/\omega_c$, ω_c being a sum of electron and hole cyclotron frequencies. In this case the scattering intensity is proportional to the probability for the pair to survive till the meeting, that is it increases exponentially with the magnetic field according to the law $\exp -t/\tau$, where τ is the carrier lifetime on the initial orbit, governed mainly by emission of optical phonons.

V. L. Gurevich *et al.* predicted collision-controlled light scattering by charge carriers in semiconductors [25].

Experimental studies of the Raman light scattering by free carriers carried out at the Ioffe Institute by B. H. Bairamov and V. V. Toropov revealed an anomalous narrowing of the spectral line with increasing carrier concentration. I. P. Ipatova *et al.* have shown [69] that the effect originates from dynamical averaging of the Doppler broadening caused by the scattering in the situation when the carrier mean free path becomes smaller than the light wavelength; so it is a solid state analog of collisional narrowing of the scattering line in gases first considered by Dicke. I. P. Ipatova, A. V. Subashiev, and V. A. Voitenko predicted a new mechanism of light scattering by free carriers: scattering by fluctuations of spin density of holes with spin $3/2$ [70].

Semimagnetic semiconductors. This is a name for semiconductor compounds containing a large amount of rare earth atoms or transition metal atoms. Solid solutions based on A_2B_6 semiconductors demonstrate strong magnetooptical effects caused by an exchange interaction of charge carriers with electrons of partially occupied d-shells of the magnetic ions. Polarization of d-electrons spins by external magnetic field leads to a giant exchange splitting of electron and hole spin sublevels exceeding by two orders of magnitude the standard Zeeman splitting. In its turn, the localized carriers produce a significant field (about 1 T) at adjacent magnetic ions, which leads to formation of magnetopolaron states.

Theory of the magnetic polaron bound on an acceptor was developed by B. L. Gelmont and I. A. Merkulov with co-workers [71]. A. V. Kavokin has shown that at some conditions a spontaneous breaking of symmetry can happen for polaronic state in a symmetrical quantum well with semimagnetic barriers, with the magnetic polaron pulled into one of the walls [72].

A detailed analysis carried out by I. A. Merkulov and K. V. Kavokin allowed them to explain anomalies in magnetic field dependence of linear polarization of luminescence and a huge number (more than 15!) of peaks in the Raman scattering spectra observed in semimagnetic quantum wells in the Foight geometry [73]. The distance between the peaks is equal to the splitting of magnetic ion spin sublevels in magnetic field. This effect, a unique optical analog of electron paramagnetic resonance allows a transparent semiclassical description taking into account that (i) the equilibrium spin of a magnetic ion is oriented by an external field in the interface plane and (ii) spin of photoexcited hole in a quantum well is almost always normal to this plane. At the moment when a localized hole is created, the total spin of ion surrounding the hole starts to be affected by the exchange field normal to this spin. Precession in the exchange field changes the projection of the ion spin on the external field direction. According to the rules of the quantum mechanics, this projection in the final state, that is after hole recombination with a secondary photon emission, should be quantized, which manifests itself as a set of peaks in the scattering spectrum.

Non-radiative recombination and thermal generation of charge carriers. As a rule, the non-radiative recombination in semiconductors occurs through a capture of carriers by impurity centers. M. Lax showed that the main mechanism of electron and hole capture by shallow donors and acceptors in Ge and Si is cascade capture via excited states accompanied by emission of low frequency acoustic phonons. V. N. Abakumov and I. N. Yassievich improved the Lax theory which allowed them to explain a lot of experimental data accumulated over many years.

Capture by deep centers is usually accompanied by excitation of local vibrations. If the temperature is not too high, this occurs due to a tunneling transition of the vibrational system between adiabatic terms corresponding to bound and free electron states.

The inverse process—thermal generation of carriers from deep centers—also occurs through tunneling. The effect of electric field on this process is of special interest. It was long believed that the main role here is played by the Frenkel–Pool effect related to a decrease by the electric field of the potential barrier for an electron to be thrown away from a charged center. V. Karpus and V. I. Perel showed that when the field is not too weak the field dependence of the generation rate is controlled by another mechanism related to a competition between the two tunneling transitions: electron tunneling from a bound state to a free one and vibrational system tunneling between the adiabatic terms. As a result it has become possible for the first time to explain the existing experimental data.

Recently, the theory of thermally activated tunnel generation was generalized by I. N. Yassievich and V. I. Perel to the case of ac electric field with a help of L. V. Keldysh's results concerning tunnel ionization of a deep center by high-frequency field. This allowed them to explain a lot of experiments carried out in recent years concerning ionization of deep centers by electromagnetic field in teraherz and far IR range and to obtain from these experiments important information about deep centers in semiconductors [74].

At high enough free carrier concentration the main contribution to the non-radiative recombination is made by Auger processes. The complicatedness of Auger recombination probability calculations is related to a necessity to calculate overlapping integrals between the Bloch amplitudes corresponding to different bands. B. L. Gelmont, Z. N. Sokolova, and I. N. Yassievich carried out such a calculation for A_3B_5 semiconductors where electron states are well described by the Kane model. In bulk semiconductors the Auger recombination is effectively suppressed because of the energy threshold related to energy and momentum conservation laws. In heterostructures the momentum component normal to interfaces is not conserved. Thus, the energy threshold is eliminated which increases the Auger recombination rate. The corresponding calculations were performed by G. G. Zegrya and V. A. Kharchenko and also by M. I. Dyakonov and V. Yu. Kachorovskii.

S. V. Kozyrev and A. Ya. Shik showed that the probability for an electron to be captured in quantum well oscillates as a function of the well width. The oscillation maxima correspond to the appearance of a resonant level on the edge of the continuous spectrum.

A review of works on the theory of non-radiative recombination in semiconductors is given in [75].

Impurities and defects. B. L. Gelmont and M. I. Dyakonov developed the theory of acceptor level in semiconductors with a degenerate valence band. Then the theory was extended to the case of gapless semiconductors where an acceptor produces a quasidiscrete level on the background of the conduction band continuous spectrum (see the review article [76]). An effective method was developed for calculation of a hole-on-acceptor wave function in **k**-representation taking into account the warped constant energy surface of the valence band and the central cell potential [77].

Yu. E. Rebane *et al.* developed the theory of electron and hole states bound at dislocations. It was found in particular that in crystals with linear topological defects there exist topological gauge fields localized at the defect cores but significantly affecting carriers in the crystal causing effects of the Aharonov–Bohm type [78].

The theory of anharmonic broadening of light absorption and scattering lines by lattice vibrations and quasilocal impurity vibrations was developed by I. P. Ipatova, A. A. Klochikhin, and A. V. Subashiev. An explanation was given to experimentally observed decrease of the integral intensity of impurity absorption line for hydrogen and deuterium in ionic crystal with increasing temperature [79]. The results are especially important in relation to the problem of crystal surface passivation by hydrogen and deuterium [80].

Crystal growth. Epitaxial growth of layers by molecular beams or from metalloorganic compounds is one of the most important technological processes in modern opto- and microelectronics. It is a matter of control of layers with a thickness of several lattice constants. Moreover, in the course of this process it appears possible to grow structures with quantum dots which are objects of the same spatial scale; this opens up completely new prospects. The problem of nucleation in the course of epitaxial growth is very important in this respect; it has been studied by R. A. Suris with co-workers. A_3B_5 semiconductors and silicon having chemical bands of tetrahedral coordination, their surfaces appear to be anisotropic. This fact leads to a strong anisotropy of the critical nucleus form and substantially affects its free energy [81], thus at high levels of oversaturation the growth can take place through alignment of adatom rows along the chemical bond direction. A new stable mode of epitaxial growth (kink-flow mode) was considered for the first time; it allows formation of quantum dots in a controlled way [82]. The evolution of the distribution function of nuclei in terms of their size has been studied theoretically [83]. It was shown that the distribution preserves a memory of its initial form in the course of epitaxial growth.

As for a study of the problem of stability of semiconductor solid solutions a significant progress is associated with papers by I. P. Ipatova, V. A. Shchukin, A. Yu. Maslov, and V. G. Malyshkin [84] (see also a review paper [85]), the problem being of fundamental importance for working out of a technology allowing spontaneous formation of nanostructures. A theory of thermodynamic and kinetic stability of epitaxial films with respect to composition fluctuations was developed. Since the difference of energies corresponding to the mixing components is small, the dominating role in the decomposition is played by the anisotropic elastic energy. A strained periodic structure of elastic domains arises in the case of the thermodynamic instability (spinodal decomposition). A surface soft mode decaying exponentially in the bulk of epitaxial film is responsible for the transition. Domains of this sort have been observed on many occasions.

Earlier it was believed that quantum dots should be ordered in a triangular lattice. However, a theoretical study by V. A. Shchukin has shown that the quadratic structure appears to be preferable. This result has been proved experimentally [86].

A program of numerical simulations was developed for selecting the optimal growth regime for GaAs in a chloride transport system [87].

Lasers. Fabricating and study of semiconductor lasers is one of the main directions of research at the Ioffe Institute. Theorists of the Institute joined in investigations in this direction already at the early stage (see a review of the earliest papers in the book by G. E. Pikus [88]).

The modern optoelectronics is unthinkable without heterolasers. The concept of double heterostructure lasers (DHS lasers) was first put forward by Zh. I. Alferov and R. F. Kazarinov in their pioneering work on this subject in the beginning of 1963 [89]. A full list of advantages offered for different devices, especially lasers and high power rectifiers—superinjection, electron and optical confinement was given by Zh. I. Alferov in 1966 [90]. At the same time the theory of superinjection was developed first for abrupt heterojunctions [91] and later for a graded heterojunction [92].

Lasers with distributed feedback are the main element in modern optical fiber communication lines. These lasers were first proposed (Zh. I. Alferov, V. M. Andreev, R. F. Kazarinov, E. L. Portnoi, and R. A. Suris, 1971) and fabricated at the Institute, where their detailed theory was developed as well [93].

A remarkable feature of the injection lasers is the possibility of direct and highly effective modulation of emitted light by pumping current. The paper [94] was the first to attract attention to the opposite effect—current modulation in a power circuit of injection laser at the difference frequency of light field beats arising when radiation with a frequency different from one of the generated light is led into the laser cavity. A theoretical analysis of the operation of injection lasers on quantum wells was done in [95] with account of the absence of the Auger-recombination threshold.

In 1971, R. F. Kazarinov and R. A. Suris formulated a pioneer idea concerning a semiconductor laser of a new type, making use of stimulated light emission in the course of sequential tunneling of electrons between layers of a narrow band semiconductor in structures with superlattices, rather than in the course of electron-hole recombination [96]. Far IR lasers based on this idea and called cascade lasers were fabricated about a quarter of century later by the group headed by F. Capasso at AT&T Bell Labs.

A new breakthrough in optoelectronics is related to injection lasers based on nanostructures with quantum dots. These lasers are expected to provide a further decrease of the threshold current and its weak temperature dependence. The low threshold current values ensure the discreteness of carrier spectrum, so the amplification spectral line must be extremely narrow while the amplification coefficient at the line maximum must be large. However, the technological scatter of the quantum dot parameters leads to inhomogeneous line broadening and to a decrease of the amplification coefficient at the maximum. The papers [97] analyze the role of these effects. Since the carrier concentration in the active region of the laser on quantum dots is small, charge neutrality breaking is to take place. As a result, the threshold current increases and starts to depend on temperature strongly. An injection laser on quantum dots was first fabricated at the Ioffe Institute (1994).

To conclude this review, which is far from being comprehensive, we list the important branches of investigations which we could not illuminate in any detail but are not able to discard: theory of electrotransport [98], a cycle of papers concerning the localization and superconductivity in quasi-1D and quasi-2D systems [99], a series of papers giving for the first time a theory of streamer breakdown in semiconductors [100], development of methods to calculate fullerenes [101], charge relaxation in 1D and 2D structures [102], theory of Zeeman effect in low-dimensional systems [103].

We are grateful to V. L. Gurevich, I. P. Ipatova, O. V. Konstantinov, I. A. Merkulov, R. A. Suris, Yu. A. Firsov, K. D. Tsendin, and I. N. Yassievich for assistance.

References

[1] G. L. Bir and G. E. Pikus, *Symmetry and Strain-induced Effects in Semiconductors* New York, Wiley, 1974.

[2] E. A. Kaner and V. G. Skobov, *Adv. Phys.* **69** 605 (1968).

[3] V. L. Gurevich, V. G. Skobov, and Yu. A. Firsov, *Zh. Teor. Eksp. Fiz.* **40** 786 (1961); Yu. M. Galperin, V. L. Gurevich, and V. I. Kozub, *Sov. Phys.: Usp.* **22** 352 (1979).

[4] A. S. Chernov and V. G. Skobov, *Phys. Lett. A* **199**, 245 (1995).

[5] V. L. Gurevich, V. B. Pevzner, and G. Iafrate, *Phys. Rev. Lett.* **77** 3881 (1996).

[6] L. E. Gurevich and B. L. Gel'mont, *Zh. Eksp. Teor. Fiz.* **51** 183 (1966).

[7] O. V. Konstantinov and V. I. Perel, *Sov. Phys.: Solid State* **6** 2691 (1965).

[8] R. F. Kazarinov, R. A. Suris, and B. I. Fuks, *Sov. Phys. Semicond.* **6** 500 (1972); **7** 102 (1973).

[9] S. I. Stepanov, V. V. Kulikov, and M. P. Petrov, *Opt. Commun.* **44** 19 (1982); M. P. Petrov, V. M. Petrov, V. V. Bryksin, I. Zouboulis, A. Gerwens, and E. Kratzig, *Opt. Lett.* **22** 1083 (1997).

[10] A. S. Furman, *Sov. Phys.: JETP* **67** 1034 (1988).

[11] V. A. Kiselev, B. V. Novikov, and A. E. Chernichenko, *Exciton spectroscopy of near-surface layers of semiconductors* (in Russian), Leningrad: Leningrad State University, 1987.

[12] E. L. Ivchenko and G. E. Pikus, in *Superlattices and other heterostructures. Symmetry and optical phenomena* Springer Series in Solid State Sci. **110**, Springer-Verlag, 1995; second ed. 1997.

[13] V. A. Kosobukin, *Sov. Phys. Solid State* **34** 1662 (1992);
V. A. Kosobukin and M. M. Moiseeva, *Phys. Solid State* **37** 2036 (1995).
[14] M. I. Dyakonov and M. S. Shur, *Phys. Rev. Lett.* **71** 2468 (1993); *Appl. Phys. Lett.* **67** (1995).
[15] Yu. A. Firsov, V. L. Gurevich, R. V. Parfeniev, and I. M. Tsidilkovskii, in *Landau Level Spectroscopy*, ed. by G. Landwehr and E. I. Rashba, Elsevier, p. 1181, 1990.
[16] M. I. Klinger, *Fiz. Tverd. Tela* **3** 1342 (1961).
[17] Yu. A. Firsov, *Polarons* (in Russian), Moscow: "Nauka", 1975.
[18] Yu. A. Firsov, *Semiconductors* **29** 515 (1995).
[19] V. V. Bryksin and Yu. A. Firsov, *Zh. Eksp. Teor. Fiz.* **61** 2373 (1971).
[20] V. L. Gurevich, V. B. Pevzner, and G. Iafrate, *Phys. Rev. Lett.* **75** 1352 (1995).
[21] O. V. Konstantinov, O. A. Mezrin, and A. Ya. Shik, *Sov. Phys. Semicond.* **17** 675 (1983).
[22] D. N. Bychkovskii and O. V. Konstantinov, *Semiconductors* **28** 715 (1994).
[23] A. P. Dmitriev and V. Yu. Kachorovskii, *Phys. Rev. B* **52** 5743 (1995).
[24] V. L. Gurevich, V. B. Pevzner , and K. Hess, *Phys. Rev. B* **51** 5219 (1995);
V. L. Gurevich, *Phys. Rev. B* **55** 4522 (1997).
[25] S. V. Gantsevich, V. L. Gurevich, and R. Katilius, *Revista del Nuovo Cimento* **2** 5 (1979).
[26] V. I. Kozub and A. M. Rudin, *Phys. Rev. B* **52** 7853 (1995);
V. L. Gurevich and A. M. Rudin, *Phys. Rev. B* **53** 10078 (1996).
[27] Yu. M. Gal'perin and V. I. Kozub, *Europhys. Lett.* **15**, 631 (1991);
V. I. Kozub and A. A. Krokhin, *J. Phys.: Cond. Matt.* **5** 9135 (1993).
[28] M. E. Raikh and I. M. Ruzin, in *Mesoscopic Phenomena in Solids*, ed. A. L. Altshuler, P. A. Lee, and R. A. Webb, Elsevier, p. 315, 1991.
[29] E. L. Ivchenko, G. E. Pikus, B. S. Razbirin, and A. N. Starukhin, *Sov. Phys. Solid State* **45** 1172 (1977).
[30] B. L. Altshuler, A. G. Aronov, and B. Z. Spivak, *Pis'ma ZhETF* **33** 101 (1981).
[31] B. L. Altshuler and A. G. Aronov, in *Electron-Electron Interaction in Disordered Systems*, ed. A. L. Efros and M. Pollak, North Holland, p. 1, 1987.
[32] W. Knap, C. Skierbiszewski, A. Zduniak, E. Litvin-Staszevska, D. Bertho, F. Kobbi, J. L. Robert, G. E. Pikus, F. G. Pikus, S. V. Iordanskii, V. Moser, K. Zekenes, and Yu. B. Lyanda-Geller, *Phys. Rev. B* **53** 3912 (1996).
[33] N. S. Averkiev, L. E. Golub, and G. E. Pikus, *Zh. Eksp. Teor. Fiz.* **113** 1429 (1998).
[34] M. I. Dyakonov, *Solid State Commun.* **92** 711 (1994);
A. Zduniak, M. I. Dyakonov, and W. Knap, *Phys. Rev. B* **56** 1996 (1997).
[35] V. V. Afonin, Yu. M. Galperin, and V. L. Gurevich, *Phys. Rev. B* **33** 8841 (1986).
[36] E. L. Ivchenko and G. E. Pikus, in *Problems of modern physics* (in Russian), Leningrad, "Nauka", p.275, 1980.
[37] A. P. Dmitriev, S. A. Emelyanov, Ya. V. Terent'ev, and I. D. Yaroshetskii, *Solid State Commun.* **72** 1149 (1989).
[38] R. I. Lyagushchenko and I. N. Yassievich, *Zh. Eksp. Teor. Fiz.* **56** 1432 (1969).
[39] A. L. Efros and B. I. Shklovskii, *Electrical Properties of Doped Semiconductors*, Springer Series in Solid State Sciences, vol. 45, Springer-Verlag, 1984.
[40] D. G. Polyakov and B. I. Shklovskii, *Phys. Rev. Lett.* **70** 3796 (1993).
[41] V. V. Bryksin, *Sov. Phys.-Solid State* **22** 1421 (1980).
[42] H. Boettger and V. V. Bryksin, *Hopping Conduction in Solids*, Akademie-Verlag, Berlin, 1985;
V. V. Bryksin, *Phil. Mag. B* **42** 297 (1980).
[43] D. I. Aladashvili, Z. A. Adamia, K. G. Lavdovskii, E. L. Levin, and B. I. Shklovskii, *Adv. in Disord. Semicond.* vol. 2, *Hopping and Related Phenomena*, ed. by H. Fritzshe and M. Pollak, World Sci., p. 283, 1990.
[44] V. L. Nguyen and B. I. Shklovskii, *Solid State Commun.* **38** 99 (1981).
[45] Yu. M. Galperin, V. L. Gurevich, and D. A. Parshin, in: *Hopping Transport in Solids*, ed. M. Pollak and B. Shklovskii, Elsevier, 1991.
[46] A. I. Gubanov, *Quantum-electronic Theory of Amorphous Semiconductors* (in Russian) Moscow, Ac. Sci. of USSR, 1963.

[47] M. I. Klinger and V. G. Karpov, *Zh. Eksp. Teor. Fiz.* **82** 1687 (1982).
[48] Yu. M. Galperin, V. G. Karpov, and V. I. Kozub, *Adv. Phys.* **38** 669 (1989).
[49] D. A. Parshin, *Phys. Solid State* **36** 991 (1994);
V. I. Kozub and A. M. Rudin, *Phys. Solid State* **38**, 189 (1996).
[50] *Electronic Phenomena in Chalcogenide Glassy Semiconductors* (in Russian), ed. K. D. Tsendin, St.-Peterburg, 1996.
[51] M. I. Dyakonov and V. I. Perel, in *Optical Orientation*, ed. F. Meier and B. P. Zakharchenya, North Holland, p. 11, 1984.
[52] V. G. Fleisher and I. A. Merkulov, *ibid*, p. 173.
[53] R. Planel, C. Benoit a la Guillaume, *ibid*, p. 353.
[54] G. E. Pikus and E. L. Ivchenko, in *Excitons* ed. E. I. Rashba and M. D. Sturge, North Holland, p. 205, 1982.
[55] E. L. Ivchenko, *Phys. Stat. Sol. (a)* **164**, 487 (1997).
[56] E. L. Ivchenko, in *Advances in Solid State Phys.* vol. 37, p. 125, 1998.
[57] D. N. Mirlin, in *Optical Orientation*, ed. F. Meier and B. P. Zakharchenya, North Holland, p. 133, 1984.
[58] V. F. Sapega, V. I. Perel, A. Yu. Dobin, D. N. Mirlin,I. A. Akimov,T. Ruf, M. Cardona, and K. Ebert, *Phys. Rev. B* **56** 6871 (1997).
[59] O. V. Konstantinov, V. I. Perel, and B. V. Tsarenkov, *Fiz. Tekhn. Poluprovod.* **3** 1039 (1969).
[60] Zh. I. Alferov, D. Z. Garbuzov, E. P. Morozov, and E. L. Portnoi, *Fiz. Tekhn. Poluprovod.* **3** 1054 (1969).
[61] L. V. Keldysh, O. V. Konstantinov , and V. I. Perel, *Fiz. Tekhn. Poluprovod.* **3** 1042 (1969).
[62] M. I. Dyakonov and V. Yu. Kachorovskii, *Sov. Phys.-Semiconductors* **20** 110 (1986).
[63] G. E. Pikus and A. N. Titkov, in *Optical Orientation*, ed. F. Meier and B. P. Zakharchenya, North Holland, p. 73, 1984.
[64] E. L. Ivchenko, I. G. Lang, and S. T. Pavlov, *Sov. Phys.-Solid State* **19** 1610 (1977).
[65] I. G. Lang, S. T. Pavlov, A. V. Prokaznikov, and A. V. Goltsev, *Phys. Stat. Sol. (b)* **127** 187 (1985).
[66] V. I. Belitskii, A. V. Goltsev, I. G. Lang, and S. T. Pavlov, *Sov. Phys. JETP* **59** 155 (1984).
[67] V. F. Sapega, T. Ruf, M. Cardona, K. Ploog, E. L. Ivchenko, and D. N. Mirlin, *Phys. Rev. B* **50** 2510 (1994).
[68] D. N. Mirlin, I. A. Merkulov, V. I. Perel, I. I. Reshina, and A. A. Sirenko, *Solid State Commun.* **82** 305 (1992).
[69] B. H. Bairamov, I. P. Ipatova, and V. A. Voitenko, *Phys. Rep.* **229** 223 (1993).
[70] B. H. Bairamov, V. A. Voitenko, and I. P. Ipatova, *Physics Uspekhi* **36** 392 (1993).
[71] Yu. F. Berkovskaya, E. M. Vakhabova, B. L. Gelmont, and I. A. Merkulov, *Sov. Phys.-JETP* **67** 750 1988;
I. A. Merkulov, K. V. Kavokin, G. Mackh, B. Kuhn-Heinrich, W. Ossau, A. Waag, G. Landwehr, and D. R. Yakovlev, *Phys. Solid State* **39** 1859 (1997).
[72] A. V. Kavokin, *Phys. Rev. B* **51** 7613 (1995).
[73] J. Stuhler, G. Schaack, M. Dahl, A. Waag, G. Landwehr, K. V. Kavokin, and I. A. Merkulov, *Phys. Rev. Lett.* **74** 2567 (1998).
[74] S. D. Ganichev, I. N. Yassievich, and W. Prette, *Phys. Solid State* **39** 1703 (1997);
S. D. Ganichev, E. Ziemann, H. Ketterl, W. Prette, I. N. Yassievich, V. I. Perel, I. Wilke, and E. E. Haller, *Phys. Rev. Lett.* **80** 2409 (1998).
[75] V. N. Abakumov, V. I. Perel, and I. N. Yassievich, *Non-radiative Recombination in Semiconductors*, North-Holland 1991; Bezyzluchatelnaya recombinatsiya v poluprovodnikakh, St. Peterburg, A. F. Ioffe Physico-Technical Institute, 1997.
[76] B. L. Gelmont, V. I. Ivanov-Omskii, and I. M. Tsidilkovskii, *Sov. Phys. Uspekhi* **19** 879 (1976).
[77] A. V. Malyshev, I. A. Merkulov, and A. V. Rodina, *Semiconductors* **30** 91 (1996).
[78] Yu. T. Rebane and W. Steeds, *Phys. Rev. Lett.* **75** 3716 (1995).
[79] A. A. Maradudin, E. W. Montroll, G. H. Wiess, and I. P. Ipatova, in *Theory of Lattice Dynamics in the Harmonic Approximation*, Academic Press, New York, p. 708, 1971.

[80] I. P. Ipatova, O. P. Chikolova-Luzina, and K. Hess, *J. Appl. Phys.* **83** 814 (1998).
[81] R. A. Suris and I. L. Aleiner, *Superlatt. Microstruct.* **10** 375 (1991).
[82] I. L. Aleiner and R. A. Suris, *Sov. Phys. Solid State* **34** 809 (1992).
[83] R. A. Suris and A. Yu. Kaminskii, *Solid State Commun.* **89** 697 (1994); A. Yu. Kaminskii and R. A. Suris, *Appl. Surf. Sci.* **104/105** 312 (1996).
[84] I. P. Ipatova, V. G. Malyshkin, and A. Shchukin, *Phil. Mag. B* **70** 557 (1994).
[85] D. Bimberg, I. P. Ipatova, P. S. Kopev, N. N. Ledentsov, I. G. Malyshkin, and V. A. Shchukin, *Physics Uspekhi* **40** 529 (1997).
[86] V. A. Shchukin, N. N. Ledentsov, P. S. Kop'ev, and D. Bimberg, *Phys. Rev. Lett.* **75** 2968 (1995)
[87] V. L. Dostov, I. P. Ipatova, A. Yu. Kulikov, and Yu. V. Zhyliaev, *Semicond. Sci. Technol.* **8** 1935 (1993).
[88] G. E. Pikus, *The Basics of Theory of Semiconductor Devices* (in Russian), Moscow: Nauka, 1965.
[89] Zh. I. Alferov and R. F. Kazarinov, *Author's Certificate* No. 181737, Application No. 950840 with priority from March 30, 1963.
[90] Zh. I. Alferov, *Sov. Phys. Semicond.* **1** 358 (1967).
[91] Zh. I. Alferov, V. B. Khalfin, and R. F. Kazarinov, *Sov. Phys. Solid State* **8** 2480 (1967).
[92] R. F. Kazarinov and R. A. Suris, *Sov. Phys. Semicond.* **9** 6 (1975).
[93] R. F. Kazarinov and R. A. Suris, *Sov. Phys. Semicond.* **6** 1184 (1972); R. F. Kazarinov, Z. N. Sokolova, and R. A. Suris, *Sov. Tech. Phys. Lett.* **1** 90 (1975).
[94] R. F. Kazarinov and R. A. Suris, *Zh. Eksp. Teor. Fiz.* **66** 1067 (1974).
[95] G. G. Zegrya and A. S. Polkovnikov, *Zh. Eksp. Teor. Fiz.* **113** 1491 (1998); *Phys. Rev. B* **58** 4039 (1998); G. G. Zegrya, A. D. Andreev, N. A. Gun'ko, and A. V. Frolushkina, *Proc. SPIE* **2399** 307 (1995).
[96] R. F. Kazarinov and R. A. Suris, *Sov. Phys. Semicond.* **5** 707 (1971); **6** 120 (1972); **7** 347 (1973).
[97] L. V. Asryan and R. A. Suris, *Semicond. Sci. Technol.* **11**, 554 (1996); *IEEE J. Select. Topics Quantum Electronics* **3**, 148 (1997); *Electron Lett.* **33** 1871 (1997).
[98] V. B. Fiks, *Ionic Conductivity in Metals and Semiconductors (Electrotransport)* (in Russian), Moscow, "Nauka", 1969.
[99] Yu. A. Firsov, V. N. Prigodin, and Chr. Seidel, *Phys. Rep.* **126** 245 (1985); Yu. A. Firsov, in *Localization and Metal-insulator Transitions*, ed. H. Fritzsche and D. Adler, New-York and London, Plenum Press, p. 477, 1985.
[100] M. I. Dyakonov and V. Yu. Kachorovskii, *Sov. Phys. JETP* **67** 1049 (1988).
[101] V. V. Rotkin and R. A. Suris, *Mol. Mat.* **4** 87 (1994); **8** 111 (1998).
[102] M. I. Dyakonov and A. S. Furman, *Sov. Phys. JETP* **65** 574 (1987).
[103] E. L. Ivchenko and A. A. Kiselev, *Sov. Phys. Semicond.* **26** 866 (1992).

Semiconductor heterostructures

Zh. I. Alferov, V. M. Andreev, and N. N. Ledentsov

1 Introduction

It is impossible to imagine now the modern solid state physics without semiconductor heterostructures. Semiconductor heterostructures and, particularly, double heterostructures including quantum wells (QWs), quantum wires (QWRs), and quantum dots (QDs) are today the subject of research of 2/3 of the semiconductor physics community.

While the possibility of controlling the type of conductivity of a semiconductor material by doping with various impurities and the idea of injecting nonequilibrium charge carriers were the seeds from which semiconductor electronics had evolved, heterostructures made it possible to solve a much more general problem of controlling the fundamental parameters of semiconductor crystals and devices: band gaps, effective masses and mobilities of charge carriers, refractive indices, electronic band structure, etc.

The development of the physics and technology of semiconductor heterostructures has resulted in remarkable changes in our everyday life. Heterostructure electronics is widely used in many areas of human civilization. It is hardly possible to imagine our life without double heterostructure (DHS) laser-based telecommunication systems, heterostructure light-emitting diodes, heterostructure bipolar transistors, and low-noise high electron mobility transistor (HEMT) for high frequency applications including, for example, satellite television. DHS laser enters now practically every house with CD-players. Heterostructure solar cells have been widely used for space and terrestrial applications—*Mir* space-station has been using AlGaAs heterostructure solar cells for nearly 10 years.

2 Historical prerequisites for the appearance of heterostructures

Systematic studies of semiconductors were started in the early 1930s at the Physico-Technical Institute under the direct leadership of its founder—Abram F. Ioffe. V. Zhuze and B. Kurchatov studied the intrinsic and impurity conductivity of semiconductors in 1932 [1]. In the same year, A. Ioffe and Ya. Frenkel created the theory of rectification in the metal-semiconductor contact based on the tunneling phenomenon [2]. In 1931 and 1936, Ya. Frenkel published his famous articles [3–5] where he predicted exciton effects, gave a name to the phenomenon, and developed a theory of excitons in semiconductors. E. Gross experimentally discovered excitons in 1951 [6, 7]. The first diffusion theory of a rectifying p-n heterojunction, which laid foundation for W. Shockley's p-n junction theory, was published by B. Davydov in 1939 [8]. By A. Ioffe's initiative, research into intermetallic compounds was started at the Physico-Technical Institute in the late 1940s. Semiconducting properties of $A_{III}B_V$ compounds were independently predicted theoretically and discovered experimentally by H. Welker [9, 10] and N. Goryunova and A. Regel at the Physico-Technical Institute [11, 12]. The high theoretical, technological, and experimental level in this area, which had already existed at the Institute by that time, served as a basis for further work.

The concept of using heterojunctions in semiconductor electronics had already been put forward at the very dawn of electronics era. Already in his first patent concerned with p-n-junction transistors W. Shockley [13] proposed a wide-gap emitter to obtain unidirectional injection. A. Gubanov in the Institute was the first to analyze theoretically current–voltage characteristics of isotype and anisotype heterojunctions [14, 15], but one of the most important theoretical investigations at this early stage of heterostructure research was done by H. Kroemer, who introduced the concept of quasi-electric and quasi-magnetic fields in a graded heterojunction and made an assumption that heterojunctions might exhibit extremely high injection efficiencies in comparison to homojunctions [16, 17]. In the same period, various concepts were developed regarding application of heterostructures in semiconductor solar cells.

A next important step was made several years later when the concept of double heterostructure lasers was formulated independently by Zh. Alferov and R. Kazarinov [18] and H. Kroemer [19]. In patent [18] the possibility of achieving high density of injected carriers and inverse population by "double" injection was outlined. It was pointed out that homojunction lasers "do not provide CW generation at elevated temperatures" and considered the possibility "to enlarge the emitting surface and use new materials in various spectral regions" to be an additional advantage of DHS lasers.

In his article, H. Kroemer proposed to use DHS for carrier confinement in the active region. He suggested that "lasing can be obtained in quite a number of indirect gap semiconductors and improved in direct gap ones with the use of a pair of heterojunction injectors".

Initially theories were developed much faster than realized experimentally. In 1966, Zh. Alferov with co-workers predicted that the density of injected carriers could be several orders of magnitude higher than the carrier density in the wide-gap emitter ("superinjection" effect) [20]. The same year, in a paper [21] submitted to a new Soviet Journal *Fizika i Tekhnika Poluprovodnikov* (English translation: Sov. Phys. Semiconductors) the authors summarized their understanding of the main advantages of the DHS for use in different devices, especially lasers and high power rectifiers:
"The regions of recombination, light emission, and population inversion coincide, being concentrated in the central layer. Owing to potential barriers at interfaces between semiconductors with forbidden bands of different width, through leakage currents of electrons and holes are completely absent, even under strong forward bias, and there is no recombination in the emitters (in contrast to p-i-n, p-n-n^+, n-p-p^+ homostructures in which recombination plays the key role).
The population inversion can be achieved for obtaining stimulated emission by purely injection method (double injection), and there is no necessity for high doping level of the central region ordegenerate doping. Because of a considerable difference between the dielectric constants, the light is completely concentrated in the central layer which acts as an effective waveguide, and thus there is no light loss in the passive regions

(emitters)."

Here are the most important features of semiconductor heterostructures were underlined at that time:

- Superinjection of carriers
- Optical confinement
- Electron confinement.

Then it only remained to find a heterostructure in which these phenomena could be realized.

At that time, general skepticism existed in respect to the possibility of creating the "ideal" heterojunction with defect-free interface and, all the more, the theoretically predicted injection properties. Even the pioneering study of the first lattice-matched epitaxial single-crystal Ge-GaAs heterojunctions by R. Anderson [22, 23] gave no proof of nonequilibrium carrier injection in heterostructures. Actual realization of the efficient wide-gap emitters was considered nearly impossible and the double heterostructure laser patent was often referred to as a "schreibtisch patent."

Mostly owing to this general skepticism, only few groups had been trying to find an "ideal couple," which was, naturally, a difficult problem. Many conditions of compatibility between thermal, electrical, and crystallochemical properties and between the crystal and band structure of the contacting materials are to be met.

A lucky combination of a number of properties, i.e., small effective mass and wide energy gap, effective radiative recombination, sharp optical absorption edge due to "direct" band structure, high electron mobility at the absolute minimum of the conduction band, and its strong reduction at the nearest minimum at the (100) point ensured for GaAs a place of honor in semiconductor physics and electronics even at that time. Since the maximum effect is achieved by using heterojunctions between a semiconductor serving as the active region and a wider band-gap material, the most promising systems considered at that time were GaP-GaAs and AlAs-GaAs. To be "compatible," materials of the "couple" must have, as the first and the most important condition, close lattice constants; therefore, heterojunctions in the system AlAs-GaAs were preferable. However, prior to starting the work on preparation and study of these heterojunctions, one had to overcome a certain psychological barrier. By that time, AlAs had long been synthesized, but many properties of this compound remained poorly studied, since AlAs was known to be chemically unstable and to decompose in moist air. The possibility of preparing stable and applicable heterojunctions in this system seemed to be not very promising.

Initially, the attempts to create DHS were related to a lattice-mismatched GaAsP system. Zh. Alferov *et al.* succeeded in fabricating by vapor phase epitaxy (VPE) first DHS lasers in this system. However, owing to lattice mismatch, the lasing occurs only at liquid nitrogen temperature, similarly to homojunction lasers [24]. Curiously, this was the first practical result obtained for a lattice-mismatched, even part relaxed system.

This experience, gained in studying the GaAsP system, was very important for understanding many specific physical properties of heterojunctions and basics of heteroepitaxy. The development of the multichamber VPE method for GaAsP system permitted to create in 1970 superlattice structures with a 200-Å period and to demonstrate the splitting of the conduction band [25].

However, by the end of 1966, Zh. Alferov and co-workers came to a conclusion that even the small lattice mismatch in $GaP_{0.15}As_{0.85}$-GaAs heterostructures gives no way of realizing the potential advantages of the DHS. At that time, D. Tret'yakov found out that nothing happened to small crystals of $Al_xGa_{1-x}As$ solid solutions of different compositions, which had been prepared two years before by cooling from a melt and put in a desk drawer by A. Borshchevskii. It immediately became clear that $Al_xGa_{1-x}As$ solid solutions are chemically stable and suitable for preparing durable heterostructures and devices. Studies of the phase diagrams and growth kinetics in this system and the development of a version of the liquid phase epitaxy (LPE) method for heterostructure growth soon resulted in the fabrication of the first lattice-matched AlGaAs heterostructure. When the first paper on this subject was published [26], the authors were lucky to be the first to find out a unique, practically ideal, lattice-matched system for GaAs; but, as it frequently happens, the same results were obtained simultaneously and independently by H. Rupprecht and J. Woodall at T. Watson IBM Research Center [27].

Further progress in the semiconductor heterostructure field was very rapid. First of all Zh. Alferov *et al.* confirmed experimentally the unique injection properties of the wide-gap emitters and the superinjection effect [28], demonstrated stimulated emission in AlGaAs DHS [29], determined the band-diagram of $Al_xGa_{1-x}As$-GaAs heterojunction, and thoroughly studied luminescence properties, carrier diffusion in a graded heterostructure, and very interesting peculiarities of current flow through the heterojunction, e.g., diagonal tunneling-recombination transitions directly between holes from the narrow-band, and electrons from the wide-band heterojunction components [30–33].

At that same time the authors created the majority of most important devices in which the main advantages of heterostructures were realized:

- Low-threshold DHS lasers operating at room temperature [34]
- High-efficiency single heterostructure (SHS) and DHS light emitting diodes (LEDs) [35]
- Heterostructure solar cells [36]
- Heterostructure bipolar transistors [37]
- Heterostructure p-n-p-n thyristor switching devices [38].

Most of these results were reproduced in other laboratories in 1–2 years, and in some cases even later.

3 Methods of heterostructure fabrication

Heterojunctions are mainly fabricated by epitaxy, producing perfect single crystal layers of the same material as that of the substrate (homoepitaxy) or some other material (heteroepitaxy). In the latter case, it is highly important to match crystal lattice parameters of the material being grown and the base material. Characteristic features making the epitaxy method widely accepted in the semiconductor technology are the following: high perfection of materials being crystallized and possibility of changing the properties of materials over wide limits and varying the layer thickness from a fraction of atomic layer to hundreds of micrometers. Three main epitaxial techniques are being widely used at the Ioffe Institute: liquid phase epitaxy, vapor phase epitaxy, and molecular beam epitaxy.

3.1 Liquid phase epitaxy

Owing to its comparative simplicity, LPE has been mainly used for fabricating heterostructures and related devices since the middle 1960s [39]. The driving force of crystallization in this technique is supersaturation of the liquid phase with the material being crystallized. Epitaxial growth of germanium from indium solution was performed in the early 1950s [40]. At the Institute, this method was most fully developed for fabricating power germanium rectifiers [41] with a breakdown voltage of 600 V and working currents of up to 500 A at a device area of up to 3 cm^2. In 1963, the LPE method was used to fabricate p-n junctions in GaAs [42] and then introduced at the Ioffe Institute for the first time to fabricate structures with heterojunctions in the system Al-Ga-As and create on their basis high-voltage diodes [26], LEDs [35], lasers [34], and solar cells [36].

Several modifications of LPE were developed for fabricating multilayer structures with heterojunctions. In one of these, employing a graphite container, multilayer structures are grown by placing alternately one melt after another over the substrate so that several layers of various composition are successively grown on it. This approach has been successfully used at the Ioffe Institute for fabricating multilayer structures based on AlGaAs/GaAs [34–36, 39], InGaAsP/GaAs [43], InGaAs/InP [44], InGaAsSb [45], and other heterojunctions. To grow laser heterostructures with quantum dots in GaInAsP/GaAs and GaInAsP/InP systems, a modified LPE method, in which layer thickness is reduced down to several lattice constants by shortening the time of contact between substrate and melt to 1–10 ms, was proposed and developed [43].

In a piston-type device first developed at the Ioffe Institute [46], melts on the substrate are changed by displacing a preceding melt by the following one which markedly reduces the number of defects at interfaces between layers of different composition and improves the planarity and quality of layers. This technique produced LEDs with efficiency as high as 45% [47]; lasers with QWs in which the threshold current density was reduced to 120 $A\,cm^{-2}$ at room temperature [48], and solar cells with efficiency $> 25\%$ at 100–1000 suns [49].

3.2 Vapor phase epitaxy

VPE is a technique for epitaxial growth of semiconductor structures including transfer of a mixture of gaseous reagents to the surface of a crystalline substrate with the subsequent growth of layers as a result of chemical reactions between the reagents and the substrate. $A_{III}B_{V}$ heterostructures are mainly grown at the Ioffe Institute by two modifications of VPE: chloride and metal-organic.

Among the specific features of the chloride technique are sufficiently high growth rates of high-purity and high-mobility undoped gallium arsenide layers $n = 10^{12}\,cm^{-3}$, $\mu_{4\,K} = 2.5 \times 10^{6}\,cm^{2}\,V^{-1}\,s^{-1}$. A chloride technique producing multilayer GaAs and GaP structures with heterojunctions was developed at the Ioffe Institute [24, 25] and used to fabricate electric energy conversion and luminescent structures: GaAs and GaP_xAs_{1-x} $(0 < x < 1)$ superlattices with separate layer thickness of ~ 10 nm on which conduction band splitting was first observed experimentally [25]; GaP rectifying structures on silicon substrates with working temperatures as high as 450 °C; GaAs diodes, transistors, and thyristors operating at currents of up to 100 A in continuous mode at reverse voltages as high as 2.5 kV and working temperatures of up to 300 °C.

In metal-organic chemical vapor deposition (MOCVD), organometal compounds are used as sources of group III elements and group V elements are delivered at the substrate in the form of hydrides with the subsequent decomposition near the substrate. The main advantage of the MOCVD technique is the relatively simple reactor design which enables obtaining QW heterostructures with abrupt heteroboundaries and steep doping profiles by rapidly changing the vapor phase composition.

Works on MOCVD were commenced at the Ioffe in the late 1970s. Metastable solid solutions in the system Ge-GaAs [50], a rather peculiar example of a crystalline material whose properties are not uniquely determined by its composition and strongly depend on growth conditions, were obtained for the first time in the world. Lasers with threshold current density as low as 80 $A\,cm^{-2}$ (300 K) were created on the basis of AlGaAs/InGaAs heterostructures with QWs [51]. Also, the concept of a laser on $In_{0.5}Ga_{0.5}As$ QDs built in the GaAs matrix in its turn sandwiched between AlGaAs layers was successfully implemented [52]. Of considerable interest are works on $A_{III}B_V$ heterostructures on Si substrates [53] and AlGaAs/GaAs solar cells [54].

3.3 Molecular beam epitaxy

Molecular beam epitaxy (MBE) is the method of epitaxial growth involving a chemical interaction of one or more molecular or atomic beams with the surface of a heated single-crystalline substrate. The source materials are placed in evaporating cells—crucibles made of a high-melting point inert material (e.g., pyrolytic boron nitride) of such dimensions and shape which ensure a uniform flow of atoms or molecules at the substrate surface. Each crucible is enveloped with a resistive heater and protective thermal shields. A substrate is mounted onto a holder ensuring its uniform heating. All MBE machines are equipped with means for obtaining and maintaining ultrahigh vacuum at a level no worse than 10^{-10} torr which is caused by the necessity for minimizing flows of uncontrollable impurities onto the surface. Reflection high-energy electron diffraction (RHEED) allows control over the growth mechanism (amorphous, polycrystalline, or epitaxial) and growth rate, reconstruction of the surface of a growing single-crystalline film, and monitoring any morphological changes occurring in epitaxy.

The results of the initial stage of MBE studies were understood as indicating that this technique is an entirely nonequilibrium process with information about it obtainable only by studying the kinetics of specific reactions.

In view of the desirable use of a broad VPE and LPE experience gained at the Ioffe Institute, the development of a thermodynamic approach to processes occurring in MBE [55, 56] was started from the very beginning of MBE studies in 1979–1980.

Gas-phase reactions were considered with account of the equilibrium constants and element activity coefficients known from LPE and VPE, with pressures corresponding to atomic and molecular flows away from the growth surface taken as equilibrium pressures [55]. This is a feature distinguishing the given approach from the thermodynamic description proposed, e.g., in [57] where the ratio of group III and IV atoms arriving at the surface was taken as equilibrium parameter.

This approach allowed describing adequately the MBE growth of binary compounds [55]. Even more dramatic was the success in applying the thermodynamics to describing the segregation of the basic components and impurities conventionally analyzed in MBE as a kinetic process. The model [58] was based on considering phase equilibria in MBE of multicomponent solid solutions and doped layers. For example, it was shown that the abrupt increase in the arsenic equilibrium pressure at the liquidus Ga-Sn-GaAs:Sn is due to the depletion of the liquid phase of gallium and, correspondingly, its enrichment with arsenic. Here, the adequate thermodynamic analysis takes into account quantitatively the appearance of segregated layers of tin and some other features. The thermodynamic approach can explain without invoking the kinetics the existence of a "forbidden" range of AlGaAs growth and suggests ways of overcoming this problem [58]. In the next stage, it was shown that the effects of impurity diffusion in MBE, e.g., the dramatic increase in the beryllium diffusion coefficient in GaAs under certain technological conditions, can also be explained quantitatively on the basis of thermodynamic concepts [59]. Also, the effects of doping [55, 60] and the influence of lattice mismatch between the substrate and the growing layer on phase equilibria were considered and the applicability of thermodynamics in this case, too, was demonstrated experimentally by the example of MBE growth of InSb on GaAs and Si [61].

The thermodynamic analysis was also extended to multicomponent solid solutions In-Ga-As-P [62].

4 Physical processes in heterostructures

4.1 Classical heterostructures

The most important results of the development of classical heterostructures can be conveniently categorized as follows.

Realization of ideal DHS leads to the appearance of unique properties and such new fundamental phenomena in the structures (Fig. 1) as:

- One-side injection
- Superinjection
- Diffusion in a built-in quasi-electric field
- Electron confinement
- Optical confinement
- Wide-gap window effect
- Diagonal tunneling through heterointerface.

Therefore, DHS structures open wide perspectives of various applications and results in diverse new classes of semiconductor devices:

- Low-threshold semiconductor lasers operating in cw mode at room temperature, lasers with distributed feedback (DFB) and distributed Bragg reflector (DBR) lasers, vertical surface-emitting lasers, IR type-II heterostructure lasers
- High effective LEDs
- Solar cells and photodetectors, based on the wide-gap window effect
- Semiconductor integrated optics, based on semiconductor DFB and DBR lasers
- Heterobipolar transistors (HBT) with wide-gap emitter
- Transistors, thyristors, dynistors with photon signal transmission
- High-power diodes and thyristors
- Infrared-to-visible converters
- Effective cold cathodes.

Fabrication of the new type of semiconductor structures and devices based on the heterostructures leads to new requirements for technology, for instance:

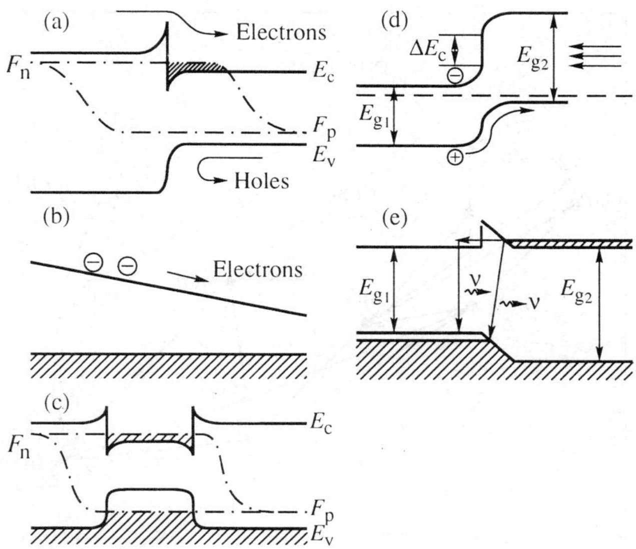

Fig 1. Fundamental physical phenomena in classical heterostructures: (a) one-side injection, (b) diffusion in built-in quasi-electric field, (c) electron and optical confinement, (d) wige-gap window effect, (e) diagonal tunneling across the heteroboundary.

- Fundamental necessity for lattice-matched structures
- Multicomponent solid solutions used for lattice-matching
- Fundamental necessity for epitaxial growth technology.

In the early stage of the development of the heterostructure physics and technology, it became clear that new lattice-matched heterostructures are to be sought in order to cover wider spectral region. The first important step in this direction was made in the Laboratory of Physics of Semiconductor Devices in 1970; it was shown that various lattice-matched heterojunctions based on quaternary III–V solid solutions can be grown, which permitted varying independently the lattice constant and forbidden band gap [63]. Later, G. Antipas with co-workers came to the same conclusions [64]. As a practical example utilizing this idea were considered various InGaAsP compositions, and soon this material was recognized as the most important for a wide variety of practical applications: photocathodes [65] and especially lasers for fiber optical communications in the infrared [66, 67] and visible spectral ranges [68–70].

After the first ideal classical heterostructures were created in the systems AlGaAs and InGaAsP the family of heterostructures grew very rapidly, and their present-day ("world map") is presented in Fig. 2.

In the early 1980s, H. Kroemer and G. Griffits published a paper [71] which stimulated strong interest in staggered line-up heterostructures (type-II heterojunctions). Spatial separation of electrons and holes at interfaces of this kind ensures tunability of their optical properties [72, 73]. Staggered band alignment furnishes an opportunity to obtain optical emission with a photon energy much lower than the forbidden band-gap width of each of the semiconductors forming a heterojunction. Fabrication of an injection laser based on type-II GaInAsSb-GaSb heterojunctions [73] opens up good prospects for creation of effective coherent light sources for the infrared. The emission in such a device is due to recombination of electrons and holes localized in self-consistent

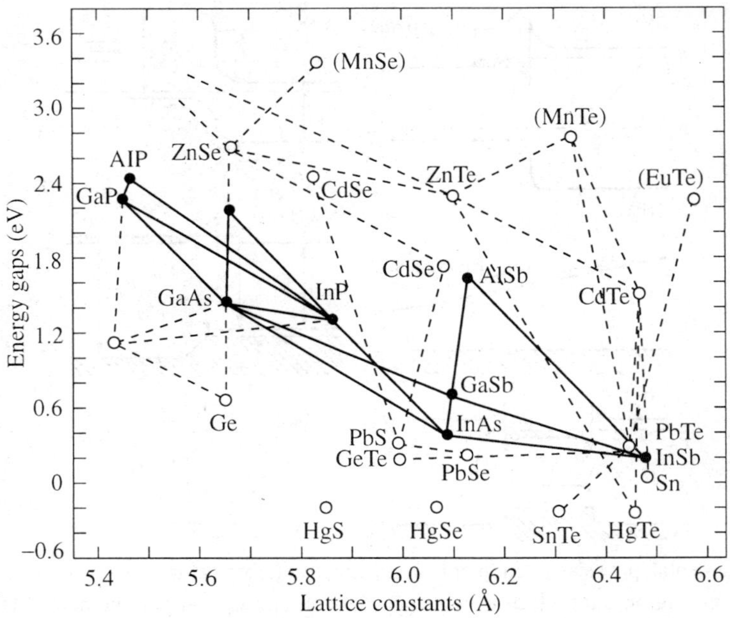

Fig 2. Dependence of the forbidden gap width on lattice parameter for Group-IV semiconductors and $A_{III}B_V$ and $A_{II}B_{VI}$ compounds. Investigated heterostructures are connected by lines (solid for $A_{III}B_V$, dashed for other compounds).

potential wells on different sides of the heterointerface. Thus, type-II heterostructures open up new possibilities both for development of the fundamental physics and for device applications which cannot be realized with type-I $A_{III}B_V$ heterostructures.

4.2 Quantum wells and superlattices

If the charge carrier motion in a solid is confined to a layer with a thickness of the order of the de Broglie wavelength of elementary excitations (or mean free path, which is shorter), then, in accordance with the quantum mechanics, quantum-size effects must be observed. In this case, a characteristic mass for quantization is the carrier (electron or hole) effective mass, rather than the free electron mass. Investigations of ultrathin layers became rather popular in the late 1950s–early 1960s (see, e.g., a review of V. Lutskii [74]). As objects of study were frequently used films of semimetals, e.g., bismuth, prepared by vacuum evaporation onto mica substrates. In thin Bi [75] and InSb films, the effective forbidden gap width increased with decreasing film thickness. In 1962, L. Keldysh studied theoretically the electron motion in a crystal in a periodic potential with a period much exceeding the lattice constant [76]. He predicted formation of minibands and occurrence of negative differential conductivity. In 1963, an idea was put forward [77, 78] of using five-layer structures with two dielectric layers and an ultrathin intermediate layer to observe the effect of resonance tunnel current rise on crossing of the quantum-size level in the ultrathin layer by the Fermi level.

The material progress in fabricating quantum-size layers and investigating the quantum-size effects is associated with the development of the concept of semiconductor heterostructures and practical realization of ideal DHS. In 1969–1970, L. Esaki and R. Tsu proposed to use multilayer periodic semiconductor heterostructures with ultrathin

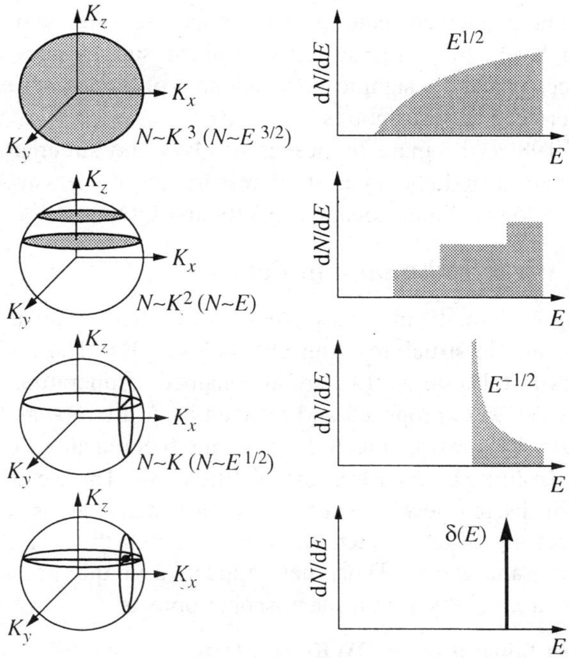

Fig 3. Density of states in structures of various dimensionality, calculated on the assumption that the carrier energy E in the band is a quadratic function of wave vector **k**.

layers (superlattices) for creating artificial materials with tunable miniband width [79]. A fundamentally important practical case of sequential tunneling in a superlattice was first considered at the Ioffe Institute by R. Kazarinov and R. Suris [80, 81] in 1970. One of the first artificial superlattices of this kind was obtained at the Ioffe Institute by Zh. Alferov *et al.* in a multichamber VPE machine of special design in the system $GaP_{0.3}As_{0.7}$-GaAs also in 1970 [25]. In 1974, L. Chang *et al.* first observed the effect of resonant tunneling in MBE-grown ultrathin semiconductor heterostructures [82].

In 1970, R. Kazarinov and R. Suris considered at the Ioffe Institute the possibility of fabricating a unipolar long-wavelength laser operating on radiative electron transitions between quantum-size levels [80]. A laser operating on this principle was fabricated by F. Capasso *et al.* (so-called cascaded laser) only a quarter of century later [83].

In 1974, R. Dingle *et al.* experimentally confirmed for the first time the manifestation of the quantum-size effect in optical spectra of heterostructures with QWs [84]. A steplike character of the absorption spectrum, corresponding to the specific density of states in QWs, was demonstrated (Fig. 3). On making the QWs thinner, the steps are shifted to higher photon energies.

Beginning with the late 1970s and early 1980s, investigations of QWs and superlattices started to dominate in the physics of semiconductors. For example, the concept of modulation-doped heterostructures [85] made possible achieving unprecedented electron and hole mobilities, which enabled progress in studying the Quantum Hall Effect and predetermined the discovery of the Fractional Quantum Hall Effect [86].

Studies of the epitaxial growth of ultrathin layers of materials with varied lattice parameter demonstrated that in the initial stages the growth is pseudomorphic with no dislocation formatted, which made possible obtaining QWs and superlattices with elastically strained layers [87]—the so-called strained superlattices [88]. The use of strained superlattices enabled a significant broadening of the variety of materials usable for synthesizing perfect heterostructures.

By the end of 1980s, the main properties of QWs and superlattices were, on the whole, well understood and the interest of researchers progressively shifted toward structures with even lower dimensionality: QWRs and QDs.

4.3 Quantum wires and quantum dots

As follows from Fig. 3, dramatic increases in density of states (singularities) are observed at certain energies only in structures with QWRs and QDs. The limiting case of size-quantization occurs in QDs or, as they are also named, "superatoms." Semiconductor QDs in glassy matrices were proposed and realized by A. Ekimov and A. Onushchenko in 1980–1981 [89]. This work initiated important theoretical investigations of QDs started at the Ioffe Institute by Al. Efros and A. Efros [90]. The electron spectrum of an ideal QD is a set of discrete levels separated by forbidden regions, which corresponds to the electron spectrum of an isolated atom even though the real QD may consist of several hundred thousand atoms. Thus, there appears a unique possibility of modeling experiments of the atomic physics on macroscopic objects.

Methods of indirect fabrication of QWRs and QDs

During a long time attempts had been made all over the world to produce QWrs and QDs and devices based on them by "conventional methods," e.g., by selective etching of structures with QDs [91], local "intermixing" of structures with QDs by ion implantation or laser annealing, growth on profiled structures [92] and cleaved surfaces, and condensation in glass matrices [89]. However, the above methods of fabricating QDs and QWrs encountered serious difficulties. For example, with lithographic methods, the characteristic size of a resist molecule and/or metal mask grain ($\sim$ 10 nm) gives no way of obtaining reproducible QDs with controllable interface abruptness at the level of one and even several monolayers (ML) including also the cases when "dry" etching techniques (e.g., ion beam etching) are used. More fundamental problems are related to degradation of lateral surfaces during ion etching [91].

Spontaneous formation of ordered nanostructures

A breakthrough in fabricating ordered arrays of uniform QWRs and QDs became possible with the use of the effects of spontaneous appearance of periodically ordered structures on the semiconductor surface and in epitaxial semiconductor films.

Ordered nanostructures may appear in closed systems, e.g., on sample annealing or upon prolonged interruption of growth. Such structures are in equilibrium, and the thermodynamic approach is used for describing them. Ordered structures may also develop in open systems during crystal growth. In a broad sense the "self-organization" of nanostructures is understood as spontaneous development of a macroscopic order in an initially random distributions [93].

Spontaneously ordered nanostructure can be conventionally divided into four large classes:

(i) Structures with periodic composition modulation in epitaxial films of semiconductor solid solutions [94].

(ii) Periodically faceted surfaces [95, 96].
(iii) Periodic structures of planar domains (e.g., islands of monolayer height) [97].
(iv) Ordered arrays of 3D coherently strained islands in heteroepitaxial lattice-mismatched systems [98].

In all of these systems, neighboring domains differ in lattice constant and/or surface structure and, consequently, domain boundaries are a source of long-range fields of elastic strain.

Both 2D and 3D islands can be used for obtaining QDs. In [99–101], buried InAs(001) layers in GaAs matrix obtained with submonolayer and monolayer depositions were studied by means of photoluminescence (PL). A conclusion was made that InAs forms microislands elongated in the $[\bar{1}10]$ direction, all having the same width [100, 101]. Later, the conclusion that islands of this kind exist in the system InAs/GaAs(001) in the case of a submonolayer deposition was confirmed by scanning tunneling microscopy [102] with the island width found to be 4 nm.

Ordered arrays of 3D coherently strained islands. In lattice-matched heteroepitaxial systems the growth mode is determined solely by the relation between the energies of two surfaces and the interface energy. If the sum of the surface energy of epitaxial layer γ_2 and the energy of interface γ_{12} is lower than the substrate surface energy, $\gamma_2 + \gamma_{12} < \gamma_1$, i.e., if the material 2 being deposited wets the substrate, then we have the Frank–van der Merve growth. Changing the $\gamma_2 + \gamma_{12}$ value may result in a transition from the Frank–van der Merve mode to an Volmer–Weber one where 3D islands are formed on a bare substrate.

In a heteroepitaxial system with lattice mismatch between the material being deposited and the substrate, the growth may initially proceed in a layer-by-layer mode. However, a thicker layer has a higher elastic energy, and the elastic energy tends to be reduced via formation of isolated islands. In these islands the elastic strains relax and, correspondingly, the elastic energy decreases. This results in a Stranski–Krastanow growth mode.

Experimental studies of coherently strained islands arrays in the systems InAs/GaAs (001) and InGaAs/GaAs(001) have shown that a narrow size distribution of islands is

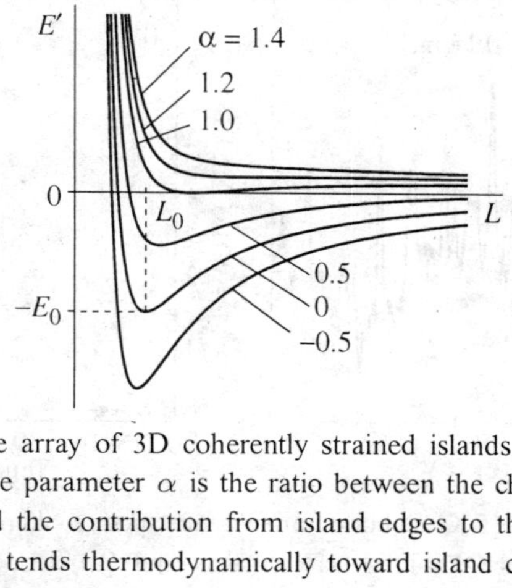

Fig 4. Energy of a sparse array of 3D coherently strained islands per unit surface area as a function of island size. The parameter α is the ratio between the change in the surface energy upon island formation and the contribution from island edges to the elastic relaxation energy. When $\alpha > 1$, the system tends thermodynamically toward island coalescence. When $\alpha < 1$, there exists an optimal island size and the system of islands is stable against coalescence.

possible [103, 104]. The characteristic size is determined by the minimum in the energy of an array of 3D coherently strained islands per unit surface area as a function of the island size (Fig. 4) [98, 105–107]. Interaction between islands via elastically strained substrate results in lateral island ordering typical of the square lattice.

Electronic structure and optical properties

Realization of ideal QDs. N. Ledentsov *et al.* demonstrated the formation of dense ($\sim 10^{11}$ cm^{-2}) arrays of dislocation-free InAs QDs in a GaAs matrix and studied the effect of growth parameters and QD composition on their optical properties [105]. The shape and size of QDs were determined for the first time, and lateral ordering of QDs into a square lattice with major axes aligned with the [001] and [010] directions was discovered. In the same work, cathodoluminescence spectra of QDs were studied with high spectral and lateral spatial resolution. It was shown that the PL line from the QD array broadened by fluctuations of shape and size in optical excitation decomposes into a set of ultranarrow lines from single QDs in the case of excitation by focused electron beam. It should be noted that the temperature-independence of the QD PL line width (contrary to two-dimensional and three-dimensional cases), first observed in [106] indicates unambiguously the atom-like nature of the electron spectrum of a QD (Fig. 5).

Absorption, photoluminescence, and luminescence excitation spectra for samples with QD arrays (Fig. 6) were also compared for the first time, and the unique features of optical properties for systems of this kind were determined [105]. For example, it was shown for the first time that the absorption peak of the ground state of a QD coincides with the PL peak for the same QD state. The authors noted that the resonant excitation of the QD ground state is only possible when the photon energy exactly corresponds to the atom-like spectrum of QD absorption. In a radiation recombination event, a QD emits a photon of exactly the same energy and, therefore, QD excitation and luminescence quanta can be generally separated in time-resolved experiments only. The given work also demonstrated for the first time that the QD excitation spectrum is modulated with an energy corresponding to the energy of LO-phonon in InAs, which

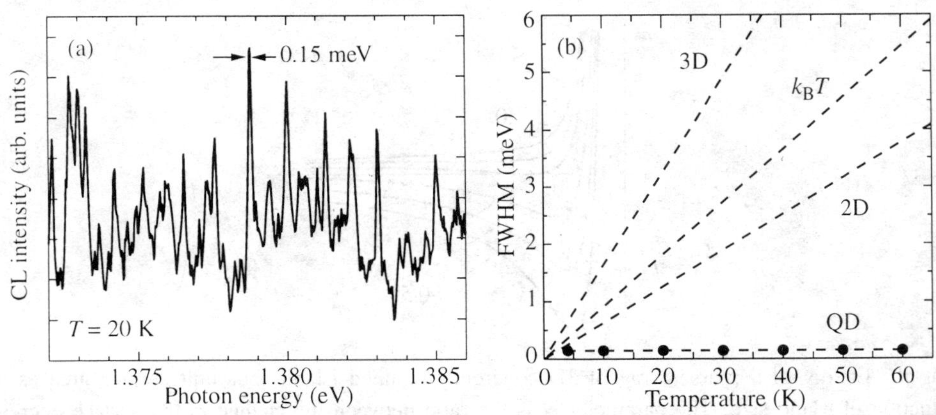

Fig 5. (a) Low-temperature (5 K) cathodoluminescence spectra taken with high spatial resolution from a structure with InAs QDs in a GaAs matrix and (b) temperature dependence of the linewidth for 3D, 2D, and isolated QD.

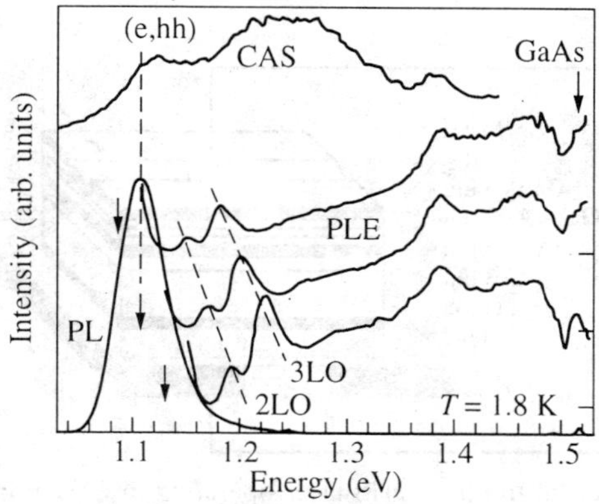

Fig 6. Spectra of calorimetric absorption, PL, and PL excitation for structures with QDs.

is due to the fact that the relaxation into the QD ground state is faster when the energy difference between the ground and excited states is a multiple of LO-phonon energy and, therefore, the possibility of radiationless recombination is lower in the excited state. Carrier capture by a QD directly from the matrix material occurs in times of order 1 ps. The time of relaxation from excited into ground state is not longer than 25–40 ps depending on the QD geometry [108]. In the case of resonant excitation [109, 110] the PL spectrum decomposes into a series of lines. The energies of the observed peaks coincide with a multiple of GaAs- and InAs-related LO-phonon energies and combination of these including GaAs bulk phonons (36.6 meV), interface modes (34 meV), phonons of wetting InAs layer (29.5 meV), and InAs QD phonons (31.9 meV).

Relationship between growth modes and QD optical properties. The effect of growth interruption on the PL from InAs-GaAs structures with 3D QDs was studied in [109]. It was shown that increasing the number of deposited InAs up to 4 ML or making longer the interruption time in growth shifts the QD PL line to longer wavelengths (1.1 μm), which agrees with the increase in characteristic QD size observed by transmission electron microscopy (TEM). These results confirmed the existence of a "limiting" or "equilibrium" QD size indicated by TEM [109]. At low growth temperature (ca 350 °C), QDs associate into elongated islands which is accompanied by the appearance of a peak at 1.8 μm (300 K) in PL spectra.

It was demonstrated that an array of ordered QDs exists only in a certain range of arsenic pressures [109].

Geometry of QDs. In the case of deposition at 460 °C, pyramidal InAs QDs with square base ($\sim$ 120 Å) and major axes aligned with directions [001] and [010] and $\sim$ 60 Å high are formed on the surface. Model calculations by molecular dynamics methods [111] indicate that an adequate cross-sectional image of a QD can be obtained by high-resolution TEM only under optimal defocusing conditions at a film thickness no more than doubled lateral QD size.

Luminescence from vertically coupled QDs. The luminescent properties of vertically correlated tunnel-coupled QDs were studied in [108, 112–114]. Increasing the number of QD deposition cycles resulted in a long-wavelength shift of both the PL line and the PL-resonant peak in the calorimetric absorption spectrum. Increasing the number

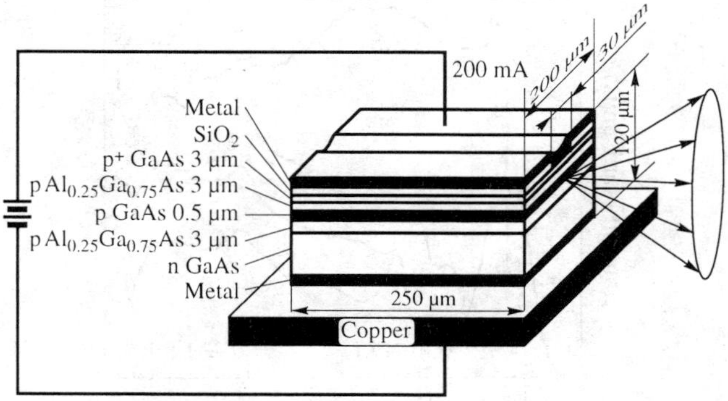

Fig 7. Schematic structure of the first DHS laser operating in CW mode at room temperature.

of cycles makes shorter the radiative recombination lifetime in QDs. For large number of coupled QD lateral association of QDs in upper rows is observed together with the appearance of the corresponding long-wavelength PL peak.

Quantum dots of type-II. QDs of type-II were first studied by F. Hatami *et al.* in GaSb-GaAs [115]. In this system, holes are localized in the GaSb region, whereas for electrons these regions present a potential barrier.

5 Application of heterostructures in electronics

5.1 Semiconductor heterostructure lasers

Lasers based on classical heterostructures

Achieving a CW operation of semiconductor lasers at room temperature in 1970 was a historical event in the vigorous development of semiconductor heterostructure application.

In 1970, the international competition became exceedingly strong. Later, one of the major competitors, I. Hayashi, who worked with M. Panish at the Bell Telephone Co. Laboratory in Murray Hill, wrote [116]: "In September, 1969, Zhores Alferov of the Ioffe Institute in Leningrad visited our laboratory. We realized he was already getting a $J_{\rm th}^{(300)} = 4.3$ kA cm^{-2} with a DHS. We had not realized that the competition was so close and redoubled our efforts... Room temperature CW operation was reported in May 1970..."

In the paper of Zh. Alferov and co-workers submitted in May 1970 [117], CW lasing was realized in stripe-geometry lasers fabricated using photolithography and mounted on copper heat sinks coated with silver (Fig. 7). The lowest threshold current density $J_{\rm th}$ at 300 K was 940 A cm^{-2} for wide lasers and 2.7 kA cm^{-2} for stripe lasers. Independently, CW lasing in DHS lasers was reported by I. Hayashi and M. Panish [118] (for broad-area lasers with diamond heat sink) in a paper submitted only a month later than [117]. Achieving a CW lasing mode at room temperature caused an outbreak of interest in physics and technology of semiconductor heterostructures.

The basic concepts of a laser with distributed feedback (DFB) were formulated in Inventor's Certificate issued in 1971 [119]. In the same year H. Kogelnik and C. Shank considered the possibility of replacing the Fabri–Perot resonator or its like in dye lasers

by periodic volume inhomogeneities [120]. It should be noted that their approach is inapplicable to semiconductor lasers, and all those who study DFB or DBR semiconductor lasers commonly use the approach formulated in [119]:
(1) Diffracting grating is created on the surface, rather than in the bulk of the waveguide layer.
(2) Interaction of waveguide modes with the surface diffraction grating not only gives rise to distributed feedback, but also produces well collimated output radiation.

A detailed analysis of the operation of a semiconductor laser with surface diffraction grating was performed in 1972 [121] where the possibility of single-mode generation was established. The first semiconductor lasers with surface diffraction grating and DFB were fabricated practically simultaneously at the Ioffe Institute [122, 123], Caltech [124], and Xerox Lab in Palo Alto [125].

On the basis of semiconductor lasers with distributed Bragg reflector was obtained a practically temperature-independent lasing wavelength [126].

QW lasers

Lasing on QWs was first obtained by J. van der Ziel *et al.*, but the generation parameters were much inferior to those of average DHS lasers [127]. Only in 1978, R. Dupuis, P. Dapkus in collaboration with N. Holonyak reported fabrication of a QW laser with parameters comparable with those of standard DHS lasers [128]. In this work, the term "quantum well" was first suggested. The actual advantage of QW lasers was demonstrated much later by W. Tsang from Bell Telephone. By significantly improving the MBE growth technology and designing an optimized structure (separate-confinement DHS lasers with graded-index waveguide: GRINSCH lasers), he could reduce the threshold current density to 160 kA cm^{-2} [129].

The development of MBE and MOCVD techniques for growth of $A_{III}B_V$ heterostructures was started at the Ioffe Institute in the late 1970s only. In the first place, it was stimulated the design and construction of the first Soviet MBE machine for domestic electronic industry. During several years, three generations of MBE machines were developed, the last of these named *Tsna* (a beautiful river not far from Ryazan', a city where the Research Technological Institute for Electronic Industry was situated: it is at this institute that the MBE machines were designed). In parallel and somewhat later, development of an MBE system was commenced at the Scientific and Engineering Association, Academy of Sciences, in Leningrad. In the late 1980s, several machines of this version were purchased by the Ioffe Institute. Both these types of MBE systems are still working at the Ioffe Institute and other laboratories of the country.

MOCVD systems were developed at the Ioffe Institute, and later, in the 1980s, a Swedish company Epiquip specially designed, with active participation of the researchers from the Ioffe two machines for the Institute.

A strong interest in low-dimensionality structures and the lack of equipment for MBE and MOCVD growth techniques had stimulated the investigations aimed at developing an LPE technique suitable for growing heterostructures with QWs.

However, till the late 1970s, it seemed impossible to grow by LPE $A_{III}B_V$ heterostructures with an active region thinner than 500 Å, because of the existence near the heteroboundaries of extended transition regions of varying chemical composition.

The situation changed after the work of N. Holonyak *et al.* appeared. They proposed to use an LPE system with rotating "boats" for growing InGaAsP superlattices [130].

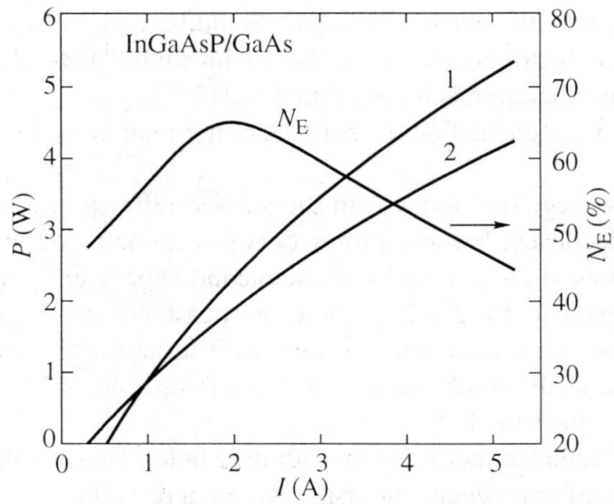

Fig 8. Watt–ampere characteristics of InGaAsP-GaAs SC DHS laser diodes with a single QW operating in CW mode. (1) Diode with weakly and strongly reflecting coatings, (2) diode with strongly reflecting coatings only. $N_E(I)$ is the dependence of energy conversion factor (efficiency) on current.

Zh. Alferov and co-workers developed a modified LPE technique with the usual translational motion of the substrate in a standard horizontal "boat" geometry for InGaAsP heterostructures [131] and a method of low-temperature LPE for AlGaAs heterostructures [132, 133]. These techniques allowed to grow practically any type of high quality heterostructures with QWs with up to 20-Å thick active region and an interface region thickness comparable with the crystal lattice parameter. Of great practical importance was the achievement with the use of the LPE technique of record-breaking threshold current densities in single-QW separate-confinement lasers based on InGaAsP/InP ($\lambda = 1.3$ and 1.55 μm) and InGaAsP/GaAs ($\lambda = 0.65–0.9\,\mu$m) heterostructures [68].

In high-power stripe-geometry InGaAs/GaAs lasers ($\lambda = 0.8\,\mu$m) (Fig. 8) it was achieved a 66% CW efficiency and 5-W radiation power at a stripe width of 100 μm and length of 1.2 mm [134, 135]. These lasers were the first in which effective cooling of a power semiconductor device was effected by recombination radiation, as proposed previously in [20]. Another important feature of InGaAsP heterostructures was their rather high stability against development of dislocations and defects (Fig. 9) [136]. These studies laid foundation for wide use of Al-free heterostructures.

The most intricate laser structure with QWs combining a single QW and short-period superlattices (SPSLs) (Fig. 10) used for creating GRIN SCH (the most preferable for obtaining the lowest threshold currents) was fabricated at the Ioffe Institute in 1988 [137].

Use of SPSL not only gives a desirable profile of graded waveguide region and creates a barrier for defect propagation into the active region, but also improves the heterointerface smoothness and makes possible growing different parts of a structure at strongly differing temperatures. Thus, both a perfect surface morphology and a high internal quantum efficiency were simultaneously achieved on a planar GaAs (001) surface. Threshold current densities as low as $J_{th} = 52\,\mathrm{A\,cm^{-2}}$ and, after certain optimization, even $J_{th} = 40\,\mathrm{A\,cm^{-2}}$ were obtained which is still a world record for injection semiconductor lasers and serves as an excellent illustration of the effective use of QWs

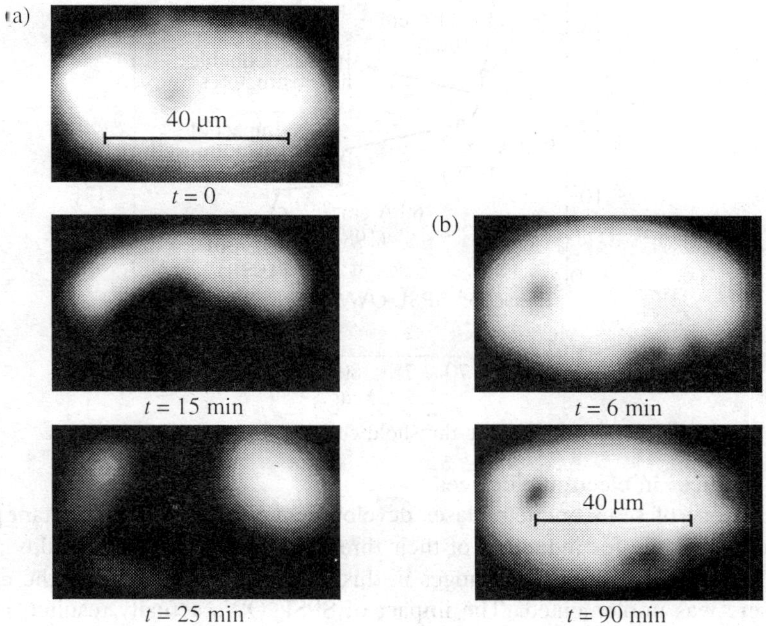

Fig 9. Temporal evolution of luminescence patterns from the active region in (a) AlGaAs-GaAs and (b) InGaAsP-GaAs SC DHS with a single QW at high excitation level. Kr^+ laser excitation beam diameter $\sim 40\,\mu$m. Excitation levels, W cm^{-2}: (a) 10^4, (b) 10^5.

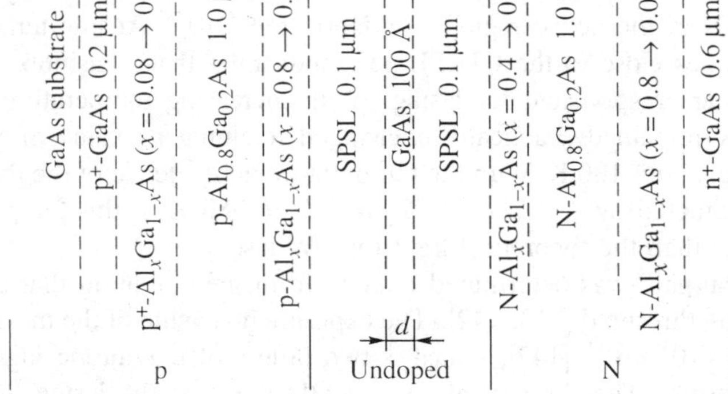

Fig 10. Structure of a SC DHS laser with QW confined by MBE-grown SPSL.

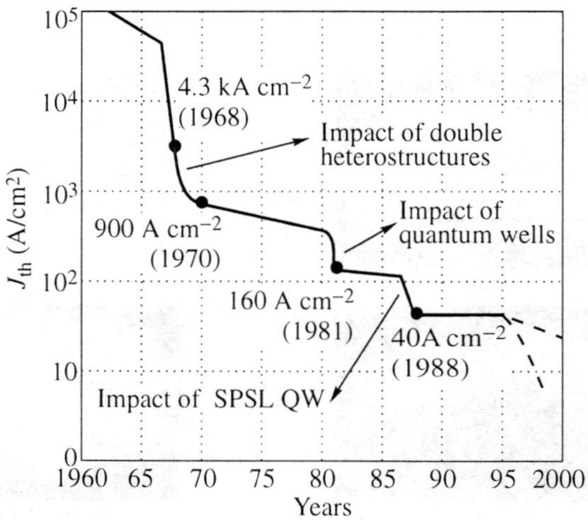

Fig 11. Evolution of the threshold current of semiconductor lasers.

and superlattices in electronic devices.

The history of semiconductor laser development is, from a certain standpoint, the history of a struggle for reduction of their threshold current which is readily illustrated by Fig. 11. The most dramatic changes in this field occurred only after the concept of DHS lasers was implemented. The impact of SPSL QWs actually resulted in that the theoretical limit was achieved for this, most important parameter. Further opportunities are associated with the use of new structures with QWRs and QDs.

QWR and QD lasers

QWR lasers. Significant advantages of QWR and QD lasers over QW devices were predicted in [138]. Historically, injection QWR lasers were first obtained by growing structures in V-grooves [139]. The energy difference between size-quantization subbands in such QWRs is comparable with kT at room temperature, in-plane quantum-size effects are difficult to observe, and no significant difference between this laser and QW laser was revealed.

QD lasers. Lasing under optical excitation via states of self-organized QD was first observed in 1993 [140]. Soon after, the first injection laser containing a single sheet of InGaAs QDs in the active region was fabricated [141]. At low temperature lasing started at energies close to the QD PL peak indicating that transitions via the ground states of QDs are responsible for lasing. With increasing temperature, the threshold current density remained practically unchanged retaining its value of 80 kA cm^{-2} to temperatures of about 180 K. With the temperature dependence of the threshold current density approximated by $J = J_0 \exp(-T/T_0)$, $T_0 = 380$ K in this temperature interval which is higher than the theoretical limit for QW lasers.

The material gain was determined directly from the condition that gain is equal to loss at the lasing threshold [113, 142]. The experimental value of the maximum "specific gain" was 1.5×10^5 cm^{-1} [142], which is two orders of magnitude higher than in the case of QW lasers. The differential gain of QD lasers at the lasing threshold is also markedly higher than in QW lasers being equal to $\sim 10^{-12}$ cm^2.

The modal gain can be enhanced by increasing the QD concentration which is achieved by using vertically correlated QDs [108, 112–114]. With these lasers the possibility of generation via the ground state of QDs up to room temperature was demonstrated [112]. It was also shown that the threshold current density decreases dramatically to ~ 90 kA cm^{-2} (300 K) on increasing the number of stacking cycles to 10 [108, 113, 114].

The threshold current density in lasers was further reduced by using AlGaAs instead of GaAs as a matrix for vertically correlated QDs [108, 114] which allowed reducing the threshold current density at room temperature to 60 A cm^{-2} [108] and increasing the differential efficiency to more than 70%. Thus, it was realized a CW mode of operation of a QD laser at room temperature with an output of 1.5 W and more. The generation went via QD states up to pumping currents exceeding 10–20-fold the threshold values, and the efficiency of the device in CW mode was as high as 42%.

Dynamic properties of QD lasers. The good dynamic characteristics of the QD laser were determined by the high differential gain coefficients [142]. Direct measurements of the cutoff frequency of a QD laser gave a value of the order of 10 GHz [143]. The experimental values of the lasing line spectral broadening factor (α-factor) were ~ 0.5, which was much lower than in the case of QW laser.

Theory of QD lasers. First detailed descriptions of a realistic QD laser with a strong nonuniform gain line broadening were given by L. Asryan and R. Suris at the Ioffe Institute [144, 145] and M. Grundmann and D. Bimberg at the Technical University in Berlin [146, 147] who showed that even with account of a substantial nonuniformity of QDs (10% size scatter) such lasers must essentially outperform QW lasers.

Surface-emitting QD lasers. Injection lasers based on vertically stacked isolated QDs were fabricated by H. Saito *et al.* The threshold current density at room temperature was 5000 A cm^{-2} [148], with lasing via excited QD states. To improve the working parameters of vertical QD lasers, the concept of vertically correlated QDs as active medium was proposed [149]. In structures with a 10-μm aperture it was realized a CW lasing mode at room temperature with a threshold current of 180 μA (180 A cm^{-2}, 300 K). The maximum efficiency was 16%. The minimum threshold current was 68 μA for 1-μm mesa-structure, which corresponds to the best values for vertical lasers based on QWs of similar geometry.

Exciton waveguides. Realization of the waveguide effect assumes the presence of a thick (with thickness comparable to the wavelength in the crystal) layer with high average refractive index. N. Ledentsov and co-workers proposed a concept of "exciton waveguide effect" in a laser in which the waveguide effect is achieved through resonant modulation of the refractive index, related to resonant exciton absorption in a structure with QDs ("submonolayer" superlattice) [150]. Lifting of the selection rules for momentum and the absence of the exciton screening effect in QDs results in that the lasing is of phononless nature. The resonant waveguide effect is realized now both in $A_{II}B_{VI}$ [150] and in $A_{III}B_V$ systems.

Self-adjusted microcavities. Since the material gain of QDs is rather high and there is no screening of exciton states in QDs, the modulation of the refractive index due to Kramers–Kronig relations, can automatically tune the vertical laser resonator mode within the gain line contour. A similar effect was observed in studying the interaction of the line of exciton absorption in a QW in the cavity mode with a pronounced spectral splitting (7 nm) of the transmission line observed [151]. However, screening of excitons at high excitation densities gives no way of realizing this effect in vertical QW lasers, and the gain lines and the resonator mode are not interrelated [152]. In [149], the effect of annealing on the properties of vertical lasers with InGaAs-GaAs QDs was studied. Annealing of QDs shifted the maximum gain to shorter wavelengths with the lasing wavelength also shifted by 6 nm (this is not characteristic of QW structures); the occurrence of the effect of self-adjustment of the resonator mode and the gain band of the vertical QD laser was demonstrated for the first time.

Infrared QD emitters. The time of phonon relaxation in a QD in transition from excited to ground state can be made much longer (up to 40 ps in the case of comparatively large QDs) as compared with the QW case where the time of subband relaxation is only $\sim$ 1 ps. This fact and the lack of lateral transport of nonequlibrium carriers in QDs are the reasons for a higher probability of interlevel carrier relaxation in QDs by emitting far infrared quanta. L. Vorob'ev *et al.* observed for the first time IR emission (10–20 μm) from QD lasers operating in the near-infrared range [153].

Prospects for development of emitters with InAs QDs in silicon. The idea of using InAs QDs in silicon matrix for fabricating high-efficiency silicon-based light-emitting devices is of much interest. Recent fabrication of 3D InAs islands in Si matrix [154] and demonstration a line of PL from silicon-overgrown InAs QDs (1.3 μm at 77 K and 1.6 μm at 300 K)indicate that this direction shows great promise.

5.2 Heterostructure light-emitting diodes

Heterostructures have a number of advantages over homo-p-n-junctions, as regards LED parameters. When the active region is made of a material with a narrower gap than that of the material through which radiation is emitted, loss of light in passive regions of the structure is reduced, and the unidirectional type of injection in heterojunctions lowers loss by carrier recombination in emitters. These advantages were first actualized in AlGaAs heterostructure LEDs in which, owing to the absence of any additional surface states at the heteroboundary, a nearly 100% internal quantum efficiency of radiative recombination was achieved [155, 156].

The effect of re-emission consisting in multiple absorption of generated light with the subsequent radiative recombination of electron-hole pairs excited by light was discovered and studied (Fig. 12) [157].

The results obtained were used to develop high-efficiency sources of spontaneous radiation on structures with a layer of a narrow-gap semiconductor sandwiched between two wide-gap layers, in which the effects of multiple reflection and re-emission of light in the active layer were used and loss of light by absorption in passive regions of the structure was reduced [47, 158–160]. These structures yielded record-breaking values of external quantum efficiency: in LEDs with semi-spherical antireflection coatings $\eta_c =$ 45% at $h\nu = 1.4-1.5$ eV, 20% at $h\nu = 1.75$ eV, and 5–10% at $h\nu = 1.8-1.87$ eV.

Long-wavelength GaInAsSb/GaSb heterostructure LEDs (λ = 1.7–2.5 μm). The 2-μm spectral region including the water absorption line ($\lambda = 1.94\,\mu$m) used in optical monitoring and absorption lines of methane CH_4 at 2.32–2.35 μm is of particular interest for LED designers.

The spontaneous light emitter developed at the Ioffe Institute for this spectral region [161, 162] was a semiconductor structure comprising an active n-GaInAsSb layer sandwiched between an n-GaSb substrate and a wide-gap p-GaAlAsSb emitter ($E_g \sim 1.2$ eV). Room temperature emission spectra of the LEDs show a single band

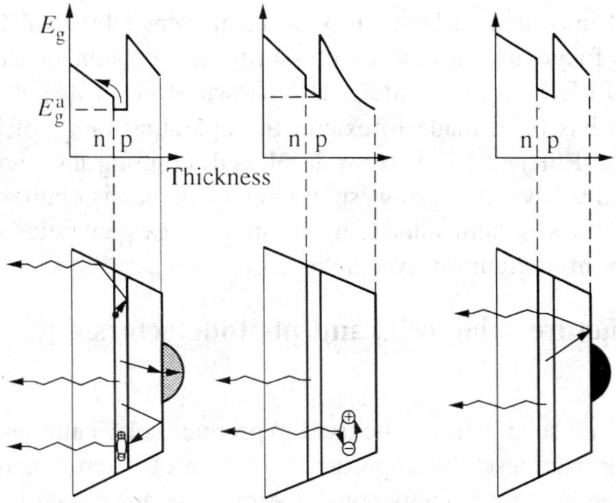

Fig 12. Forbidden gap width in AlGaAs/GaAs heterostructure LEDs of three types and schematic of light reflection and "re-emission" processes.

whose peak wavelength practically linearly depends on the composition of the solid solution in the active region. The external quantum efficiency of the LEDs is the highest ($\eta = 4\%$) at 2.0–2.2 μm.

Uncooled long wavelength non-isoperiodic InAsSb/InAsSbP heterostructure LEDs (λ = 3–6 μm). The requirements to air purity in industrial areas becoming more and more stringent, efforts of quite a number of researchers are aimed at developing photon-coupled pairs "LED-photodetector" operating in the mid-infrared spectral range (3–6 μm) and sensitive to major air contaminants, such as C_nH_m, SO_2, NH_3, CO_2, CO, etc., whose absorption lines lie in this spectral range.

LEDs for the 3–6 μm spectral range were developed using heterostructures based on InAsSb narrow-gap solid solutions [163, 164]. One of the main difficulties in developing long-wavelength LEDs consists in the lack of an isoperiodic substrate for InAsSb solid solutions, which predetermines the use of strained non-isoperiodic heterostructures in the system InAs/InAsSb/InAsSbP for creating long-wavelength LEDs. Figure 13(a) shows a double InAsSb/InAsSbP heterostructure grown on an InAs substrate [163]. Emission spectra of the LEDs ($\lambda = 3–4\,\mu$m) are shown in Fig. 13(c).

With the use of gradient structures, structurally perfect InAsSbP (Fig. 13(b)) and

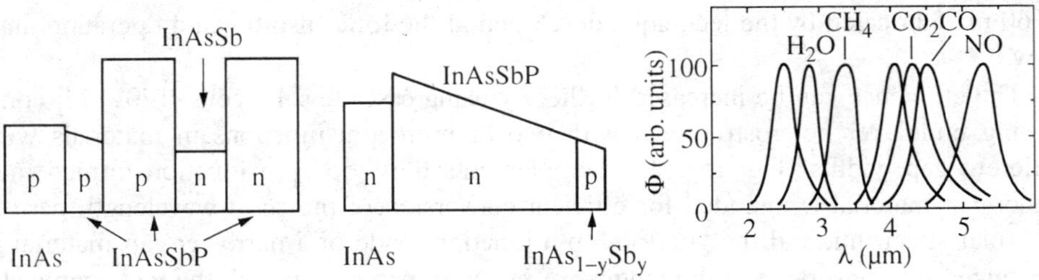

Fig 13. Band diagrams for (a) heterostructure LED ($\lambda = 3–4\,\mu$m), (b) graded bandgap LED ($\lambda = 4–6\,\mu$m), and (c) emission spectra for long-wavelength LEDs.

InGaAs layers and uncooled LEDs with $\lambda > 4\,\mu$m were fabricated for the first time. These LEDs have found use in devices measuring the content of carbon mono- and dioxide. Spectra of LEDs of this kind are also shown in Fig. 13(c).

Recently, effort has been made to extend the operating range of InAsSb LEDs to longer wavelengths. Efficient LEDs were developed, emitting light with $\lambda = 5-6\,\mu$m (300 K). These LEDs have the narrowest, among spontaneous emitters, emission band ($< 0.6\,\mu$m), which makes them suitable for creating a new generation of nitrogen oxide (NO) sensors with an operating wavelength $\lambda_{max} = 5.3\,\mu$m.

5.3 Heterostructure solar cells and photodetectors

Solar cells

$A_{III}B_V$ heterostructure solar cells can be used to produce solar batteries with higher efficiency and radiation hardness. An important advantage of heterostructure photoelectric converters is their ability to efficiently convert strongly concentrated (up to 1000–2000-fold) sunlight, which opens up possibilities of reducing significantly (in proportion to the extent of concentration) the area and cost of solar cells and, as a result, making cheaper the "solar" electric power.

n-GaAs–p-AlGaAs heterostructure solar cells were first proposed and fabricated at the Ioffe Institute [165]. The use of a wide-gap "window" made of a thin layer of an AlGaAs solid solution (Fig. 14(a)), which is practically fully transparent for the sunlight, ensures passivation of the surface of the photoactive region [166, 167] and gives efficiencies close to the limiting theoretical values [168]: 24.6% for 100 suns in space and 27.5–28% for 100–200 suns in Earth conditions.

In recent years MOCVD has been widely used for fabricating AlGaAs/GaAs heterostructures for solar cells. Of particular interest is a MOCVD-grown heterostructure solar cell with a built-in Bragg mirror. In a structure developed at the Ioffe [169], the backside potential barrier is replaced with a multilayer dielectric mirror comprising 24 alternating GaAs (60 nm) and AlAs (70 nm) layers. The reflectivity of such a mirror is about 95% in the spectral range 750–900 nm. As a result, that part of solar radiation which is not absorbed in the base layer is reflected into the active region (Fig. 14(d)), so that the collection efficiency for carriers generated by light remains high at thinner base region and shorter diffusion lengths, which increases the radiation hardness of solar cells [169].

Owing to their higher efficiency and improved radiation hardness, AlGaAs/GaAs heterostructure solar cells are widely used in space. In the solar battery of the command module of *Mir* space-station were installed heterostructure solar cells with a total area of 60 m^2 fabricated by the technique developed at the Ioffe Institute and operating until now.

The efficiency can be increased further by using cascade solar cells [170–174] comprising multilayer heterostructures with two or more p-n junctions in materials with different gap widths (Fig. 15). In such elements the "top" p-n junction formed in a wider-gap material is intended for efficient conversion of the short-wavelength part of the solar spectrum, and the "bottom" p-n junction made of a narrower-gap material is optimized for conversion of the long-wave radiation passing through the wide-gap cell.

The increase of efficiency in cascade cells was demonstrated at the Ioffe Institute on a mechanically stacked cell [171]. In this design, GaAs is used as a material for the

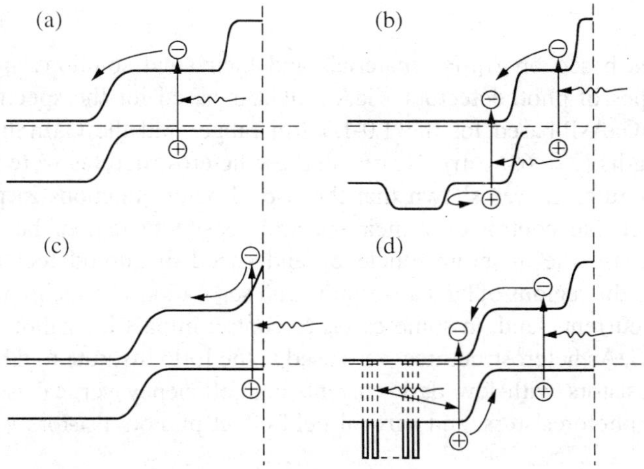

Fig 14. Band diagrams for AlGaAs/GaAs heterostructure solar cells: (a) structure with p-n junction in GaAs and frontal wide-gap p-AlGaAs "window;" (b) structure with backside potential barrier in n-region; (c) structure with an AlGaAs frontal layer of varied composition; (d) structure with a Bragg "mirror" in n-region.

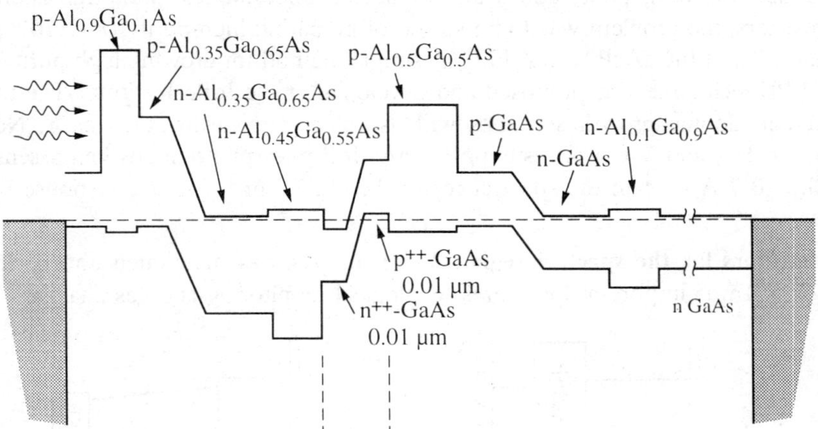

Fig 15. Band diagram for a monolithic cascade solar cell with AlGaAs/GaAs heterojunctions, fabricated by low-temperature LPE.

wide-gap cell and GaSb or InGaAs as a material for the narrow-gap one. The GaAs cell is made transparent for infrared radiation with wavelength exceeding 0.9 μm, and the narrow-gap cell ensures efficient conversion of the long-wavelength part of solar radiation in the spectral range 0.9–1.8 μm. The efficiency of such solar cells is 32–33% for AM1.5, 50–100 suns.

Of considerable interest is a monolithic cascade solar cell (Fig. 15) in which the "top" and "bottom" cells are connected by a tunnel p^+-n^+ junction providing for low ohmic loss. Cascade cells of this kind were fabricated using AlGaInP/GaAs [172, 173] and AlGaAs/GaAs [174] heterojunctions. The highest efficiencies achieved in these elements are 30–31% [172, 173] for direct and concentrated AM1.5 sunlight.

Photodetectors

Heterostructures based on $A_{III}B_V$ materials and their solid solutions are widely used now in new types of photodetectors: GaAs/AlGaAs-based for the spectral region 0.3–0.9 μm, InP/InGaAsP-based for the 1.0-1.6 μm range, and the GaInSbAs system for longer wavelengths (2.0–6.0 μm). Nearly all these heterostructures were first fabricated at the Ioffe Institute. It was shown that the use of heterojunctions in photodetectors not only does enable control over their spectral sensitivity region, but also makes it possible to increase the quantum efficiency and speed of photodetectors by bringing into coincidence the regions of light absorption and separation of nonequilibrium carriers, to reduce dark currents, and, in some cases, to control impact ionization processes [37, 175]. AlGaAs-GaAs heterostructures were used at the Ioffe Institute to fabricate first fast heterophototransistors with low dark current, high-efficiency vertical isotype N-n^0-n^+ heterostructure photoresistors, and vertical field effect phototransistors with a response time of $\sim 10^{-10}$ s [175, 176].

InGaAsP/InP heterostructure photodetectors. In the early 1980s, in connection with the development of a glass fiber with low loss at 1.3–1.55 μm, intensive work was commenced all over the world on creating photodetectors for this spectral region based on InGaAsP solid solutions. To create fast and low-noise high-performance light-sensitive devices, such as p-i-n photodiodes and avalanche photodiodes, phototransistors, and photoresistors, the problem was to be solved of preparing high-purity ($n < 10^{16}$ cm^{-3}) epitaxial InP and InGaAsP layers [177]. A simple method for growing high-purity layers by the LPE technique was proposed and developed at the Ioffe Institute. The method consisted in doping growth solutions with isovalent rare-earths, such as Y, Nd, Gd, Td, Dy, Er, Ho, and Yb. The resulting InGaAs/InP p-i-n photodiodes had a sensitivity $S_\lambda = 0.5{-}0.7$ A W^{-1} in the spectral region 1.3–1.55 μm [178] at a response time of ≤ 50 ns.

Photodetectors for the spectral region 2–5 μm. As was mentioned above, spectral region 2–5 μm is important for express ecological monitoring and gas analysis.

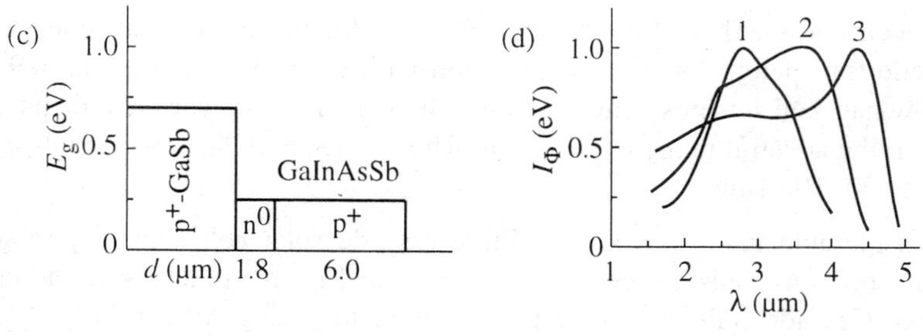

Fig 16. (a)–(c) Schematic of photodiode heterostructures and (d) spectral characteristics for indicated photodiodes at room temperature: curve 1 for (a), curve 2 for (b), and curve 3 for (c).

Uncooled GaInAsSb/GaAlAsSb photodiodes with a quantum efficiency of $\sim 0.9-1.0$ A W^{-1} and response time of ~ 0.5 ns operating in the spectral range 1.2–2.4 μm and also fast p-i-n heterophotodiodes having lightly doped GaInAsSb active layer, which enabled achieving as short a photoresponse time as 100–200 ps, were first fabricated at the Ioffe Institute by LPE onto GaSb substrates [179]. For the first time, uncooled and slightly cooled long-wavelength InAs/InAsSbP, InAsSb/InAs, and GaSb/InGaAsSb heterostructure photodiodes were developed (Fig. 16) for the spectral range 3–5 μm [180, 181] with quantum efficiency of 1.2 A W^{-1} at a wavelength $\lambda = 3\,\mu$m and response time of $\sim$ 50 ns. Also for the first time was discovered the phenomenon of "resonance" impact ionization in GaSb and InAs-like materials in which the forbidden gap width is close to the valence band spin-orbit splitting [182]. As a result, long-wavelength avalanche photodiodes with low level of excess noise were created on the basis of materials with high ratio of impact ionization coefficients for holes and electrons ($\geq 20-60$). Fast avalanche photodiodes of this kind for the spectral range 2–4 μm were developed using InAs, InAsSb, and GaAlAsSb/GaSb [182], and also GaInAsSb/GaAlAsSb heterostructures with separated regions of multiplication and absorption [183].

5.4 Devices for superhigh-speed electronics

Pulsed semiconductor devices. In the middle 1970s, an LPE technique for preparing undoped epitaxial GaAs layers and fabricating on their basis high-voltage p-n junctions with a breakdown voltage as high as 1.8 kV was developed at the Ioffe Institute. A strong modulation of the resistance of the high-resistivity base region by the intrinsic recombination radiation was revealed and means were found of enhancing this effect [184]. The development of a technique for growth of undoped epitaxial GaAs layers enabled proceeding to fabrication of more intricate structures—pulsed thyristors and transistors.

The high internal quantum efficiency of radiative recombination and the possibility of controlling in the course of epitaxial growth the forbidden gap widths of layers constituting the p-n-p-n structure made it possible to propose and study thyristors in which the turn-on voltage was controlled by intermediate conversion of the electric control signal into an optical one with the subsequent reverse conversion into an electric signal at the collector junction [185–187].

A significant increase in the turn-on voltage (to 1.5 kV) and speed was achieved in structures named photon-injection switches (PIS) [188]. A schematic of a PIS structure is given in Fig. 17. A characteristic feature of this structure is that, depending on the connection type, PIS can operate both as a fast pulsed thyristor and as a transistor. In the developed devices, comparatively slow diffusion-drift processes of nonequilibrium carrier transport in base regions were replaced by practically inertialess photon transport, which increased by more than an order of magnitude of the operating speed. In a PIS operating in the thyristor mode, the time of anode current rise is not longer than 100 ps, and the switching is stable to within 10–15 ps [187, 189]. In a structure connected as a transistor, the role of the base current is played by the photon flux from the LED part absorbed in the collector region of the n-n-p-n$^+$ transistor operating as a phototransistor. Despite the multistage nature of the conversion, the amplification coefficient in a circuit with common base is as high as 20–50. By now, transistors have been developed with collector-emitter voltage capacity of 500 V, pulse currents of up to 20 A at $\beta = 10$, and signal rise and decay times of $\sim$ 10 ns.

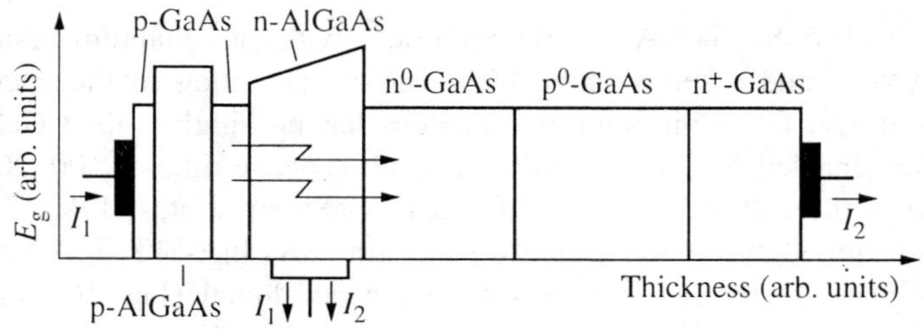

Fig 17. Forbidden gap profile along the structure of a photon-injection switch. Currents are shown for transistor connection: I_1—control current, I_2—output current. In thyristor connection, the current I_2 flows through the whole structure.

Heterobipolar transistors. As already noted, practical interest in heterojunctions arose in 1951, when, to increase the efficiency of the emitter junction at high injection levels, W. Shockley proposed a transistor with emitter having wider-gap emitter than the base region. However, only after a technique for fabrication of GaAs-AlGaAs heterostructures was developed in the early 1970s, were the first wide-gap-emitter HBTs outperforming silicon transistors created at the Ioffe Institute. It was shown that the wide-gap emitter in a HBT not only prevents the amplification coefficient from falling at high injection levels, but also enables the whole set of parameters (especially the operation speed) of the transistors to be improved through the possible free choice of doping levels of the base and emitter regions. The possibility of controlling the flows of majority and minority carriers in heterostructures let to actualize a transistor with wide-gap emitter and collector, which furnished an opportunity to drastically improve the operating speed and reduce the saturation voltage of an HBT operating in a switching mode [190, 191].

5.5 Microwave devices based on resonance-tunnel heterostructures and superlattices

Of growing interest for microwave applications are short-period superlattices in which ultrahigh operating frequencies have been achieved (up to several terahertz with frequency multiplication) and the active medium volumes are so large, as compared with double-barrier structures, that considerable microwave radiation power can be obtained. Here, much success has been achieved by researchers of the Ioffe Institute in cooperation with physicists from the University of Regensburg, Germany [192]. The key factor was the use of GaAs-AlAs and InGaAs-InAlAs superlattices with ultrathin (about monolayer thick) barriers. Extremely good results have been obtained with GaAs-AlAs superlattices with submonolayer AlAs barriers which are in fact structures with arrays of "anti"-quantum AlAs dots in a GaAs matrix.

6 Conclusion

Recently, impressing results have been obtained for short-wave radiation sources based on $A_{II}B_{VI}$ selenides and $A_{III}B_V$ nitrides. The use of heterostructure-related concepts and growth techniques developed for $A_{III}B_V$-based QDs and superlattices predetermined to a great extent the success of this research. A natural and predictable tendency is to apply heterostructure-related concepts and technological procedures to the new materials. The recently developed various $A_{III}B_V$, $A_{II}B_{VI}$, and $A_{IV}B_{VI}$ heterostructures substantiate this statement.

The classical heterostructures, QDs, and superlattices are already fairly perfect, and one can use quite a number of their unique properties. At the same time, structures with QWRs and QDs are still very young, and breathtaking discoveries and new unexpected applications are in store in this field. It is possible to say already now that ordered equilibrium QD arrays can be used in quite a number of devices: lasers, optical modulators, detectors and emitters for the far infrared, etc. Resonant tunneling via semiconductor atoms introduced into wider bandgap layers can improve device characteristics significantly. In a wider sense, QD structures are going to be developed both extensively and intensively. "Extensively" means new material systems capable of covering new regions of the energy spectrum. The system studied to the greatest extent—InGaAs-GaAs—has already been used to improve significantly characteristics of semiconductor lasers. It is likely that the problems related to the operation life of "green" and "blue" semiconductor lasers and even more general problems pertaining to fabrication of defect-free structures based on wide bandgap $A_{II}B_{VI}$ semiconductors and $A_{III}N$ nitrides can be solved by using QD structures in these systems.

"Intensively" assumes an understanding of the fact that the degree of ordering depends to a great extent on the exceedingly complicated growth conditions, material constants, and specific values of the surface free energy. The way to resonant-tunnel and single-electron devices and structures is an in-depth thorough study and evaluation of these parameters in order to achieve the maximum possible degree of ordering. On the whole, it is necessary to find more powerful mechanisms of self-organization to create ordered QD arrays. Correlated arrays of self-organized QWRs and QDs are rather promising for transverse Esaki–Tsu superlattices. Vertically coupled dots can be understood as a one-dimensional superlattice—an entirely novel object of study.

It is hardly possible to cover in a single paper even the main directions of the modern physics and technology of semiconductor heterostructures. Their number is much greater and the development of these directions will determine not only the prospects of solid state physics, but, in a sense, the future of the mankind.

The authors are grateful to D. Bimberg, V. Korol'kov, P. Kop'ev, A. Gorelenok, Yu. Zhilyaev, M. Mikhailova, V. Ustinov, A. Tsatsul'nikov, and Yu. Yakovlev for helpful discussions.

References

[1] V. P. Zhuze and B. V. Kurchatov, *Zh. Eksp. Teor. Fiz.* (JETP) **2** 309 (1932).
[2] Ya. I. Frenkel and A. F. Ioffe, *Phys. Zs. SU* Bd. 1, H. 1, 60 (1932).
[3] Ya. I. Frenkel, *Phys. Rev.* **37** 17 (1931).
[4] Ya. I. Frenkel, *Phys. Rev.* **37** 1276 (1931).
[5] Ya. I. Frenkel, *Phys. Rev.* **37** 17 (1931).
[6] E. F. Gross and N. A. Karriev, *Dok. Akad. Nauk* **84** 261 (1952).
[7] E. F. Gross and N. A. Karriev, *Dok. Akad. Nauk* **84** 471 (1952).
[8] B. I. Davydov, *Zh. Eksp. Teor. Fiz.* **9** 451 (1939).
[9] N. H. Welker, *Zs. Naturforsch.* **7a** 744 (1952).
[10] N. H. Welker, *Zs. Naturforsch.* **8a** 248 (1953).
[11] N. A. Goryunova, *Thesis, Leningrad State University–Physico-Technical Inst.* 1951.
[12] A. I. Blum, N. P. Mokrovskii and A. R. Regel, *Proc. VII Conf. on Semiconductors* (Kiev, 1950) [*Izv. AN SSSR. Ser. Phys.* **XVI** 139 (1952)].

[13] W. Shokley, *US Patent 2569347*, September 25 (1951).
[14] A. I. Gubanov, *Zh. Tekh. Fiz.* **20** 1287 (1950).
[15] A. I. Gubanov, *Zh. Tekh. Fiz.* **21** 304 (1951).
[16] H. Kroemer, *Proc. JRE* **45** 1535 (1957).
[17] H. Kroemer, *RCA Rev.* **28** 332 (1957).
[18] Zh. I. Alferov and R. F. Kazarinov, *Author's Certificate* No. 181737, Application No. 950840 with priority from March 30, 1963.
[19] H. Kroemer, *Proc. IEEE* **51** 1782 (1963).
[20] Zh. I. Alferov, V. B. Khalfin, and R. F. Kazarinov, *Sov. Phys.-Solid State* **8** 2480 (1967).
[21] Zh. I. Alferov, *Sov. Phys. Semicond.* **1** 358 (1967).
[22] L. Andreson, *IBM J. Res. Develop.* **4** 283 (1960).
[23] L. Andreson, *Solid State Electron.* **5** 341 (1962).
[24] Zh. I. Alferov, D. Z. Garbuzov, V. S. Grigor'ev, *et al.*, *Sov. Phys.-Solid State* **9** 208 (1967).
[25] Zh. I. Alferov, Yu. V. Zhilyaev, and Yu. V. Shmartsev, *Sov. Phys. Semicond.* **5** 174 (1971).
[26] Zh. I. Alferov, V. M. Andreev, V. I. Korol'kov, *et al.*, *Sov. Phys. Semicond.* **1** 1313 (1968).
[27] H. S. Rupprecht, I. M. Woodall, and G. I. Pettit, *Appl. Phys. Lett.* **11** 81 (1967).
[28] Zh. I. Alferov, V. M. Andreev, V. I. Korol'kov, *et al.*, *Sov. Phys. Semicond.* **2** 843 (1969).
[29] Zh. I. Alferov, V. M. Andreev, V. I. Korol'kov, *et al.*, *Sov. Phys. Semicond.* **2** 1289 (1969).
[30] Zh. I. Alferov, V. M. Andreev, V. I. Korol'kov, *et al.*, *Proc. IX ICPS*, Moscow, 23-29 July 1968, "Nauka", Leningrad, **1** 534 (1969).
[31] Zh. I. Alferov, *Proc. Int. Conf. on Luminescemce* (Newark, Delaware USA, August 25-29), 1969.
[32] Zh. I. Alferov, V. M. Andreev, V. I. Korol'kov, *et al.*, *Sov. Phys. Semicond.* **3** 460 (1969).
[33] Zh. I. Alferov, *J. Luminesc.* **1** 2, 869 (1970).
[34] Zh. I. Alferov, V. M. Andreev, E. L. Portnoi, *et al.*, *Sov. Phys. Semicond.* **3** 1107 (1970).
[35] Zh. I. Alferov, V. M. Andreev, V. I. Korol'kov, *et al.*, *Sov. Phys. Semicond.* **3** 785 (1969).
[36] Zh. I. Alferov, V. M. Andreev, M. V. Kagan, *et al.*, *Sov. Phys. Semicond.* **4** 2047 (1971).
[37] Zh. I. Alferov, F. A. Ahmedov, V. I. Korol'kov, and V. G. Nikitin, *Sov. Phys. Semicond.* **7** 780 (1973).
[38] Zh. I. Alferov, V. M. Andreev, V. I. Korol'kov, *et al.*, *Sov. Phys. Semicond.* **4** 481 (1970).
[39] V. M. Andreev, L. M. Dolginov, and D. N. Tret'akov, *Liquid Phase Epitaxy in the Semiconductor Devices Technology* ed. Zh. I. Alferov, Moscow, Izd. "Sov. Radio" 328p., 1975.
[40] R. R. Low, C. W. Mueller, J. I. Pankove, *et al.*, *Proc. IRE* **40** 1352 (1952).
[41] Zh. I. Alferov, *Thesis, Leningrad, Physico-Technical Inst.* 1961.
[42] H. Nelson, *RCA Rev.* **24** 603 (1963).
[43] Zh. I. Alferov, N. Yu. Antonishkis, I. N. Arsent'ev, *et al.*, *Sov. Phys. Semicond.* **21** 162 (1987).
[44] Zh. I. Alferov, A. T. Gorelenok, and D. Z. Garbuzov, *Sov. Techn. Phys. Lett.* **8** 113 (1982).
[45] A. A. Andaspaeva, A. P. Guseynov, and A. N. Baranov, *Fiz. Tekhn. Poluprovodn.* **24** 1708 (1990).
[46] Zh. I. Alferov, V. M. Andreev, S. G. Konnikov, *et al.*, *Kristall und Technik* **11** 1013 (1976).
[47] Zh. I. Alferov, V. M. Andreev, D. Z. Garbuzov, *et al.*, *Pisma Zh. Tekhn. Fiz.* **48** 809 (1978).
[48] V. M. Andreev, A. B. Kazantsev, V. R. Larionov, *et al.*, *AIP Conf. Proc.* Ed: Zh. I. Alferov, **240** 24 (1991).
[49] V. M. Andreev, L. B. Karlina, A. B. Kazansev, *et al.*, *Proc. of the First World Conf. on Photovoltaic Energy Conversion*, Hawaii, USA, 1713 (1994).
[50] Zh. I. Alferov, M. Z. Zhingarev, S. G. Konnikov, *et al.*, *Sov. Phys. Semicond.* **16** 532 (1982).
[51] N. N. Ledentsov, D. Bimberg, Yu. M. Shernyakov, *et al.*, *Appl. Phys. Lett.* **70** 2888 (1997).
[52] Zh. I. Alferov, N. Yu. Gordeev, S. V. Zaitsev, *et al.*, *Sov. Phys. Semicond.* **30** 197 (1996).
[53] D. A. Vinokurov, V. M. Lantratov, M. A. Sinitsin, *et al.*, *Sov. Phys. Semicond.* **25** (1991).
[54] V. M. Andreev, V. V. Komin, I. V. Kochnev, *et al.*, *Proc. Fist World Conf. on Photovoltaic Energy Conversion*, Hawaii, vol. II, p. 1894, 1994.
[55] P. S. Kop'ev and N. N. Ledentsov, *Sov. Phys. Semicond.* **22** 1093 (1988) and references in this paper.

[56] R. Heckingbottom, C. J. Todd, and G. J. Davies, *J. Electrochem. Soc.* **127** 444 (1980).
[57] R. Heckingbottom and G. Davies, *J. Cryst. Growth* **50** 644 (1980).
[58] S. V. Ivanov, P. S. Kop'ev, and N. N. Ledentsov, *J. Cryst. Growth* **104** 345 (1990).
[59] S. V. Ivanov, P. S. Kop'ev, and N. N. Ledentsov, *J. Cryst. Growth* **108** 661 (1990).
[60] S. V. Ivanov, P. S. Kop'ev, and N. N. Ledentsov, *J. Cryst. Growth* **111** 151 (1991).
[61] S. V. Ivanov, A. A. Boudza, R. N. Kutt, *et al.*, *J. Cryst. Growth* **156** 191 (1995).
[62] A. Yu. Egorov, A. R. Kovsh, A. E. Zhukov, *et al.*, *Phys. Semicond.* **31** 989 (1997).
[63] Zh. I. Alferov, V. M. Andreev, S. G. Konnikov, *et al.*, *Proc. Int. Conf. on Physics and Chemistry of Semiconductor Heterojunctions and Layer Structures* (Budapest, October 1970) [*Academiai Kiado*, **1**, 93 (1971)].
[64] G. A. Antipas, R. L. Moon, and J. W. James, *Conf. Ser. IOP* **17** 48 (1973).
[65] L. James, G. Antipas, and R. Moon, *Appl. Phys. Lett.* **22** 270 (1973).
[66] A. P. Bogatov, L. M. Dolginov, and L. V. Drughinina, *Kvant. Electron.* **1** 2294 (1974).
[67] J. J. Hsieh, *Appl. Phys. Lett.* **28** 283 (1976).
[68] Zh. I. Alferov, I. N. Arsent'ev, and D. Z. Garbuzov, *Sov. Techn. Phys. Lett.* **1** 147 (1975).
[69] Zh. I. Alferov, F. A. Ahmedov, and V. I. Korol'kov, *Sov. Techn. Phys. Lett.* **1** 191 (1975).
[70] W. R. Hitchens, N. Holonyak Jr., and P. D. Wright, *Appl. Phys. Lett.* **27** 245 (1975).
[71] H. Kroemer and G. Griffiths, *IEEE Electron. Dev. Lett.* **EDL-4** 20 (1983).
[72] Zh. I. Alferov, D. Z. Garbuzov, E. P. Morozov, and E. L. Portnoy, *Sov. Phys. Semicond.* **3** 885 (1969).
[73] A. N. Baranov, B. E. Dzhurtanov, A. N. Imenkov, *et al.*, *Sov. Phys. Semicond.* **20** 1385 (1986).
[74] V. N. Lutskii, *Phys. Stat. Sol. (a)* **1** 199 (1970).
[75] V. N. Lutskii and L. A. Kulik, *Pisma Zh. Eksp. Teor. Fiz.* **8** 3 (1968).
[76] L. V. Keldysh, *Fiz. Tverd. Tela* **4** 2265 (1962).
[77] R. Davies and H. Hosack, *J. Appl. Phys.* **33** 864 (1963).
[78] L. V. Iogansen, *Zh. Eksp. Teor. Fiz.* **45** 207 (1963).
[79] L. Esaki and R. Tsu, *IBM J. Res. Develop.* **14** 61 (1970).
[80] R. F. Kazarinov and R. A. Suris, *Sov. Phys. Semicond.* **5** 707 (1971).
[81] R. F. Kazarinov and R. A. Suris, *Sov. Phys. Semicond.* **5** 120 (1972).
[82] L. L. Chang, L. Esaki, and R. Tsu, *Appl. Phys. Lett.* **24** 593 (1974).
[83] J. Faist, F. Capasso, D. L. Sivco, *et al.*, *Science* **264** 553 (1994).
[84] R. Dingle, W. Wiegmann, and C. H. Henry, *Phys. Rev. Lett.* **33** 827 (1974).
[85] R. Dingle, H. L. Störmer, A. C. Gossard, and W. Wiegmann, *Appl. Phys. Lett.* **33** 665 (1978).
[86] D. C. Tsui, H. L. Störmer, and A. C. Gossard, *Phys. Rev. Lett.* **48** 1559 (1982).
[87] J. W. Matthews and A. E. Blakeslee, *J. Cryst. Growth* **32** 265 (1976).
[88] G. C. Osbourn, *J. Appl. Phys.* **53** 1586 (1982).
[89] A. I. Ekimov and A. A. Onushenko, *Pisma Zh. Eksp. Teor. Fiz.* **34** 363 (1981).
[90] Al. L. Efros and A. L. Efros, *Sov. Phys. Semicond.* **16** 772 (1982).
[91] S. P. Beamount, *Low-Dimensional Structures in Semiconductors*, ed. A. R. Peaker and H. G. Grimmeiss, Plenum Press, New York, p. 109, 1991.
[92] E. Kapon, M. C. Tamargo, and D. M. Hwang, *Appl. Phys. Lett.* **50** 347 (1987).
[93] H. Haken, *Synergetics*, Springer, Berlin-Heidelberg, 1977.
[94] I. P. Ipatova, V. G. Malyshkin, and V. A. Shchukin, *J. Appl. Phys.* **74** 7198 (1993).
[95] V. I. Marchenko, *JETP* **54** 605 (1981).
[96] A. F. Andreev and Yu. A. Kosevich, *Zh. Eksp. Teor. Fiz.* **81** 1435 (1981).
[97] V. I. Marchenko and A. Ya. Parshin, *Zh. Eksp. Teor. Fiz.* **79** 257 (1980).
[98] V. A. Shchukin, N. N. Ledentsov, P. S. Kop'ev, and D. Bimberg, *Phys. Rev. Lett.* **75** 2968 (1995).
[99] Zh. I. Alferov, D. Bimberg, A. Yu. Egorov, *et al.*, *Usp. Fiz. Nauk* **165** 224 (1995).
[100] P. D. Wang, N. N. Ledentsov, C. M. Sotomayor Torres, *et al.*, *Appl. Phys. Lett.* **64** 1526 (1994).
[101] P. D. Wang, N. N. Ledentsov, C. M. Sotomayor Torres, *et al.*, *Appl. Phys. Lett.* **66** 112 (1995).
[102] V. Bressler-Hill, A. Lorke, S. Varma, *et al.*, *Phys. Rev. B* **50** 8479 (1994).

[103] J. M. Moison, F. Houzay, F. Barthe, *et al.*, *Appl. Phys. Lett.* **64** 196 (1994).
[104] D. Leonard, M. Krishnamurthy, C. M. Reaves, *et al.*, *Appl. Phys. Lett.* **63** 3203 (1993).
[105] N. N. Ledentsov, M. Grundmann, N. Kirstaedter, *et al.*, *Proc. 22nd Int. Conf. Phys. Semicond.*, Vancouver, Canada, August 1994, ed. D. J. Lockwood, World Scientific, Singapore, Vol. 3, p. 1855, 1994.
[106] M. Grundmann, J. Christen, N. N. Ledentsov, *et al.*, *Phys. Rev. Lett.* **74** 4043 (1995).
[107] N. N. Ledentsov, V. M. Ustinov, S. V. Ivanov, *et al.*, *Usp. Fiz. Nauk* **166** 423 (1996).
[108] N. N. Ledentsov, *Proc. of the 23rd Int. Conf. on the Physics of Semiconductors*, Berlin, Germany, July 21–26, 1996, ed. M. Scheffler and R. Zimmermann (World Scientific, Singapoure), v. 1, p. 19, 1996.
[109] N. N. Ledentsov, M. Grundmann, N. Kirstaedter, *et al.*, *Proc. of the 7th Int. Conf. on Modulated Semiconductor Structures*, Madrid, 10–14 July, 1995, *Solid State Electron.* **40** 785 (1996).
[110] R. Heitz, M. Grundmann, N. N. Ledentsov, *et al.*, *Appl. Phys. Lett.* **68** 361 (1996).
[111] S. Ruvimov, Z. Liliental-Weber, N. N. Ledentsov, *et al.*, *Mat. Res. Soc. Symp. Proc.* Vol. 421, eds. by: R. J. Shul, S. J. Pearton, F. Ren, C.-S. Wu, Pittsburgh, PA, p. 383, 1996.
[112] Zh. I. Alferov, N. A. Bert, A. Yu. Egorov, *et al.*, *Phys. Semicond.* **30** 194 (1996).
[113] N. N. Ledentsov, J. Bohrer, D. Bimberg, *et al.*, *Mat. Res. Soc. Symp. Proc.* Vol. 421, eds.: R. J. Shul, S. J. Pearton, F. Ren, C.-S. Wu, Pittsburgh, p. 133, 1996.
[114] N. N. Ledentsov, V. A. Shchukin, M. Grundmann, *et al.*, *Phys. Rev. B* **54** 8743 (1996).
[115] F. Hatami, N. N. Ledentsov, M. Grundmann, *et al.*, *Appl. Phys. Lett.* **67** 656 (1995).
[116] I. Hayashi, *IEEE Trans. Electron. Dev.* **ED-31** 1630 (1984).
[117] Zh. I. Alferov, V. M. Andreev, D. Z. Garbuzov, and Yu. V. Zhilyaev, *Sov. Phys. Semicond.* **4** 1573 (1971).
[118] I. Hayashi, M. B. Panish, P. W. Foy, *et al.*, *Appl. Phys. Lett.* **17** 109 (1970).
[119] Zh. I. Alferov, V. M. Andreev, R. F. Kazarinov, *et al.*, *Author's Certificate No. 392875*, Application No. 1677436 with priority from July 19, 1971.
[120] H. Kogelnik and C. V. Shank, *Appl. Phys. Lett.* **18** 152 (1971).
[121] R. F. Kazarinov and R. A. Suris, *Sov. Phys. Semicond.* **6** 1184 (1973).
[122] Zh. I. Alferov, S. A. Gurevich, R. F. Kazarinov, *et al.*, *Sov. Phys. Semicond.* **8** 541 (1974).
[123] Zh. I. Alferov, S. A. Gurevich, N. V. Klepnikova, *et al.*, *Sov. Techn. Phys. Lett.* **1** 286 (1975).
[124] M. Nakamura, A. Yariv, H. W. Yen, S. Somekh, *et al.*, *Appl. Phys. Lett.* **22** 315 (1973).
[125] D. R. Scifres, R. D. Burnham, W. Streifer, *et al.*, *Appl. Phys. Lett.* **25** 203 (1974).
[126] S. A. Gurevich, S. Yu. Karpov, E. L. Portnoy, *et al.*, *Pisma Zh. Tekhn. Fiz.* **12** 268 (1986).
[127] J. P. van der Ziel, R. Dingle, R. C. Miller, *et al.*, *Appl. Phys. Lett.* **26** 463 (1975).
[128] R. D. Dupuis, P. D. Dapkus, N. Holonyak Jr., *et al.*, *Appl. Phys. Lett.* **32** 295 (1978).
[129] W. T. Tsang, *Appl. Phys. Lett.* **40** 217 (1982).
[130] E. Rezek, H. Shichijo, B. A. Vojak, and N. Holonyak Jr., *Appl. Phys. Lett.* **31** 534 (1977).
[131] Zh. I. Alferov, D. Z. Garbuzov, I. N. Arsent'ev, *et al.*, *Sov. Phys. Semicond.* **19** 679 (1985).
[132] Zh. I. Alferov, V. M. Andreev, A. A. Vodnev, *et al.*, *Sov. Techn. Phys. Lett.* **12** 450 (1986).
[133] Zh. I. Alferov, D. Z. Garbuzov, S. V. Zaitsev, *et al.*, *Sov. Phys. Semicond.* **21** 914 (1987).
[134] Zh. I. Alferov, N. Yu. Antonishkis, I. N. Arsent'ev, *et al.*, *Sov. Phys. Semicond.* **22** 650 (1988).
[135] D. Z. Garbuzov, *et al.*, *Technical Digest CLEO*, paper THU44, 396 (1988).
[136] D. Z. Garbuzov, *et al.*, *Conf. Digest 12th Int. Semicond. Laser Conf.* (Davos, Switzerland, 1990) p. 238.
[137] Zh. I. Alferov, A. M. Vasiliev, S. V. Ivanov, *et al.*, *Pisma Zh. Tekhn. Fiz.* **14** 1803 (1988).
[138] Y. Arakawa and H. Sakaki, *Appl. Phys. Lett.* **40** 939 (1982).
[139] S. Simhony, E. Kapon, T. Colas, *et al.*, *Appl. Phys. Lett.* **59** 2225 (1991).
[140] N. N. Ledentsov, V. M. Ustinov, A. Yu. Egorov, *et al.*, *Phys. Semicond.* **28** 832 (1994).
[141] N. Kirstaedter, N. N. Ledentsov, M. Grundmann, *et al.*, *Electron. Lett.* **30** 1416 (1994).
[142] N. Kirstaedter, O. G. Schmidt, N. N. Ledentsov, *et al.*, *Appl. Phys. Lett.* **69** 1226 (1996).
[143] D. Bimberg, N. Kirstaedter, N. N. Ledentsov, *et al.*, *IEEE J. of Selected Topics in Quantum Electronics* **3** 196 (1997).

[144] L. V. Asryan and R. A. Suris, *Semicond. Sci. Technol.* **11** 554 (1996).
[145] L. V. Asryan and R. A. Suris, *IEEE Journal of Selected Topics in Quantum Electronics* **3** 148 (1997).
[146] M. Grundmann and D. Bimberg, *Jpn. J. Appl. Phys.* **36** Part I, 4181 (1997).
[147] M. Grundmann and D Bimberg, *Phys. Rev. B* **55** 9740 (1997).
[148] H. Saito, K. Nishi, I. Ogura, S. Sugou, and Y. Sugimoto, *Appl. Phys. Lett.* **69** 3140 (1996).
[149] J. A. Lott, N. N. Ledentsov, V. M. Ustinov, *et al.*, *Electronics Lett.* **33** 1150 (1997).
[150] N. N. Ledentsov, I. L. Krestnikov, M. V. Maximov, *et al.*, *Appl. Phys. Lett.* **69** 1343 (1996), ibid. **70** 2766 (1997).
[151] L. A. Graha, Q. Deng, D. G. Deppe, and D. L. Huffaker, *Appl. Phys. Lett.* **70** 814 (1997).
[152] C. H. McMahon, J. W. Bae, C. S. Menonui, *et al.*, *Appl. Phys. Lett.* **66**, 2171 (1995).
[153] L. E. Vorob'ev, D. A. Firsov, V. A. Shal'gin, *et al.*, *Pisma Zh. Tekhn. Fiz.* **67** 256 (1998).
[154] G. E. Cirlin, V. N. Petrov, V. G. Dubrovskii, *et al.*, *Pisma Zh. Tekhn. Fiz.* **24** 10 (1988).
[155] Zh. I. Alferov, V. M. Andreev, D. Z. Garbuzov, and V. D. Rumyantsev, *Sov. Phys. Semicond.* **9** 462 (1975).
[156] A. Abdulaev, V. G. Agafonov, V. M. Andreev, *et al.*, *Sov. Phys. Semicond.* **11** 272 (1977).
[157] Zh. I. Alferov, V. M. Andreev, D. Z. Garbuzov, *et al.*, *Sov. Phys. Semicond.* **8** 2350 (1974).
[158] Zh. I. Alferov, V. M. Andreev, D. Z. Garbuzov, *et al.*, *Sov. Techn. Phys. Lett.* **45** 368 (1975).
[159] Zh. I. Alferov, V. M. Andreev, D. Z. Garbuzov, *et al.*, *Sov. Techn. Phys. Lett.* **3** 657 (1977).
[160] Zh. I. Alferov, V. M. Andreev, D. Z. Garbuzov, *et al.*, *Sov. Techn. Phys. Lett.* **4** 241 (1978).
[161] A. A. Andaspaeva, A. P. Guseynov, A. N. Baranov, *et al.*, *Pisma Zh. Tekhn. Fiz.* **15** 71 (1989).
[162] A. A. Popov, V. V. Sherstnev, and Yu. P. Yakovlev, *Techn. Phys. Lett.* **23** 701 (1997).
[163] A. A. Popov, M. V. Stepanov, V. V. Sherstnev, and Yu. P. Yakovlev, *Techn. Phys. Lett.* **23** 828 (1997).
[164] M. Aidaraliev, N. V. Zotova, S. A. Karandashov, *et al.*, *Phys. Stat. Sol. (a)* **115** k117, (1989).
[165] Zh. I. Alferov, V. M. Andreev, N. S. Zimogorova, and D. N. Tretyakov, *Sov. Phys. Semicond.* **3** 1633 (1969).
[166] V. M. Andreev, T. M. Golovner, M. B. Kagan, *et al.*, *Sov. Phys. Semicond.* **3** 2289 (1973).
[167] Zh. I. Alferov, V. M. Andreev, D. Z. Garbuzov, *et al.*, *Sov. Phys. Semicond.* **11** 1765 (1977).
[168] V. M. Andreev, V. P. Khvostikov, E. V. Paleeva, *et al.*, *Proc. of the 25th IEEE Photovoltaic Specialists Conf.* Washington, USA, p. 143, 1996
[169] V. M. Andreev, V. S. Kalinovsky, and V. V. Komin, *et al.*, *Proc. of the 4th European Space Power Conf.* Poitiers, France, 1995.
[170] L. M. Fraas, J. E. Avery, J. Martin, *et al.*, *IEEE Trans. Electron Dev.* **37** 444 (1990).
[171] V. M. Andreev, L. B. Karlina, V. P. Khvostikov, *et al.*, *Proc. of the 13th European Photovoltaic Solar Energy Conf.* Nice, France, p. 329, 1995.
[172] K. A. Bertness, S. R. Kurtz, D. J. Friedman, *et al.*, *Appl. Phys. Lett.* **65** 989 (1994).
[173] M. Yamaguchi and S. Wakamatsu, *Proc. of the 25th IEEE Photovoltaic Specialists Conf.*, Washington, USA, p. 9, 1996.
[174] V. M. Andreev, V. P. Khvostikov, E. V. Paleeva, *et al.*, *Proc. of the 23d Int. Symp. of Compound Semiconductors* St. Petersburg, Russia, p. 425, 1996.
[175] M. Z. Zhingarev, V. I. Korol'kov, M. P. Mikhailova, *et al.*, *Sov. Techn. Phys. Lett.* **5** 355 (1979).
[176] L. A. Volkov, V. G. Danil'chenko, and V. I. Korol'kov, *Sov. Techn. Phys. Lett.* **14** 682 (1988).
[177] A. T. Gorelenok, *Photodetectors and Photoconverters* Leningrad, Nauka p. 37–63, (in Russian) 1986.
[178] V. M. Andreev, A. T. Gorelenok, M. Z. Zhingarev, *et al.*, *Pisma Zh. Tekhn. Fiz.* **55** 668 (1985).
[179] I. A. Andreev, M. A. Afrailov, A. N. Baranov, *et al.*, *Sov. Techn. Phys. Lett.* **15** 15 (1989).
[180] I. A. Andreev, A. N. Baranov, M. P. Mikhailova, *et al.*, *Sov. Techn. Phys. Lett.* **18** 50 (1992).

[181] M. P. Mikhailova, S. V. Slobodchikov, N. D. Stoyanov, *et al.*, *Sov. Techn. Phys. Lett.* **22** 642 (1996).
[182] A. P. Dmitriev, M. P. Mikhailova, and I. N. Yassievich, *Phys. Stat. Sol. (b)* **140** 9 (1987).
[183] I. A. Andreev, M. A. Afrailov, A. N. Baranov, *et al.*, *Sov. Techn. Phys. Lett.* **14** 435 (1988).
[184] Zh. I. Alferov, Ya. V. Bergmann, V. I. Korol'kov, *et al.*, *Sov. Phys. Semicond.* **12** 38 (1978).
[185] Zh. I. Alferov, V. I. Korol'kov, N. Rahimov, *et al.*, *Sov. Phys. Semicond.* **12** 42 (1978).
[186] V. I. Korol'kov, V. S. Yuferev, and A. A. Yakovenko, *Sov. Phys. Semicond.* **14** 1005 (1978).
[187] V. I. Korol'kov, A. S. Prokhorenko, and A. V. Porkov, *Sov. Phys. Semicond.* **18** 311 (1992).
[188] Yu. M. Zadiranov, V. I. Korol'kov, S. I. Ponomarev, *et al.*, *Sov. Techn. Phys. Lett.* **32** 466 (1987).
[189] Zh. I. Alferov, V. M. Efanov, Yu. M. Zadiranov, *et al.*, *Sov. Techn. Phys. Lett.* **12** 529 (1986).
[190] Zh. I. Alferov, F. A. Akhmedov, and V. I. Korol'kov, *Sov. Techn. Phys. Lett.* **1**, 769 (1975).
[191] Zh. I. Alferov, G. D. Andreev, V. I. Korol'kov, *et al.*, *Sov. Phys. Semicond.* **7** 621 (1973).
[192] E. Schomburg, J. Grenzer, K. Hofbeck, *et al.*, *IEE J. Selected Topics in Quantum. Electron.* **2** 724 (1996).

Optical spectroscopy of semiconductors and dielectrics

B. P. Zakharchenya, A. A. Kaplyanskii, and S. A. Permogorov

Introduction

Spectroscopical studies of solids began at the Physico-Technical Institute at the times when Academician Abram Fedorovich Ioffe, the founder of the Institute, was still alive. They were initiated by Evgeny Fedorovich Gross whom Ioffe had invited to join the Institute soon after the end of the World War II in order to organize a laboratory for optical research. E. F. Gross had been widely known for his outstanding experiments on the Raman scattering in crystals. The most significant had been his famous observation of the Brillouin scattering in quartz crystals. At the pre-laser epoch this experiment was extremely difficult to perform, and the observation of the Reyleigh line fine structure required a 100-hours exposition. This was done in 1930.

E. F. Gross and his post-graduate assistant Nuri Kariyev, following the ideas of S. M. Ryvkin and V. P. Zhuze who had suggested the participation of a quasiparticle "exciton" in the photoconductivity of crystalline cuprous oxide, found in this crystal an unusual, amazingly beautiful phenomenon—the optical spectrum of the exciton. This spectrum, consisting of narrow lines, is shown in Fig. 1. It is the absorption spectrum of a thin Cu_2O plate cooled to liquid helium temperature. By its hydrogen-like convergence to the continuous absorption edge, it wonderfully resembles the Lyman and Balmer series, so well known to us since the days of school courses... At first, the phenomenon was discovered at liquid nitrogen temperature with six exciton lines had been observed. The experimental results were published in 1952 [1], although the experiment itself had been carried out one year before. E. F. Gross interpreted that spectrum with confidence as the optical spectrum of the exciton, a quasiparticle consisting of two light particles, electron and hole, bound by Coulomb interaction in a medium possessing a fairly high dielectric constant. It should be recalled that such model of exciton, well fitting in the usual Bloch band scheme of semiconductor, was proposed by Wannier and Mott. But still earlier, in 1931, the concept of the exciton as an excited state migrating in a crystal from cell to cell owing to translation symmetry, had been introduced into the solid state physics by a prominent theorist; Institute's staff member, Yakov Il'ich Frenkel.

The observation of an atomic-like spectrum in a semiconductor was absolutely unexpected event for the physicists of that period, among whom soon formed a strong group of opponents, advocating a more trivial interpretation of the phenomenon in terms of impurities. As for E. F. Gross, he was firm in his conviction that the observed spectrum is of excitonic nature. In short time a scientific school was created by Gross to perform

Fig 1. Photograph of absorption spectrum of a Cu_2O crystal platelet at $T = 4.2$ K. Numbers $n = 2, \ldots 7$ indicate transitions into exciton states with corresponding principal quantum numbers.

the following tasks: search for an analogous phenomenon in other semiconductor crystals; studies of the effect of external electric and magnetic fields on exciton spectra and the effect of elastic deformation-tensor fields on narrow lines of excitons; investigations of the emission spectra of excitons; study of the dynamics of excitons and the spatial dispersion effects; investigation of optical orientation and alignment of excitons, electrons, and nuclei. Recently the spectroscopy of excitons in low-dimensional nanostructures has been extensively developed.

At the Institute, the elaborate spectroscopical methods developed for semiconductors has been successfully applied to other objects: dielectric crystals and, most recently, fullerenes. In a brief review we cannot show the whole diversity of optical and spectroscopic investigations being performed at our institute; suffice it to say that they all originate from the E. F. Gross' initial observation of the exciton spectrum.

1 Optics and spectroscopy of semiconductors

1.1 Electro- and magneto-optics of semiconductors

First experiments on the effect of electric fields on exciton lines in Cu_2O were carried out as early as 1953 [2]. In these experiments were observed the Stark effect in excitons, flaring-up of new forbidden lines, and ionization of exciton states, accompanied by a shift of the absorption edge corresponding to the Keldysh–Franz effect. Magneto-optical phenomena were investigated in more detail. The starting point was the observation of a Zeeman splitting of a very narrow quadrupolar $n = 1$ exciton line in Cu_2O [2] and the observation of so-called diamagnetic shift of exciton lines, with $n > 2$ (here n is the principal quantum number) [3]. The discovery of this relativistic effect caused by acceleration of the orbital motion of light particles in a magnetic field, was possible owing to the gigantic size of highly excited excitonic states. These effects formed the basis for the magneto-optics of semiconductors. Next, again in Cu_2O crystal, oscillations of magneto-absorption were discovered, which is the main magneto-optical effect in semiconductors [4]. Practically at the same time an analogous phenomenon was found in InSb crystals by Burstein and Zverdling and in Ge by B. Lax *et al.* In these works the effect was interpreted as due to optical transitions between Landau sublevels in valence and conduction bands. By contrast, in a work carried out at the Institute, the oscillations of magneto-absorption were attributed to transitions into quasi-one-dimensional excitonic states formed by electrons and holes in diamagnetic Landau subbands participating in the optical transition. Such excitons were named diamagnetic, and their existence was proved by numerous experiments [5]. In the foreign literature they were called magnetoexcitons.

The experiments performed at the laboratory of R. Seisyan revealed new magneto-optical effects in InAs, InSb, InP, CdTe, ZnTe, ZnSe, CdSe, and CdS crystals, in a PbTe semimetallic compound [6] and in CdHgTe and InGaAs solid solutions [7]. The same group is engaged in magneto-spectroscopic investigations of low-dimensional structures. For instance, magneto-optical experiments provide strong arguments in favor of the concept of a "Coulomb well" [8], which is formed as a result of the Coulomb attraction of light holes by an electron localized within the boundaries of a quantum well. The magneto-optical experiments have proved to be a key method in investigation of such artificial localized states as above-barrier excitons formed by Bragg localization of the electron wave function within the background of the continuum of states [9].

1.2 Exciton motion and spatial dispersion

An important property of excitons as intrinsic excitations of crystals is their free motion. First experimental evidences of exciton motion were obtained in studying the effect of magnetic field inversion on the exciton absorption in CdS crystals. Propagation of an exciton orthogonally to the magnetic field [10] produces an additional electric field in the exciton coordinate framework, which in its turn shifts the exciton levels, depending on the mutual orientation of the magnetic field and exciton wave vector.

The most direct evidence of free exciton motion was obtained in a study of phonon-assisted exciton luminescence [11]. The process of radiative exciton recombination has an annihilative nature, and in this sense the exciton is more similar to positronium than to the hydrogen atom [12]. The whole amount of exciton energy—internal and kinetic—must be transferred to the emitted photon and phonons. As a result, the distribution of exciton kinetic energies can be directly visualized in the emission spectrum.

In the case of exciton system in thermal equilibrium, the shape of the phonon-assisted emission band is Maxwellian. For non-equilibrium energy distributions, the band shape provides information on the electronic temperature and can be used as an internal thermometer. A detailed analysis of the emission band shapes for processes with emission of different number of LO phonons [13] has also demonstrated the dependence of the phonon-assisted emission band shape on the magnitude of the wave vector transferred from an exciton to phonons. This dependence allows one to detect the elastic scattering of excitons by impurities or defects.

The established properties of exciton-phonon luminescence have a universal character and are widely used for characterization of new materials and structures.

Very important new results have been obtained in a study of the excitation spectra of exciton luminescence. It was found that under optical excitation, even in the region of fundamental absorption, the most efficient generation of excitons takes place through the excitation of correlated electron–hole pairs which can be considered as "hot" excitons with kinetic energies far exceeding the exciton binding energy [14]. For such pairs the energy relaxation with emission of LO phonons is so fast that the pairs have no time to separate into carriers but form normal exciton states with kinetic energies less than the energy of LO phonon. Further thermalization of these states involves acoustic phonons. It has been shown that relaxation of hot exciton is the main mechanism of optical excitation of excitons in semiconductor crystals and heterostructures.

In quantum-mechanical approach, the relaxation of hot excitons should be considered as a process of multiphonon resonant scattering of the initial photon through real intermediate states into a final state which may be either a mechanical exciton or an emitted photon. Owing to the continuum nature of the energy spectrum of exciton intermediate states, the processes of hot luminescence and resonant Raman scattering in the region of fundamental absorption are indistinguishable. More rigorously, the optical processes in the region of exciton resonance should be considered as resonant secondary emission [15] with properties intermediate between those of luminescence and light scattering. In particular, the Raman scattering through real intermediate states attains a measurable decay time. Magneto-optical studies of resonant Raman scattering in CdS has allowed measuring the exciton relaxation times in the picosecond range [16].

Scientists of the Ioffe Institute made a significant contribution to the studies of exciton polaritons. The existence of the exciton dipole moment causes splitting of the longitudinal and transverse exciton states, which results in a characteristic "anticrossing" of dispersion curves in the region of exciton resonance. In this region the intrinsic excitations of semiconductors have mixed exciton–photon properties and are called "polaritons".

In 1971, the energy dependence of inelastic polariton scattering on LO phonons was measured in the region of $n = 1$ exciton resonance in CdS crystals [17]. An analysis of these data provided for the first time the polariton dispersion curve. A peculiar property of exciton resonance is the existence of so-called additional waves in the resonant region. It appears that at a given energy several polariton states with quite different wave vectors can simultaneously interact with the incident light. The interference pattern of additional waves was observed for the first time in transmission and reflection spectra of ultra-thin CdSe platelets [18].

The interference of additional waves is a clear manifestation of the spatial dispersion of exciton states, which arises owing to the spatial nonlocality of the dielectric constant. The first experimental proof of the spatial dispersion effect on exciton resonances was obtained in a study of optical anisotropy of the $n = 1$ quadrupole exciton transition in cubic Cu_2O crystals [19].

As an effect opposite to the free exciton motion, the exciton localization in disordered systems, in particular, in semiconductor solid solutions, has been studied. Recently, the semiconductor solid solutions have attracted great interest owing to their wide use in modern thin-layer heterostructures. Exciton localization in solid solutions is caused by the tailing of carrier- and exciton-band edges due to compositional fluctuations. A systematic study of a wide class of II–VI solid solutions by modern spectroscopical methods [20] has shown that the optical absorption and emission in solid solutions, as well as the dynamic characteristics of emission, are mostly governed by the properties of localized exciton states.

1.3 Optical orientation in semiconductors

The observation of quasi-atomic exciton spectra led to the application of some methods of optical spectroscopy of isolated atoms to a study of quantum phenomena in semiconductors. It was natural that after the observation of Stark and Zeeman effects there appeared the method of optical orientation of spins, based on the conservation of angular momentum and first applied by Kastler to the optics of isolated atom.

Lampel [21] and Parsons [22] were the first to succeed in experimenting with semiconductors in that direction. Parsons performed optical orientation of electron spins and its detection by means of polarized luminescence.

At the Ioffe the investigations on optical orientation started in the beginning of the 1970s and have been carried on intensively by a number of experimental and theoretical groups. The method of optical detection of nuclear magnetic resonance by measuring the degree of luminescence polarization in weak external magnetic fields was first developed in these investigations. The early experiments were performed for GaAlAs by Berkovits, Safarov and Ekimov and theoretically explained by Perel' and Dyakonov. The observed effect is due to the contact Fermi interaction of the nuclear spins with the spins of electrons localized in s-states. The magnetic field of polarized nuclei has a reciprocal effect on the polarization of the electrons responsible for its generation. As a result, the degree of polarization of luminescence, resulting from the recombination of polarized carriers, changes. A thorough study of electron-nuclear interaction in semiconductors has led to observation of a deep cooling (down to 10^{-6} K) of the nuclear spin system of semiconductors by light, including the effect of so-called resonance cooling, and also to the discovery of quadrupole interaction between semiconductor lattice nuclei and of multispin resonances.

For the first time the optical alignment of excitons was observed. The methods of optical orientation and depolarization of luminescence by an external magnetic field (the Hanle effect, also adopted from spectroscopy of isolated atom) are widely used for investigating different mechanisms of spin relaxation, diffusion of spin-labelled electrons, and for detecting the spin resonance in semiconductors.

All the above-mentioned physical phenomena and many others are treated in a collective monograph “Optical Orientation”, edited by Zakharchenya and Meier [23].

At present, the optical orientation methods are successfully used for studying quantum-size structures, semimagnetic semiconductors, and semiconductor-magnetic metal heterostructures. The ideas of optical orientation have greatly influenced the studies of hot luminescence of electrons, discovered at the Ioffe Institute in 1976 [24].

1.4 Photoluminescence of hot electrons

In studying the spectral and polarization properties of photoluminescence of hot electrons in GaAs crystals (denoted further as hot photoluminescence, or HPL) a phenomenon of optical alignment of the momenta of photoexcited electrons was discovered. In effect, the interband absorption of light in semiconductors with a diamond-like structure produces a strongly anisotropic velocity distribution of photoexcited carriers. Under excitation with linearly polarized light the recombination radiation of such hot electrons is linearly polarized. In external magnetic field a depolarization of HPL due to cyclotron motion of electrons has been observed.

Based on these two effects, a new method for studying of fast relaxation processes through HPL depolarization was developed and widely used. In particular, for the first time the characteristic times of optical phonon emission by hot electrons, as short as 140 and 100 femtoseconds, were found for GaAs and InP crystals, respectively. Times of $\Gamma - L$ and $\Gamma - X$ intervalley scattering and times of energy and momentum relaxation via scattering on neutral acceptors, and those of scattering between valence subbands were also determined.

The discrete nature of hot electron energy distribution function, due to emission of LO phonons, manifests itself in HPL spectra of moderately doped crystals (Fig. 2). The distribution function of hot electrons under optical pumping was reconstructed for the first time from HPL spectra in a wide range of energies with due account of the energy dependence of the transition probability. Methods of the HPL spectroscopy were successfully used for studying band structure away from the Brillouin-zone center for

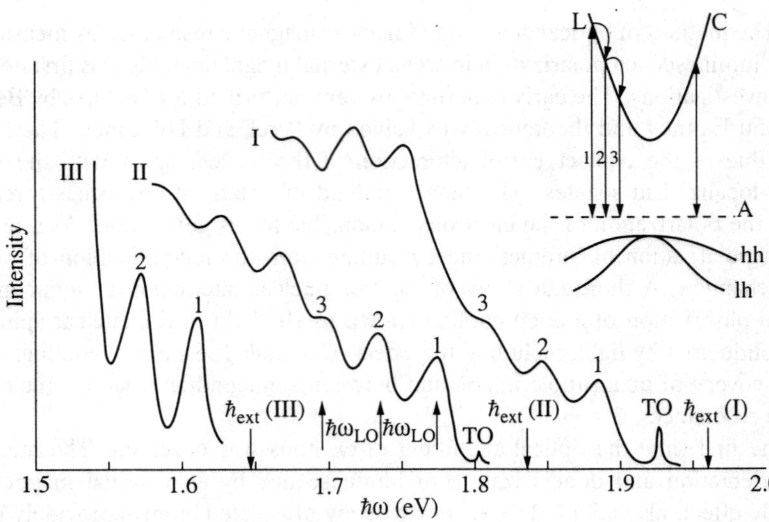

Fig 2. Spectra of HPL in acceptor doped GaAs at $T = 2$ K taken with different excitation energies (I–III). Inset illustrates the origin of emission bands.

a number of semiconductors. So, the positions of the side valleys were determined for GaAs and InP with high precision from the HPL spectra. HPL studies under uniaxial stress provided an unambiguous determination of the side valleys' symmetry and gave reliable values of the shear deformation potentials for the L- and X-valleys. Luttinger parameters were determined for InP with good accuracy by two independent methods (from the position of the zero-phonon peak in HPL spectrum and from the anisotropy of linear polarization).

Circular polarization of the HPL in external magnetic field was studied in crystals doped with different acceptor impurities, both shallow (Zn,Ge) and deep (Mn). The circular polarization in this case was due solely to magnetization of acceptor-bound equilibrium holes. The results obtained for Mn-doped GaAs were explained by antiferromagnetic interaction of localized holes with the 3d core. This conclusion was later confirmed by other authors through ESR studies.

The alignment of two-dimensional electrons was observed in a study of HPL in multiple quantum wells (MQW). Contrary to the three-dimensional case, where the initial polarization is almost independent of the kinetic energy of photoexcited electrons, the polarization in quantum wells (QW), vanishes in agreement with theory, with decreasing of kinetic energy.

A study of electron–phonon interaction in GaAs/AlAs and GaAs/GaAlAs MQWs was performed. The energies and type of phonons (interface or confined phonons) which dominate the energy relaxation were determined from the energy losses in HPL spectra. The scattering rate was evaluated from measurements of magnetic depolarization. For GaAs/AlAs MQW the scattering rate changes from 11 to 6 reciprocal picoseconds for QW widths in the range 4–14 nm in fair agreement with the calculations performed within the framework of the dielectric continuum model (DCM). These results confirmed the applicability of the DCM to analysis of electron–phonon interaction in QWs.

Comparative studies of scattering rates of hot electrons by neutral acceptors in 3D (thick films) and 2D (QWs) cases were performed. It was found that in QWs the scattering rate was reduced by half owing to a decrease of the density of final electron states.

The HPL polarization in superlattices (SL) proved to be very sensitive to the ratio of energies of electron motion along the miniband and in the lateral direction (in QW plane). The width of the SL miniband was determined from the energy dependence of the degree of polarization. Polarization measurements in magnetic field demonstrated a sharp anisotropy of momentum distribution of photoexcited electrons. The polarization characteristics of HPL undergo radical change in the course of the 2D-to-3D transition, i.e., from a QW to a SL with a wide miniband. The upper limit for the applicability of tight-binding approximation in the theory of SL was established.

1.5 Raman scattering

Lately a series of Raman scattering (RS) studies of MQWs was performed. Spin-flip RS for a MQW doped with acceptor impurities (spin-flip of an acceptor-bound hole) and an undoped MQW (momentum-flip of a localized exciton) were observed for the first time. The g-factors for the acceptor-bound hole and the electron and hole in the exciton, as well as the g-factor anisotropy and dependence on the wall thickness were determined.

A new type of RS by acoustic phonons in SL was found. It is due to fluctuations of QW widths and, consequently, of electron confined-state energies, i.e., their inhomogeneous broadening. Actually, this type of RS is related to individual wells and has a form of a continuum. It coexists with RS from SL represented by narrow peaks due to the so-called folded phonons. Both components of RS are clearly seen in Fig. 3. The ratio of the intensities of the two components of RS and its temperature dependence is governed by the ratio of homogeneous and inhomogeneous widths of the intermediate states.

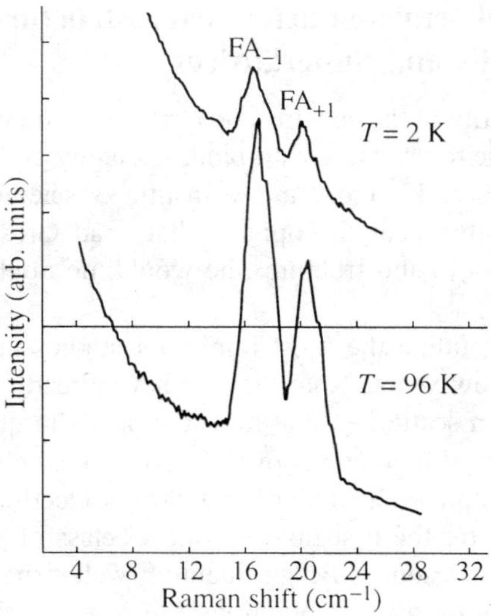

Fig 3. Low-energy Raman spectrum of GaAs/AlGaAs MQWs. Sharp peaks are due to folded acoustic phonons.

Wide potentialities of the HPL spectroscopy are employed in a number of laboratories in USA, Germany, and France. The phenomenon of optical alignment manifests itself in a number of optical effects, such as induced birefringence and dichroism. The above-mentioned observation of acoustic phonon scattering initiated work on magneto-Raman acoustic spectroscopy. The investigations described in sections 1.4 and 1.5 are discussed in more detail in [25–28].

1.6 New methods of optical detection of microwave resonances in semiconductors

A new phenomenon, the effect of cyclotron resonance (CR) on the intensity of luminescence in semiconductors, has been found. For the first time the CR was detected by monitoring the recombination luminescence of electron–hole drops in ultra-pure germanium crystals. The developed technique of optical detection of CR is widely used now for investigation and local diagnostics of semiconductor materials and low-dimensional structures [29]. Multiquantum resonances, which were optically detected in semiconductors, superlattices, and in AgCl, allowed studying EPR with an absorption of up to 7 microwave quanta, i.e., detecting EPR at low and high frequencies simultaneously. The developed technique was used successfully to determine the fine structure of exciton energy levels in various semiconductors and in AgCl [30].

A significant achievement in the field of EPR in semiconductors was the development by L. Vlasenko of a new high-sensitive method for the spectroscopy of deep centers. This method is based on the known effect of spin-dependent recombination, i.e., a resonant change of the recombination rate of photoexcited carriers and, consequently, change of photoconductivity of samples under EPR saturation conditions. In this case photoconductivity is detected by monitoring the Q-factor of the EPR spectrometer cavity. This method was successfully applied to silicon and silicon-based devices. New paramagnetic recombination centers were found and investigated by this method [31, 32].

1.7 Spectroscopy of semiconductor heterostructures (Quantum Wells and Superlattices)

The overwhelming majority of the spectroscopic studies of quantum-size structures rests upon the existence of excitonic states with binding energy and oscillator strength much higher than those in the bulk. From many Institute's scientists engaged in studying this problem, one can often hear the opinion that "had Gross lived till the advent of the exciton spectroscopy of nanostructures, he would, no doubt, have become a Nobel Prize winner".

In what follows, we outline the most important studies in this field carried out at the Institute in the last few years. Note that we have already mentioned the works on hot luminescence, Raman scattering, and magneto-optics in quantum-size structures.

Following the theoretical predictions of E. L. Ivchenko, coherent interaction of excitons in neighboring quantum wells under the action of electromagnetic field have been observed experimentally for the first time in a novel class of periodical semiconductor quantum well structures—resonant Bragg quantum well structures. In such structures with a period d satisfying the Bragg conditions $K_z d = \pi$ (where K_z is the z component of the wave vector) the exciton resonance reflectivity and the exciton line broadening increase along with the number of wells in a structure [33, 34].

By analyzing the exciton-reflection and four-wave mixing spectra in quasi-one-dimensional semiconductor clusters in the GaAs/AlAs system (grown by MBE on [311] GaAs substrates at P. S. Kop'ev's laboratory), it has been shown that the excitonic transitions in such clusters are characterized by high oscillator strength and high nonlinear susceptibility and spatial anisotropy [35, 36].

Three possible mechanisms of photoinduced modification of exciton-resonance refractive index, related to modification of the excitonic parameters (resonance frequency of transitions, oscillator strength, and damping), have been studied experimentally in quantum well structures. Diffraction of light on a photoinduced spatial modulation of the refractive index caused by exciton damping modulation has been found [37, 38].

The dependence of exciton oscillator strength on quantum well width has been measured in semimagnetic quantum-well structures CdTe/CdMnTe. This dependence is controlled by the overlap of the electron and hole wave functions. Its character indicates the influence of the Coulomb potential of a quantized electron on the hole wave function. A type I-to-type II transition has been observed in the dependence of the exciton oscillator strength on magnetic field [39, 40].

By studying magnetic-field-induced polarized luminescence and optical orientation the following effects have been found in $A_{III}B_V$ heterostructures: (i) dependence of the hole spin relaxation time on the superlattice period related to heavy-light hole mixing [41]; (ii) anisotropy of electron g-factor in superlattices [42]. In type I and type II GaAs/AlAs superlattices the constants of exciton-exchange and anisotropic-exchange splitting have been determined using the resonant optical orientation and optical alignment of exciton dipole momenta in magnetic fields. The contributions of free and localized excitons have been separated using the effect of optical alignment [43].

In modulation-doped quantum-well structures with low-density 2DEG a phenomenon of combined exciton-cyclotron resonance has been found in magnetic fields. The effect involves photogeneration of excitons with simultaneous excitation of a free electron between Landau levels [44].

Methods based on optical detection of EPR and level anticrossing have been developed to study the exciton fine structure in GaAs/AlAs type I and type II superlattices. On the basis of these studies the possibility of local diagnostics of superlattice parameters was shown, and the properties of two interfaces in a quantum well were investigated separately [45]. Magneto-optical studies of hybrid ferromagnetic-semiconductor nanostructures were performed for the first time. Optically oriented carriers in a semiconductor were used to detect magnetization in a ferromagnetic film. A coercive force modification of the ferromagnetic interface by optical orientation has been found [46].

The fine structure of localized excitons and exciton-impurity complexes has been studied theoretically and experimentally in superlattices and quantum dots. Polarization spectroscopy was used successfully to measure the small exchange splitting between the optically active and inactive exciton states in the case when these states were not resolved spectroscopically, as well as to measure all the basic parameters characterizing the fine structure of excitons and their spin dynamics [47]

Resonant cooling of the nuclear spin system in quantum wells due to modulation of the spin polarization of 2D electrons with NMR frequency has been discovered. The anisotropy of the electron g-factor in isolated quantum wells has been observed and investigated. A simple and reliable optical method has been developed for determinating the g-factor sign in isolated quantum wells of various width. A simple and reliable optical method has also been developed for determinating the *g*-factor sign of electrons in quantum wells, using the Hanle effect and quantum beats of electron spin [48].

In studying the optical orientation in doped quantum dots obtained by MOS-hydride epitaxy, the effects due to nuclear polarization are particularly well pronounced. In InP/InGaP quantum dots nuclear magnetic resonance was optically detected and the hyperfine structure of exciton investigated (the Overhauser effect). In semimagnetic cubic CdMnTe semiconductors, the dependence of the magnetic-field-induced radiative linear polarization on the field direction has been discovered, despite the fact that all the effects in this object are usually described in a spherical approximation. Similar experiments on quantum wells have revealed a sharp anisotropy of the hole *g*-factor in the quantum-well plane, which is due to D_{2d} symmetry of the quantum well (the quantum-mechanical mixing of heavy- and light-hole states) [49].

1.8 Investigation of high density excitons[1]

At a sufficiently high density of excitons, formation of exciton molecules (biexcitons) and even of exciton liquid is possible. This follows from a clear analogy between excitons and positronium atoms. The expectation of the exciton liquid formation follows from the theorem of statistical physics which states that if molecules or atoms are subject to attraction, no matter how weak, a liquid must form, provided the temperature is low enough. The exciton science gives answer to the question as to whether the condensed state would be an insulator or a metal and what properties the exciton molecules would have. The structure of the exciton molecule depends mainly on the number of valleys in the conduction and valence bands. The anisotropy of electron and hole effective masses is also important. In the simplest case, in which both the conduction and valence bands have only one valley and the electron and hole effective masses are isotropic, only biexcitons and an insulating excitonic liquid can exist. For multivalley semiconductors, such as Ge, Si, GaP, AlAs, AlSb, AlN, and many solid solutions of these semiconductors, it has been shown by both experimental and theoretical investigations that: (i) there may exist at least three (or more) kinds of exciton molecules, such as biexcitons, triexcitons and tetraexcitons; (ii) the exciton condensate is always metallic, and, furthermore, the metallic liquid has abnormally high density. The reason for such a behavior is rather simple. In multivalley semiconductors there are many different electrons and holes occupying the 1s-state, because the additional quantum number is the number of the valley. For example, in silicon 12 electrons are in 6 different valleys, but all these electrons may be in the 1s-state (spin↑ and spin↓ in each valley). The number of different holes in the exciton molecule is four ($\pm 3/2$ spins and $\pm 1/2$ spins). Note that in the case of a positronium-like band structure the exciton molecule may have 2 electrons and 2 holes in the 1s-state. In silicon the following exciton molecules may exist: biexcitons, triexcitons, and tetraexcitons. The molecules which consist of three and four excitons are stronger bound and have smaller radius as compared with the exciton radius. At present the question is discussed as to whether there exist some new strongly bound (in terms of energy per exciton) and, therefore, stable exciton molecules containing 11 or 12 excitons. These molecules have 4 holes in the 1s-state and 7 or 8 holes in the 2p-state. All electrons in such a molecule occupy the 1s-state.

The physical properties of the exciton liquid had been practically completely investi-

[1] A. A. Rogachev [50–52]

gated before 1976 by Soviet physicists under the guidance of Academician L. V. Keldysh. For the present discussion it is essential to mention that the exciton liquid density in Ge and Si is 10 times higher than in such metallic liquids as sodium or potassium (with the effective Bohr radius taken as unit length). Both the density and the binding energy in the exciton molecules are smaller than in the exciton liquid. The reason is the same as the reason why smaller drops are always weaker bound than bigger ones.

Of current interest is a study of quantum dots or, in other words, three-dimensional small-size quantum wells, in multivalley semiconductors. There are two different kinds of quantum dots, viz. isoelectronic impurities and artificially grown narrow quantum wells. The laboratory of Electronic Semiconductors is carrying out the investigations of isoelectronic impurities in such multivalley semiconductors as GaP, AlN, and their solid solutions. The peculiarity of recombination in multivalley semiconductors is the possibility that a center captures 4 electrons and 4 holes, or even more, simultaneously. The luminescence spectra of such semiconductors contain more than ten different lines, and their identification needs creation of a theoretical scheme of classification. If a strong electric field is applied to the semiconductor surface, it is screened by charge carriers which form a self-consistent quantum well near the surface. The charge carriers filling this quantum well are in equilibrium with excitons and exciton molecules. In the multivalley semiconductors, the surface system of excitons contains bound excitons, bound biexcitons, bound triexcitons, bound tetraexcitons and other surface exciton molecules. This leads to the appearance of exciton luminescence spectrum consisting of many narrow and wide lines. Investigation of exciton luminescence spectra is an effective means to study the surface of semiconductors [50–52].

2 Optical spectroscopy of dielectrics

2.1 Study of dielectrics with impurity centers and defects

This study dealt primarily with ionic crystals (halogenides, oxides) activated by iron-group impurity ions (unfilled $3d^n$ shell) and $4f^n$ rare-earth ions. Interest in the properties of these crystals was spurred in the early 1960s by the development of the solid-state ruby laser (Al_2O_3:Cr^{3+}). The optical absorption and luminescence spectra of impurity ions obtained within the broad spectral-transparency range of crystalline hosts contain usually narrow, purely electronic (zero-phonon) transitions, which makes crystals doped with $3d^n$ or $4f^n$ ions extremely convenient subjects for application of sophisticated spectroscopic techniques. The first studies of activated crystals were carried out in cooperation with the group headed by P. P. Feofilov (State Optical Institute), which was engaged in intense research of these crystals already in the pre-laser times.

Development of new methods of research

Piezospectroscopy. In order to answer the fundamental question on the nature of impurity centers or defects in a lattice, one has to have information on the local center symmetry, which is directly related to their structure. A new efficient piezospectroscopic method of determining the local symmetry of centers, based on the effect of reversible spectral-line splitting of local centers under directed elastic strain discovered at the Ioffe Institute, [53] was developed. The point symmetry group of a center is a subgroup of the symmetry of the whole crystal, so that centers with a lower symmetry can occupy in the lattice several equivalent positions differing only in spatial orientation ("orientational degeneracy"). The physical equivalence of differently oriented centers breaks down, however, under uniaxial compression of the lattice, which should result in different shifts of their spectra and, hence, in reversible splitting of the crystal's spectral lines. The number and polarization of the splitting components depend on the symmetry of the center and on the electronic degeneracy of its levels, which permits one to reliably extract the most important characteristics of centers from the observed splitting pattern. This method was applied successfully to investigation of the structure of impurity centers in many crystals, namely, oxides and fluorides, activated by $3d^n$- and $4f^n$-group ions, centers in the diamond, aggregate color centers in alkali halide crystals and so on. The piezospectroscopic method is enjoying presently widespread use in probing local centers in both dielectrics and semiconductors.

Besides directed strain, an external action which can be applied only to solids, one employed widely in spectroscopic studies of the symmetry and structure of local centers in crystals also such external factors traditional for optical spectroscopy as the magnetic field (the Zeeman effect) [54] and the electric field (the Stark effect) [55]. Application of an electric field was found to be particularly advantageous in studies of the linear Stark effect, observed with centers possessing a constant dipole moment, whose symmetry group does not contain inversion operations. The linear-in-field Stark-induced shifts of optical transition frequencies of inversion-invariant centers in centrosymmetric crystals give rise to a characteristic symmetric "pseudo-Stark" splitting [56] in spectra, which is entirely due to the difference between the line shifts in differently oriented centers and was shown [55] to yield valuable information on the symmetry of a center.

ODMR. Optical detection of magnetic resonance makes use of the effect of the change in optical absorption or luminescence of a crystal at the instant of resonant absorption of microwave power by local centers in transitions between Zeeman-split sublevels in a magnetic field. New efficient ODMR arrangements were developed [57] at the Ioffe Institute which are based on (i) a change in the intensity of recombination tunneling afterglow in the EPR of local centers involved in radiative recombination, and (ii) effect of EPR in the ground and excited states of a center on the photostimulated luminescence associated with the centers. These methods are particularly convenient to employ in studying the properties of crystalline phosphors, where photostimulated luminescence is used to visualize ionizing radiation (x-ray screens). The observed phenomenon of cross relaxation in the Zeeman manifold of interacting centers underlies the proposed method of EPR detection of defects from the magnetic circular dichroism in the absorption of color centers which form in the vicinity of these defects and serve as detectors of their EPR. The ODMR method was employed to study short-lived excited states of many intrinsic and extrinsic defects in alkali-halide and other ionic crystals. Information was obtained on hyperfine interactions in triplet states of ns^2 ions and excitons, the Jahn–Teller effect in excited state, and on the color-center wave function in relaxed excited states.

Optical detection of non-equilibrium high-frequency acoustic phonons. The luminescent properties of dielectric crystals with impurity ions enjoy important applications in the terahertz acoustic range (10^{12} Hz) in solids, which is inaccessible for coherent ultrasound techniques. Optical detection of non-equilibrium terahertz phonons injected into a crystal at a low temperature (e.g., by heat pulses generated by a strip heater) is based on the interaction of phonons with excited impurity ions, where non-equilibrium phonons induce resonant transitions between electronic sublevels of the radiative state of an ion, which is detected from the change in the ion luminescence signal [58]. The optical detection methods were further refined in a series of studies performed at the Ioffe Institute [59]. Ballistic phonon beams in crystals were detected for the first time. These experiments revealed terahertz-phonon focusing caused by the elastic anisotropy of the crystal, anharmonic decay of terahertz phonons, and interaction of colliding ballistic phonon beams. The cross section of the phonon resonant interaction with the electronic levels of an impurity center was found to depend on the acoustic mode (LA, TA, direction of propagation), which permitted one to formulate general symmetry-based selection rules for one-phonon transitions between levels. A comprehensive study was made of spatial imprisonment of non-equilibrium phonons due to multiple resonant scattering from impurity ions. Optical methods of studying the behavior of phonons in disordered crystals and glasses were developed. Recent experiments [60] made on solid highly-porous aluminum oxide whose framework is made up of nanometer-scale spherical particles revealed effective excitation under optical pumping of non-equilibrium terahertz vibrations corresponding to the lowest quantum-confined acoustic Lamb modes of spherical nanoparticles, and they were found to have very long (ms-scale) decay times.

Dynamical processes in excited states of doped dielectrics

The various physical phenomena observed under optical excitation of impurity ions in dielectric hosts attract considerable interest of researchers and find broad application (quantum electronics, optical information recording etc.). A totally new phenomenon observed under optical pumping of concentrated ruby crystals (Al_2O_3:Cr^{3+}) stands out from among the results obtained in this area at the Ioffe Institute [61]. As established from the pseudo-Stark splitting of Cr^{3+} lines, pumping within the Cr^{3+} absorption bands at low temperatures ($T < 150$ K) gives rise to formation of stable domains with a strong ($\sim 10^5$ V/cm) electric field of opposite signs, which is parallel to the trigonal axis of the crystal. Phenomenological theory [62] relates this phenomenon in ruby to the N-shaped field dependence of photocurrent having a part with absolute negative resistance (current flowing against the field), which gives rise to an electrical instability and spontaneous charge separation in an optically excited sample. Direct measurements of the photocurrent in ruby did indeed detect negative (i.e., against the field) sign of the photocurrent within a certain interval of applied fields ("absolute negative conductivity"). Experiments on resonant excitation of photocurrent within the narrow Cr^{3+} spectral lines showed that the photocurrent in ruby originates from directed hopping transport of "excess" charge over the system of Cr^{3+} ions excited into the ^2E metastable state. Taking into account a number of microscopic factors reveals that the spatial distribution of excited Cr^{3+} ions surrounding the excess charges is asymmetric, thus causing directed charge hopping over these ions against the field applied to the sample.

Observation of the anomalous photoelectric properties of ruby stimulated interest in the poorly known phototransport charge processes, which were found to be essential for many applications of doped dielectrics. These studies are carried out in cooperation with German and American scientists. The position of the electronic impurity levels in the band gap of dielectrics was derived from the impurity photoionization thresholds, which is essential, for instance, for using crystals as solid-state lasers [63]. Studies of photoionization-induced persistent hole burning in the inhomogeneously broadened spectral-line profile of impurity ions has recently revealed "off-resonance" burning of extremely wide holes in the vicinity of the laser pump frequency. This phenomenon is connected with a strong linear time-varying Stark shift of ion transition frequencies in the variable internal electric fields created under photoinduced charge transfer between the impurities generating these fields [64].

2.2 Spectroscopic studies of phase transitions

Soft-mode spectroscopy

Investigation of vibrational spectra, which provide information on lattice dynamics, permitted one to observe for the first time and to study in considerable detail structural phase transitions (PT) in mercurous halogenides Hg_2X_2 (X = Cl, Br, I), a new class of dielectric crystals [65]. These tetragonal crystals possess unique and potentially valuable anisotropic physical properties, namely, a extremely high optical birefrigence ($n_e - n_0 \gg 1$) and a extremely low sound velocity ($v \approx 250$ m/s in Hg_2I_2). Purely spectroscopic measurements (Raman scattering, as well as IR absorption) established unambiguously the mechanism of the low-temperature transition from tetragonal to rhombic phase in Hg_2Cl_2 ($T_c = 185$ K) and Hg_2Br_2 ($T_c = 143$ K) as an improper ferroelastic transformation, which involves doubling of the cell volume and is induced by the soft acoustic (TA) mode from the Brillouin zone BZ boundary (the X point) of the paraphase. A soft mode from the BZ boundary, which is forbidden in first-order spectra of the paraphase, was observed in ferrophase Raman spectra (allowed at the PT due to the $X-\Gamma$ Umklapp process in the BZ), namely, in second-order scattering (overtone) and first-order defect-induced scattering. Spectra of the third compound, Hg_2I_2, exhibited under the deepest cooling only a softening of the TA(X) mode characteristic of a virtual ferroelastic, while a real PT in Hg_2I_2 was initiated recently only at a high pressure [66]. On the whole, crystals of the Hg_2X_2 group (X = Cl, Br, I, mixed compounds) do not have equals in the diversity and distinctness of spectroscopic manifestations of PTs, and can be considered as model objects for studying general properties of PTs in crystals by optical vibrational spectroscopy.

Spectra of impurity probe ions

A new, fairly general type of impurity-ion spectral splitting under a ferroelectric phase transition was revealed, namely, pseudo-Stark splitting of spectral lines of an impurity ion in the internal electric field of the polar ferrophase [67]. The magnitude of the splitting is proportional to the transition order parameter, which permits one to investigate the temperature behavior of the order parameter using the probe-ion spectrum.

2.3 Investigation of new materials

Photonic crystals

Intense studies of the optical properties of synthetic opals, porous a-SiO_2 materials whose framework forms a three-dimensional cubic lattice of closely packed identical spheres, a few fractions of a micrometer in size, have been started in recent years at the Ioffe Institute. Opals were shown for the first time [68] to be very convenient subjects for studying the behavior of electromagnetic waves in the visible range in the so-called "photonic crystals", [69] i.e., in media with three-dimensional periodic modulation of the refractive index. By analogy with the behavior of electrons in a periodic crystal lattice, one discusses presently the existence in photonic crystals of band gaps, i.e., spectral intervals with zero density of photonic states, where light waves undergoing Bragg reflection from "atomic planes" of the three-dimensional refractive-index lattice cannot propagate freely in the medium. One may expect in such regions observation of very unusual properties of light (suppression of spontaneous emission, light localization, and so on). Investigation of the spectra of light transmission and reflection along different directions in the opal lattice has revealed a strong anisotropy of the photonic band gap ("pseudoallowed band"), for which the light frequencies forbidden for all directions of propagation are not observed but the density of states in the band is at the same time strongly suppressed [70]. A possibility was demonstrated [68, 70] of an efficient variation of the properties of the opal as a photonic crystal by filling the empty opal voids with semiconducting materials (CdS) having a large refractive index.

Matrix-isolated fullerene molecules

The unusual properties of fullerenes, the recently discovered (1985) carbon modification, have spurred considerable interest in investigation of these systems, in particular, of their electronic and vibrational spectra. Until quite recently, spectral bands measured in the optical electronic spectra of the fullerene molecules C_{60} and C_{70} were fairly broad ($>$ 10 meV), which precluded obtaining accurate information on the structure and properties of electronic states in the fullerenes. The pioneering studies performed at the Ioffe Institute [71, 72] demonstrated that introducing fullerene molecules into crystalline dielectric hosts of some organic compounds gives rise to the appearance of a rich narrow-line optical spectrum produced by purely electronic transitions between the ground and excited singlet and triplet electronic states of molecules and by their vibrational satellites. This permitted one to determine with a high accuracy the energies of the molecular vibronic levels and the singlet-triplet splitting of the lower excited electronic state of the molecule (for C_{70}). The narrow-line spectra of the nonspherical fullerene molecules C_{70} embedded in anisotropic crystalline hosts were found to be polarized. The Zeeman effect in the fullerene molecule was observed for the first time, which made it possible to directly prove the triplet character of the lowest electronic state in the C_{70} molecule and to determine the electronic g factor of the molecule in this state. The proposed approach has proved to be extremely efficient in probing the electronic spectra of the various fullerene molecules, as well as of a number of their derivatives.

Epitaxial dielectrics

Preparation and investigation of perfect dielectric films on the surface of semiconductors have a considerable application potential. At the Ioffe Institute, the rare-earth luminescent probe method was used successfully for the first time to measure elastic strains and crystalline perfection of thin (nm-scale) single-crystal layers of Ca, Sr, and Cd fluorides grown by MBE on the surface of semiconductors (Si, GaAs) [73]. One observed specific features in the strain-induced inhomogeneous broadening of impurity lines in the spectra of thin CaF_2:Sm layers, as well as in narrow-hole burning in line profiles, which are associated with reversible charge transport between the dielectric layers and the semiconductor substrate.

3 Conclusion

We have attempted to depict a panorama of optical and spectroscopic investigations carried on at the Ioffe Institute, by briefly describing but part of these works being performed at the laboratories of Zakharchenya, Kaplyanskii, Rogachev, Baranov, Seisyan and Titkov. These research groups are also engaged very successfully in some other studies to which we could not dedicate here any special paragraphs for lack of space in the present brief survey. Such are the studies on Raman spectrography of pure and doped semiconductors and, lately, of the structures with quantum dots (B. Bairamov), the optical polarization spectroscopy of surface (A. Titkov), the study of the possibility of creating lasers on Si-Er systems (M. Bresler, O. Gusev), the study of semimagnetic semiconductors by optical methods (D. Yakovlev, Yu. Kusrayev), the optical study of interaction of non-equilibrium acoustic phonons with excitons and carriers in bulk and spatially confines semiconductors (A. Akimov [74]). Very important investigations on the spectrography of narrow-band and amorphous semiconductors and on the laser spectrography of shallow impurities have been performed at Ivanov-Omskii's laboratory.

The research in optics and spectroscopy carried on at PTI exceeds in its scope by far the analogous investigations at any of scientific institutions in Russia or abroad. The high quality of the research has always been ensured by collaboration of experimentalists with such prominent theorists as V. Perel', G. Pikus, A. Aronov, E. Ivchenko, M. Dyakonov, I. Merkulov, etc.

Diverse as these works are, they all have grown out of the exciton studies by E. F. Gross and his school.

References

[1] E. F. Gross and N. A. Karryev, *Dokl. Akad. Nauk SSSR* **84** 471 (1952).

[2] E. F. Gross and B. P. Zakharchenya, *Dokl. Akad. Nauk SSSR* **97** 57 (1954).

[3] E. F. Gross and B. P. Zakharchenya, *J. phys. et radium*, **1** 68–71 (1957).

[4] E. F. Gross, B. P. Zakharchenya, and P. P. Pavinskii, *Zh. Tekhn. Fiz.* **27** 2177 (1957).

[5] B. P. Zakharchenya and R. P. Seisyan, *Usp. Fiz. Nauk* **97** 194 (1969);
R. P. Seisyan and B. P. Zakharchenya, in *Landau Level Spectroscopy*, ed. E. I. Rashba and G. Landwehr, North Holland, Amsterdam, Ch. 7, p. 345–444, 1991.

[6] K. I. Geiman, S. I. Kokhanovskii, R. P. Seisyan, V. A. Yukish, and Al. L. Efros, *Sov. Phys. Solid State* **28** 476 (1986).

[7] V. I. Ivanov-Omski, S. I. Kokhanovski, R. P. Seisyan, V. A. Smirnov, V. Yukish, and Al. L. Efros, *Solid State Commun.* **46** 26 (1983);
S. I. Kokhanovski, Yu. M. Makushenko, R. P. Sesyan, Al. L. Efros, T. V. Yazeva, and M. A. Abdullaev, *Sov. Phys. Semicond.* **25** 298 (1991).

[8] S. I. Kokhanovski, K. Moumanis, M. E. Sasin, R. P. Seisyan, A. V. Kavokin, H. Gibbs, and G. Khitrova, *Phys. Status Solidi (a)* **164** 67 (1997).

[9] M. R. Vladimirova, A. V. Kavokin, M. A. Kaliteevskii, M. E. Sasin, R. P. Seisyan, and V. M. Ustinov, *Compound Semiconductors*, ed. M. S. Shur and R. A. Suris, Inst. of Physics Conf. Series, No 155, p. 157, 1997.

[10] E. F. Gross, B. P. Zakharchenya, and O. V. Konstantinov, *Fiz. Tverd. Tela* **3** 305 (1961).

[11] E. F. Gross, S. A. Permogorov, and B. S. Razbirin, *J. Phys. Chem. Sol.* **27** 1647–1651 (1966).

[12] E. F. Gross, S. A. Permogorov, and B. S. Razbirin, *Sov. Phys. Usp.* **14** 104 (1971).

[13] S. A. Permogorov, *Excitons*, ed. E. I. Rashba, and M. D. Sturge, North-Holland, p. 177–202, 1982.

[14] S. A. Permogorov, *Phys. Status Solidi (b)* **68** 9–42 (1975).

[15] A. A. Klochikhin, S. A. Permogorov, and A. N. Reznitsky, *JETP* **44** 1176 (1976).

[16] S. A. Permogorov and V. V. Travnikov, *Solid State Commun.* **29** 615–619 (1979).

[17] E. F. Gross, S. A. Permogorov, V. V. Travnikov, and A. V. Sel'kin, *Sov. Phys. Solid State* **13** 580 (1971).

[18] V. A. Kiselev, B. S. Razbirirn, and I. N. Ural'tsev, *JETP Lett.* **18** 296 (1973).

[19] E. F. Gross and A. A. Kaplyanskii, *Dokl. Akad. Nauk* **132** 98 (1960).

[20] S. Permogorov and A. Reznitsky, *J. Lumin.* **52** 201–223 (1992).

[21] G. Lampel, *Phys. Rev. Lett.* **20** 491 (1968).

[22] R. R. Parsons, *Phys. Rev. Lett.* **23** 1132 (1969).

[23] F. Meier and B. Zakharchenya, *Optical Orientation. Modern Problems in Condensed Matter Sciences* vol. 8, North-Holland, 1984.

[24] B. P. Zakharchenya, D. N. Mirlin, and V. I. Zemskii, *Zh. Eksp. Teor. Fiz.* **70** 1092 (1976).

[25] B. P. Zakharchenya, D. N. Mirlin, V. I. Perel, and I. I. Reshina, *Sov. Phys. Usp.* **25** 143 (1982).

[26] M. A. Alekseev, I. Ya. Karlik, D. N. Mirlin, and V. F. Sapega, *Sov. Phys. Semicond.* **23** 479 (1989).

[27] D. N. Mirlin and V. I. Perel, in *Spectroscopy of Nonequilibrium Electrons and Phonons*, ed. C. V. Shank and B. P. Zakharchenya, North-Holland, p.?, 1992.

[28] D. N. Mirlin and A. V. Rodnina, *Phys. Solid State* **38** 1747 (1996).

[29] P. G. Baranov, Yu. P. Veshchunov, R. A. Zhitnikov, N. G. Romanov, and Yu. G. Shreter, *JETP Lett.* **26** 249 (1977).

[30] N. G. Romanov and P. G. Baranov, *Semicond. Sci. Technol.* **9** 1080 (1994).

[31] L. S. Vlasenko, M. P. Vlasenko, V. N. Lomasov, and V. A. Khramtsov, *Sov. Phys. JETP* **64** 612 (1986).

[32] L. S. Vlasenko, Yu. V. Martynov, T. Gregorkievich, and C. A. T. Ammerlaan, *Phys. Rev. B* **52** 1144 (1995).

[33] V. P. Kochereshko, G. R. Pozina, E. L. Ivchenko, D. R. Yakovlev, A. Waag, W. Ossau, G. Landwehr, R. Hellman, and E. O. Gubel, *Proc. 22 Int. Conf. Physics of Semiconductors*, Vancouver, Canada, Ed. by D. J. Lockwood, World Scientific, p. 1372–1375, 1994.

[34] D. R. Yakovlev, G. R. Pozina, V. P. Kochereshko, A. Waag, W. Ossau, and G. Landwehr, *JETP Lett.* **61** 628 (1995).

[35] E. L. Ivchebko, A. V. Kavokin, V. P. Kochereshko, P. S. Kop'ev, and N. N. Ledentsov, *Superlattices and Microstructures*, **12** No 3, 317 (1992).

[36] S. T. Cundiff, E. O. Gubel, V. P. Kochereshko, D. R. Yakovlev, P. S. Kop'ev, and N. N. Ledentsov, *Solid State Commun.* **92** 467–471 (1994).

[37] E. L. Ivchenko, V. P. Kochereshko, I. N. Uraltsev, and D. R. Yakovlev, *Phys. Status Solidi (b)* **161** 217–221 (1990).

[38] E. V. Goncharova, V. P. Kochereshko, M. A. Yakobson, J. Cibert, and Le Si Dang, *JETP Lett.* **61** 894 (1995).

[39] A. V. Kavokin, V. P. Kochereshko, G. R. Pozina, I. N. Uraltsev, D. R. Yakovlev, G. Landwehr, R. N. Bicknell-Tassius, and A. Waag, *Phys. Rev. B* **46** 9788–9791 (1992).

[40] E. L. Ivchenko, A. V. Kavokin, V. P. Kochereshko, G. R. Pozina, I. N. Uraltsev, D. R. Yakovlev, R. N. Bicknell-Tassius, A. Waag, and G. Landwehr, *Phys. Rev. B* **46** 7713–7722 (1992).

[41] V. P. Kochereshko and I. N. Uraltsev, *Semiconductors and Insulaters: Optical and Spectroscopic Research*, ed. Yu. I. Koptev, Nova Science Publ. Inc., N.Y., v. 15, p. 1–19, 1992.

[42] E. L. Ivchenko, V. P. Kochereshko, I. N. Uraltsev, and D. R. Yakovlev, *High magnetic fields in semiconductor physics III: Quantum Hall effect and magnetooptics*, ed. G. Landwehr, Springer Series on Solid State Sciences, p. 533–536, 1992.

[43] E. L. Ivchenko, V. P. Kochereshko, A. Yu. Naumov, I. N. Uraltsev, and P. Lavallard, *Superlattice and Microstructures* **10** 497–500 (1991).

[44] D. R. Yakovlev, V. P. Kochereshko, R. A. Suris, H. Schenk, W. Ossau, A. Waag, G. Landwehr, P. C. M. Christiansen, and J. C. Maan, *Phys. Rev. Lett.* **79** 3974–3977 (1997).

[45] P. G. Baranov and N. G. Romanov, *Proc. 22th Conf. on the Physics of Semiconductors* Ed. D. J. Lockwood, World Scientific, vol. 2, p. 1400–1403, 1994.

[46] R. I. Dzhioev, B. P. Zakharchenya, and V. L. Korenev, *Phys. Solid State* **37** 1929 (1995).

[47] R. I. Dzhioev, B. P. Zakharchenya, E. L. Ivchenko, V. L. Korenev G. Kusraev, N. N. Ledentsov, V. M. Ustinov, A. E. Zhukov, and A. F. Tsatsul'nikov, *JETP Lett.* **65** 804 (1997); H. M. Gibbs, E. L. Ivchenko, G. Khitrova, V. L. Korenev, M. N. Tkachuk, and B. P. Zakharchenya, *Phys. Rev. B* **56** 20, 13405 (1997).
[48] V. K. Kalevich, B. P. Zakharchenya, K. V. Kavokin, A. V. Petrov, P. Le Jeune, D. Marie, D. Robaut, T. Amand, J. Barrau, and M. Brousseau, *Phys. Solid State* **39** 681 (1997).
[49] A. V. Kondakov, Yu. G. Kusrayev, K. V. Kavokin, I. A. Merkulov, and B. P. Zakharchenya, *Compound Semiconductors*, ed. M. S. Shur and R. A. Suris, Inst. Phys. Conf. Series, No 155, p. 747, 1997.
[50] A. A. Rogachev, in *Handbook on Semiconductors*, Completely Revised and Enlarged Edition, Series Editor T. S. Moss, ed. P. T. Landsberg, Elsevier Science Publishers B.V., vol. 1, Ch. 9, p. 449–487, 1992.
[51] A. A. Rogachev, *Annales de Physique*, Colloque C2, supplement au No 3, vol. 20, C2, p. 389-392, Juin 1995.
[52] A. A. Rogachev, *Fiz. Tverd. Tela* **40** No 5, 141-143 (1998).
[53] A. A. Kaplyanskii, *Opt. Spectrosc. (USSR)* **10** 165 (1961); **16** 602 (1964); **16** 1031 (1964).
[54] B. P. Zakharchenya, L. M. Kanskaya, and I. B. Rusanov, *Opt. Spectrosc. (USSR)* **19** 306 (1965).
[55] A. A. Kaplyanskii and V. N. Medvedev, *Opt. Spectrosc. (USSR)* **23** 743 (1967).
[56] W. Kaiser, S. Sugano, and D. L. Wood, *Phys. Rev. Lett.* **6** 605 (1961).
[57] P. G. Baranov and N. G. Romanov, *Appl. Magnetic Resonace* **2** 361 (1991).
[58] K. F. Renk and J. Deisenhofer, *Phys. Rev. Lett.* **26** 764 (1971).
[59] *Nonequilibrium Phonons in Nonmetallic Crystals* ed. W. Eisenmonger and A. A. Kaplyanskii, North Holland, p. 373–454, 1986.
[60] S. P. Feofilov, A. A. Kaplyanskii, and R. I. Zakharchenya, *J. Lumin.* **66-67** 349–357 (1996).
[61] A. A. Kaplyanskii, *J. Lumin.* **48-49** 1–10 (1991).
[62] M. I. Dyakonov, *JETP Lett.* **39** 185 (1984).
[63] S. A. Basun, T. Danger, A. A. Kaplyanskii, D. S. McClure, K. Peterman, and W. C. Wong, *Phys. Rev. B* **54** 6141 (1996).
[64] S. A. Basun, M. Rankas, U. Happek, A. A. Kaplyanskii, J. Rennie, J. C. Vial, W. M. Yen, and R. S. Meltzer, *Phys. Rev. B* **56** 12992 (1997).
[65] A. A. Kaplyanskii, M. F. Limonov, and Y. F. Markov, *Jnd. J. of Pure and Appl. Phys.* **26**252 (1988).
[66] A. A. Kaplyanskii, Yu. F. Markov, and A. Sh. Turaev, *Phys. Solid State* **36** 1497 (1994).
[67] S. A. Basun, A. A. Kaplyanskii, and S. P. Feofilov, *Phys. Solid State* **34** 1807 (1992).
[68] V. N. Astratov, V. N. Bogomolov, A. A. Kaplyanskii, A. V. Prokofiev, L. A. Samoilovich, and Yu. A. Vlasov, *Nuovo Cimento* **D17** 1349 (1995).
[69] E. Eblonovitch, *J. Modern Optics* **41** 173 (1994).
[70] Yu. A. Vlasov, V. N. Astratov, O. Z. Karimov, A. A. Kaplyanskii, V. N. Bogomolov, and A. V. Prokofiev, *Phys. Rev. B* **55** 13357 (1997).
[71] B. S. Razbirin, A. N. Starukhin, A. V. Chugreev, Yu. S. Grushko, and S. N. Kolesnik, *JETP Lett.* **60** 451 (1994).
[72] A. N. Starukhin, B. S. Razbirin, A. V. Chugreev, D. K. Nelson, Yu. S. Grushko, S. N. Kolesnik et. al., *J. Lumin.* **72-74** 457 (1997).
[73] N. S. Sokolov, J. C. Alvarez, and N. L. Yakovlev, *Appl. Surface Science* **60/61** 421–425 (1992).
[74] A. V. Akimov, E. S. Moskalenko, L. J. Challis, and A. A. Kaplyanskii, *Physica B* **219/220** 9 (1996).

Power semiconductor electronics

I. V. Grekhov

1 Introduction

Practically the whole amount of electric power is produced at thermal, nuclear, and hydroelectric power plants in the form of a relatively low ac voltage of industrial frequency (50–60 Hz) which is transformed into high voltage, transported to consumption sites, and there again transformed into low voltage. Nearly half the electric power is consumed in this form, while the other is subjected to more complicated conversion by means of power semiconductor electronics. The scale of conversion is, naturally, enormous. For example, in 1988 the total output of all electric power plants of the USSR was $\sim$ 350 million kW, whereas the installed power of semiconductor converters was $\sim$ 150 million kW, the figures of the same order of magnitude. Estimates show that the optimal saturation of the energetics with power electronics could save no less than 10% of the power produced, which is approximately equal to the contribution from all nuclear power plants.

The highest economy resources lie in the domain of low- and medium-power consumers, especially in the field of electric drives which consume more than 50% of the electric power produced.

Vast potential fields of economy are reactive power compensation in transmission lines and local power networks, variable speed drives with energy recuperation in electric transport, electronization of domestic appliances, control in electrometallurgy, etc.

In any developed country the power electronics is maintained by a large specialized field of industry; in the USSR about 20 large plants had worked in this field.

The development of the power electronics in this country originated from works performed at the Ioffe Institute in the laboratory of Academician V. M. Tuchkevich. It is at this Institute that the first power germanium and silicon diodes in this country, first germanium transistor, first power silicon thyristor, and first high-power devices based on avalanche breakdown were created. The original technology of industrial production of power silicon devices developed in cooperation with the *Elektrovypryamitel'* plant allowed fast organization of their large-scale production, which laid foundation for a new industry of power semiconductor devices.

The Ioffe Institute was always in close contact with this industry, assisting in its development; many important, fundamentally new projects were initiated by the Institute, such as, e.g., semiconductor converters for dc transmission lines and dc links, converters for industrial isotope separation systems, and many other things. During the last decade new principles of high power commutation by semiconductor devices have been developed which drastically, by orders of magnitude, increased the peak pulse power of these devices and gave birth to a new field of development—pulsed power semiconductor electronics.

Also, the world's first GaAs-based power devices were created at the Ioffe Institute. These works, performed in the laboratory of Academician Zh. I. Alferov, laid foundation for one more avenue of development in power electronics—creation of devices based on wide-bandgap materials for special fields of application.

2 Silicon devices of power electronics

Any power semiconductor converter is a complex electrotechnical installation made up of hundreds of components whose technical level determines the level of the converter as a whole. However, the key role is played by the technical level of the main active elements—power semiconductor devices.

Silicon has already been the basis material for power semiconductor devices for three decades, and there seems to be no real alternative to it, at least for the nearest future. Hence we consider here only silicon devices; a brief review of the state of affairs in the field of power devices based on other materials can be found, e.g., in [1, 2].

The preferable ranges of power and frequency of device application as of 1990 and the prospects for 2000 are presented in Fig. 1 [3]; the present-day state of affairs is close to this prognosis.

The devices dominating in the range of extremely high power and low frequencies are *high-voltage thyristors* mainly used in converter substations of dc transmission lines, dc links between power systems, static compensators, and other power installations; another possible consumer in near future is thermonuclear engineering.

The semiconductor structure of a power thyristor consists of four layers of silicon, $n^+p'np^+$, in a single crystal. These layers form three pn junctions. The outer junctions (emitters) are forward-biased, and the central one (collector) blocks the voltage applied to the device. The switching is initiated by passing a current pulse through the control circuit along the thin base p′-layer; this current causes electron injection from the n^+ emitter. Owing to the large resistance of the p′ layer, the injection is localized in a

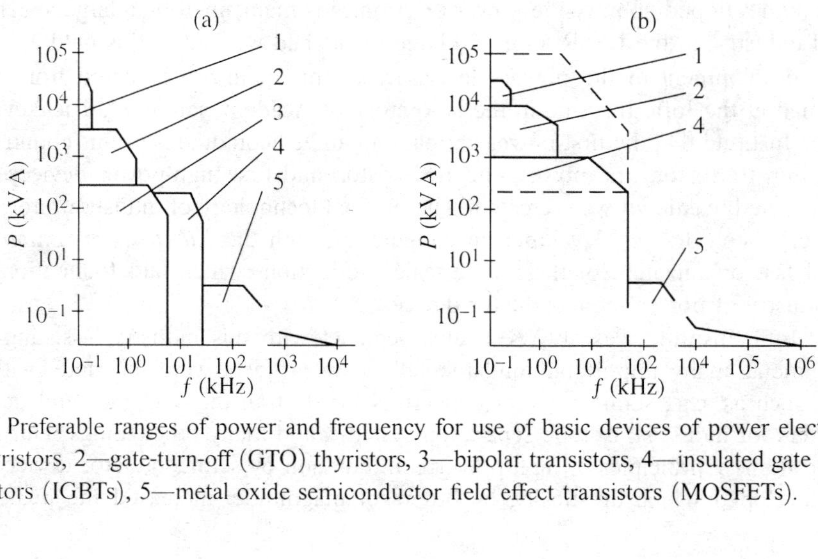

Fig 1. Preferable ranges of power and frequency for use of basic devices of power electronics. 1—thyristors, 2—gate-turn-off (GTO) thyristors, 3—bipolar transistors, 4—insulated gate bipolar transistors (IGBTs), 5—metal oxide semiconductor field effect transistors (MOSFETs).

narrow (of order of a hundred micrometers) region along the boundary of n^+p' layers, and it is there that the switching occurs to produce a plasma channel. The on-state spreads from the channel outwards at a relatively low velocity (0.1–0.001 mm/μs) over the whole device area.

As a rule, the working currents of transmission lines fall within the range of several kiloamperes, and working voltages in the range from hundreds of kilovolt to megavolts. For this reason, the development of superpower thyristors has taken a line increasing both the working current and the working voltage. The current is mainly increased by making larger the diameter of the semiconductor element, and the voltage by improving the purity and increasing the resistivity of silicon. Both these directions constitute a serious challenge to device designers and single-crystal silicon manufacturers. These tasks are tackled with fairly successfully; high purity material obtained by float zone melting with a wafer diameter of the order of 125 mm is now, in principle, commercially available. A great stride forward was a change to silicon doped by transmutation method in nuclear reactors, including those installed at nuclear power plants; this doping method affords highly reproducible and uniform resistivity of silicon. The high-quality material enables creating thyristors with very good parameters; for example, a thyristor produced by SPCO on silicon with a wafer diameter of 125 mm has an average working current exceeding 5 kA at a voltage of 5 kV, and a thyristor produced on a 53-mm wafer has a working voltage of 9 kV. We have shown that, using silicon with a resistivity of $\sim 2\,\mathrm{k\Omega\,cm}$, pn junctions with a breakdown voltage of 25–30 kV can be fabricated, and even this value is no limit [1]. However, at such voltages the space charge region (SCR) width in a reverse-biased collector is several millimeters, which necessitates making the n-base layer very thick. When forward current is passed through the device, this layer must be filled with an electron-hole plasma so as to reduce its resistance to an acceptable value. Since plasma injection occurs from outer, heavily doped layers, the diffusion length L of electrons and holes in it must be comparable with the base layer thickness $L = (D\tau)^{1/2} \approx (0.3{-}0.5)W_n$, where τ is the carrier lifetime in the plasma. During turn-off of the thyristor, plasma disappears, mainly via recombination, and, therefore, high-voltage devices have long turn-off time; for example, a thyristor working at 9 kV has a turn-off time of $\sim 10^{-3}$ s. This fact, and also the heat removal deteriorating with increasing wafer thickness, restrict the possibilities of markedly increasing the working voltage further. As for the working current, it can be, in principle, increased substantially, since the maximum diameter of float-zone silicon will soon be increased to 150 mm and then, possibly, even to 200 mm.

The main devices intended for the power range up to several megawatts and frequencies up to a kilohertz are *gate-turn-off* (GTO) thyristors. In contrast to the conventional thyristors, GTO devices are turned off not by changing the voltage polarity in the main circuit, but by passing a turn-off (negative) pulse of current in the control circuit. Owing to a large tangential resistance of the p′ base and a comparatively low breakdown voltage of the n^+ emitter, the turn-off process occurs at a small (of order of hundreds of micrometers) distance from the edge of the n^+ layer. Therefore, the n^+ emitter of a power GTO consists of a great number (several thousand) of separate 200–300-μm wide cells (Fig. 2). Among the fields of application of these devices are traction drives, ac drives for ventilators, pumps, compressors, etc. Parameters of GTO are very high. For example, a power GTO produced by SPCO with a semiconductor element 100 mm in diameter has the maximum turn-off current of 3 kA (average working current 800 A),

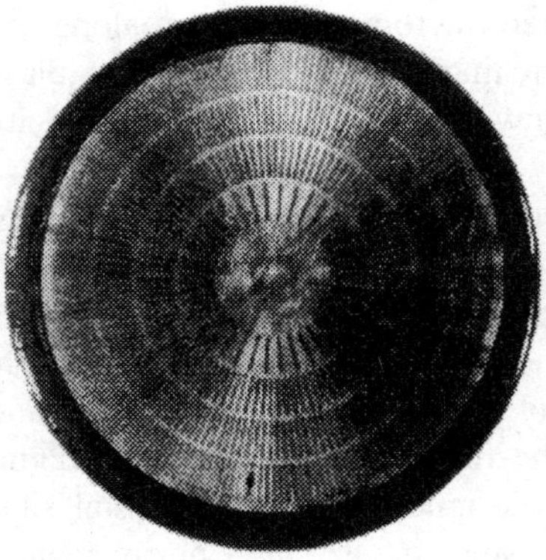

Fig 2. Design of a semiconductor element of GTO thyristor, wafer diameter 76 mm.

blocking voltage of 6 kV, and turn-off time of 35 μs; thus, the GTO is a much faster device than the conventional thyristor. However, the switching losses in the GTO are very high, which limits the maximum frequency to several kHz, and the amplitude of the turn-off current is rather high, $\sim$ 20% of the working current. Both these disadvantages stem from the fact that in the GTO similarly to the conventional device, the base n-region is filled in the on-state by high-density electron-hole plasma via injection of minority carriers from the outer n^+ and p^+ layers of the $n^+p'np^+$ structure. Such double-side injection furnishes an opportunity to drastically reduce the resistance of the wide n-base, so that the voltage drop in the on-state is low even in very high-voltage devices. In turn-off, the whole amount of plasma is to be removed from base layers; in GTO, plasma is mainly removed by a current pulse passed through the control circuit. However, only that part of the plasma can be removed fast which is situated in the near-collector region. The rest of the plasma filament comparatively slowly diffuses from the p^+-emitter to the collector and maintains current through the device under the conditions of increasing collector voltage. It is this effect that causes large switching losses in the GTO thyristor and makes necessary using an additional power RC-circuit—snubber—which delays voltage build-up. The situation can be improved essentially by using the method of turning-off the GTO thyristor by an extremely strong current pulse with a high rise rate (4 kA/μs), recently developed by ABB Co. In this case a strictly simultaneous and fast turn-off of all emitter cells can be achieved without snubber. The main problem is to design a control circuit with an extremely low inductance, since the maximum voltage in it must not exceed the breakdown voltage of the n^+p'-emitter ($\sim$ 20 V).

In the range of medium powers and frequencies (hundreds of kW, tens of kHz) are clearly visible (Fig. 1) the results of dramatic events of the last decade, related to the intrusion of microelectronics into the power electronics. The classical bipolar transistor, which had dominated this field beginning with the late 1970s and was previously the most widely used device of power electronics, started to come out of use, replaced by *insulated gate bipolar transistor* (IGBT). The design of an elementary cell of the device is shown in Fig. 3. Positive bias applied to the gate 1 forms an inversion layer 3 in the p-base of the bipolar transistor, shorting the n^+-emitter and the collector. The electron current flowing in this circuit causes injection of holes from the p^+-emitter, which pass through the $n'n$-layer and arrive at the p-base of the bipolar n^+pn transistor to cause

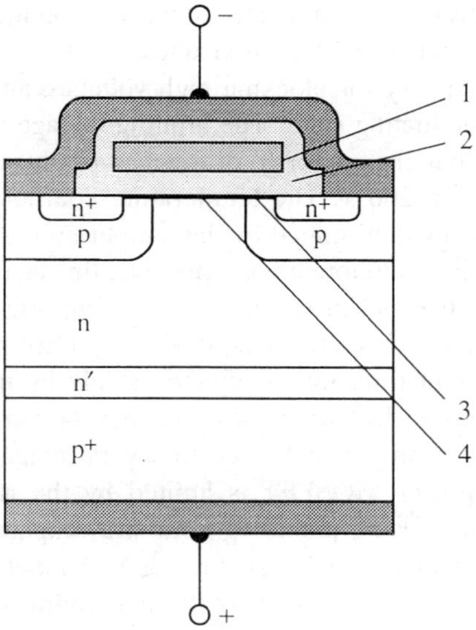

Fig. 3. Design of elementary cell of an IGBT. 1—gate, 2—gate oxide, 3—inversion n-layer in the p-base of the bipolar transistor, 4—enriched layer in n-base.

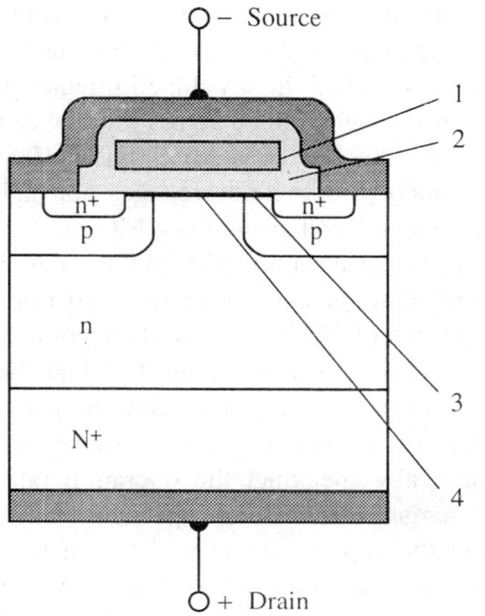

Fig. 4. Design of elementary cell of a MOSFET. 1—polysilicon gate, 2—gate oxide, 3—inversion n-layer, 4—enriched n-layer in the n-drain.

injection of electrons from the n^+-emitter. Thus, the hole flux is in fact the control current of the bipolar transistor through which most part of the power current flows. For turn-off, a negative bias is applied to the gate 1, the inversion layer disappears, and the control current of the bipolar transistor breaks. This results in an abrupt drop of the power current, and the device turns off. The elementary cell diameter is $\sim 20\ \mu$m, and the power device consists of several hundred thousand of such cells on a single chip. This is a typical rather large integrated circuit. A fundamental advantage of the IGBT is the low power expenditure in the control circuit for turning on and off, and also the high speed. A disadvantage is the high (as compared with the GTO) voltage drop in the on-state and relatively high cost, determined by the complex fabrication technology. The limiting voltage of the present-day IGBT is ~ 1.6 kV, the current 200–300 A. Intensive work on improving the IGBT is currently in progress in all countries; it is believed that by 2000 this device will completely replace bipolar transistors in the medium power range, except those fields of application where the cost is the key parameter.

High-power MOSFETs are highly promising and rapidly developing domain of semiconductor device engineering. The MOSFET, similarly to IGBT, is in fact a power integrated circuit consisting of several hundred thousand microtransistors with a characteristic dimension of 20 μm. The design of such a microtransistor is shown in Fig. 4. At a zero bias on the gate 1 of the pn junction blocks the external voltage. At a positive bias on the gate, electrons injected from the n^+-layer form a thin inversion n-channel 3 with high conductivity between the n^+-source and the n-region of the drain. A layer 4 enriched with electrons is also formed at the interface between the oxide and the n-material; the electron current from this layer flows through the lightly doped n-region of the drain to the N^+-substrate. The resistance of the device in conducting state is mainly determined by the thickness and doping level of the n-region, since no conductivity modulation occurs here. It is for this reason that MOSFETs are exclusively low-voltage devices: a thick lightly doped n-region is necessary for blocking high voltages, and this means a high voltage across the device in conducting state. The limiting voltage which can be blocked by a MOSFET is 1000 V, but the majority of devices produced are intended to operate at voltages not higher than 200 V. The temperature coefficient of the residual voltage across a MOSFETs is mainly determined by the temperature dependence of electron mobility in the n-region, being therefore always positive; this facilitates parallel connection of devices and rules out thermal instability. Among important advantages of MOSFETs are their good frequency characteristics and low control power expenditure. Owing to the fact that the current transport is effected solely by majority carriers, the switching time is determined by drift processes and may be very fast (tens of nanoseconds); the working frequency range is extended to several magahertz. Generally speaking, the operation rate of power MOSFET is limited by the rate of recharging of the total capacitance in the gate circuit, including both stray capacitance and the capacitance related to the formation of a current-conducting n-channel. The limiting working current of the device is in principle limited by the maximum dimensions of the chips which are determined by the admissible percentage of rejected chips at the existing level of the technological process. The present-day state of the problem was considered in detail in [4]. A description of the physical processes in occurring MOSFETs and problems related to the design of devices of this kind can be found, e.g., in [5].

Power monolithic and hybrid integrated circuits. The main line of development of present-day converters of low and medium power consists in integrating, either in a crystal (power up to $\sim$ 1 kW) or in a package (up to $\sim$ 30 kW), a power device and the information part of the converter—systems for control, protection, and diagnostics. This allows reducing dramatically the dimensions and cost of converters and increasing their reliability. MOSFETs are commonly used as power devices at $\sim$ 200 V, and IGBTs, at higher voltages. The use of power field-controlled devices is of fundamental importance, since they are much more readily coupled to digital control systems and have low switching losses.

A comprehensive and skillful presentation of the physical and technological foundations of modern power semiconductor devices can be found, e.g., in [6].

3 Pulsed power electronics

This field of science and technology, constituting a major part of power electronics, had been developing for a comparatively long time exclusively in the USSR and the US in certain secrecy, since very high-power electric pulses (from mega- to terawatts) of micro-, nano-, and subnanosecond duration were mainly used in major military projects. Then the situation changed dramatically, and the experience gained in this field started to be used in various industrial technologies. It was found that while in the stage of physical experiments most problems related to fast commutation of exceedingly high power could be solved with conventional hydrogen thyratrons, vacuum switches, and dischargers, for the majority of industrial applications these devices are extremely inconvenient because of the short operation life, complicated synchronization, and a number of other disadvantages. At the same time common silicon devices power-electronics described above also cannot be used because of rather low switching power.

As already discussed above, the turn-on process in the highest-power silicon switching—thyristor—starts in a very narrow region (~ 0.1 mm) near the gate electrode and rather slowly (0.1–0.01 mm/μs) spreads over the device area. Therefore, even in the microsecond range the thyristor can switch relatively low power only. More modern devices—IGBTs and MOSFETs—turn on simultaneously in hundreds of thousands of elementary cells, and do this rather fast—in tens of nanoseconds, and even several nanoseconds. However, because of the very small area ($\sim 10^{-6}$ cm^{-2}) of the conducting channel in each cell the total conducting area is small, and a parallel or in-series connection of a great number of devices is technically complicated and unacceptable economically, owing to the high cost of a single device.

All conventional devices of power electronics, in which the filling of the SCR of the pn junction is effected by injection from the ends, cannot be used in principle to switch high power in the subnano- and picosecond ranges. Indeed, the SCR width in a silicon pn junction with a breakdown voltage of, e.g., 2.5 kV is $\sim 4.2 \times 10^{-2}$ cm. Carriers, even those moving at the maximum possible (saturated) velocity $V_s = 10^7$ cm/s, can only pass this region in 2 ns. This means that novel physical principles are necessary for subnano- and picosecond switching.

Pulsed power semiconductor electronics became reality after new principles of high power switching by semiconductor devices [7] in the micro-, nano-, and subnanosecond ranges, which enabled a dramatic—by orders of magnitude—increase in their pulse power, were developed at the Ioffe Institute.

3.1 Microsecond range

To increase the pulse power of thyristor-type devices, it was proposed to initiate the turn-on simultaneously over the whole device area using a thin layer of electron-hole plasma formed over the entire plane of the collector pn junction. Several ways of creating such a layer have been studied—pulsed avalanche breakdown, impact ionization in a high-power ultra-high frequency field, and ionization by a pulse of coherent or incoherent light. All of these produced on the whole satisfactory results, but were too complicated in technical regard. The most realistic method is the reverse-injection control [8] when the plasma layer is created by a short-time voltage reversal in the power circuit. This device (Fig. 5), named reversely switching dynistor (RSD) consists of several tens of thousands of alternating thyristor and transistor sections with a characteristic size smaller than the thickness of the wide n-base of a four-layer structure. The central pn junction common to all sections—the collector—is reverse-biased and blocks the external voltage. On changing the polarity the low-voltage n^+p-emitter of transistor sections breaks down, and a current pulse flows through them, accompanied by plasma injection into the n-region in the form of hundreds of thousands of cones overlapping at the collector and forming a rather uniform plasma layer over the entire device area. At that time a saturable core choke connected in series with the RSD is in unsaturated state and separates the power and control circuits. Then the choke is saturated, the polarity of the voltage across the structure again changes its sign, and the external field extracts holes into the p-base, causing counter injection of electrons from the n^+-emitter with the subsequent switching of $n^+p'np^+$ structures simultaneously over the entire device area.

A commutation oscillogram for a device with $S = 20\,\text{cm}^2$ is shown in Fig. 5(d).

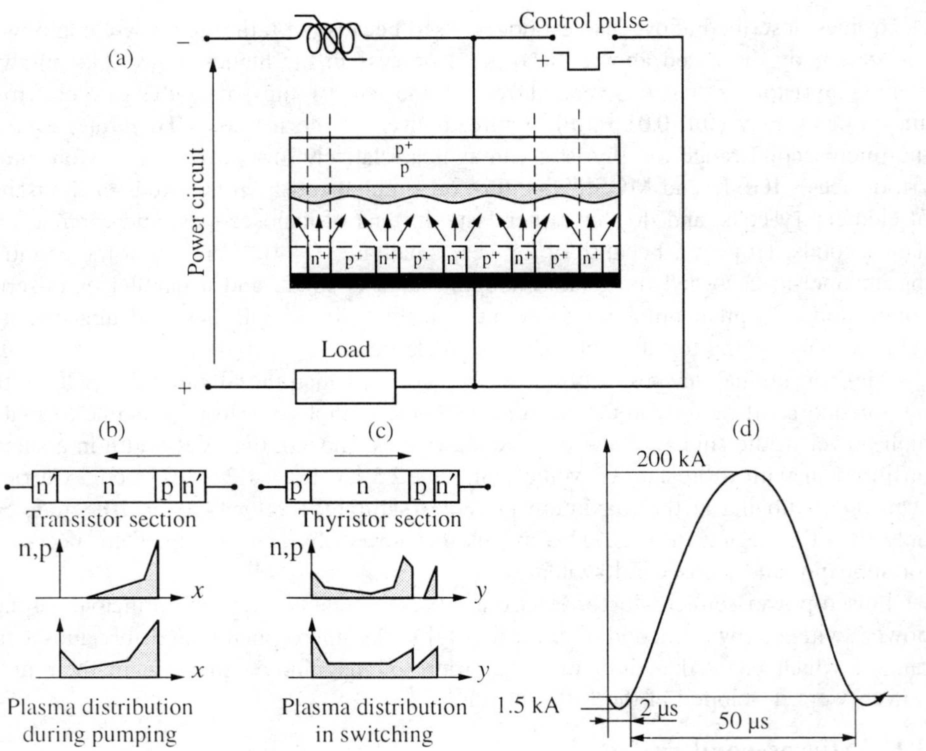

Fig 5. (a) Design of semiconductor RSD structure. (b) Plasma distribution in the n-base of a transistor section, with control current flowing (pumping). (c) Plasma distribution in bases of a thyristor section in switching. (d) Commutation oscillogram for an RSD with working area $S = 20$ cm^2 (wafer diameter 56 mm).

After a control current pulse (1.5 kA, 2 μs, 2×10^{-3} C) the device commutes a current of amplitude $\sim$ 300 kA and $dI/dt \approx 2 \times 10^4$ A/μs. The delay time between the control and power pulses is zero for fundamental reasons. Therefore, devices connected in-series will be turned-on strictly simultaneously by a single control pulse passed through the whole assembly. These two specific features—high pulse power and simplicity of in-series connection—enable constructing switches and pulse generators of practically unlimited power on the basis of RSD. Figure 6 shows, an example of RSD switch designed to operate at a current of 250 kA, 25 kV voltage, and pulse duration of 100 μs in the mode of rarely repeated pulses; with forced cooling the switch operates in the frequency mode with exceedingly long working life. It was shown recently that 3 RSD-stacks in parallel (80-mm silicon wafer) are capable to switch 0.5 MA at 0.5 ms pulse width. RSD switches of this kind can be used in pumping systems of ultra-high-power lasers, electromagnetic launchers, various technologies using pulsed magnetic fields, etc.

Owing to the high pulse power and low switching losses in RSD, these devices have been used to create a set of effective and reliable generators of high-power high-voltage (30–50 kV) pulses operating in a frequency range 0.1–1.0 kHz for a number of industrial technologies—purification of industrial and agricultural waste waters by high-

Fig 6. Power RSD switch. Working voltage 25 kV, current 250 kA at pulse duration 100 μms (in rare pulse mode).

power electric discharge, pulsed power supply systems of electrostatic precipitators, pumping of technological lasers, etc., and also high-power high-frequency ($\sim$ 50 kHz) generators for electrothermics.

Now work is in progress on creating special fast RSD for high-power converter technology. The advantage of devices must be the greater, the higher the converter power. The possible power and frequency range of these devices is shown by broken curve in Fig. 1(b).

3.2 Nanosecond range

One of the most promising methods of formation of high-power pulses with nanosecond rise time is to accumulate energy in an inductance and then break the current through it by means of an in-series connected semiconductor diode for which this current is a reverse current. Preliminarily, a forward current pulse is passed through the diode to create a plasma supply in the base region, ensuring the possibility of reverse current flow. The energy stored in the inductance and passed into the load in the form of a short high-power pulse depends on the total amount of plasma which can be extracted from the diode by the reverse current.

In the early 1980s, it was shown at the Ioffe Institute that, with an initially markedly nonuniform plasma distribution, the current break in a $p^+p'nn^+$ diode occurs very fast, in several nanoseconds [9]. Such a distribution was created by a very short (hundreds of nanoseconds) pulse of forward current. In this case, a diffusion layer of high-density plasma ($\sim 10^{17}$ cm^{-3}) is formed in a narrow (10–20 μm) region at the p^+p' junction, wherefrom a plasma wave of low concentration ($\sim 10^{15}$ cm^{-3}) is carried away, reaching the n^+-layer in tens of nanoseconds under conditions of bipolar drift. When a reverse current is passed, an identical reverse wave is formed whose back front approaches the p′n junction simultaneously with the complete exhaustion of the diffusion plasma layer. After that an SCR spreading into the completely plasma-free n-region starts to form. The relation between the resistivity of the n-region and the reverse current density is chosen in such a way that the field in the quasi-neutral part exceeds 10^4 V/cm. Under these conditions the SCR extends at a saturation velocity $V_s = 10^7$ cm/s, and the current break in several nanoseconds. Diodes operating in this mode were named drift step recovery diodes (DSRD). DSRDs which can block voltages of, e.g., 2 kV operate at a reverse current density of $\sim 10^2$ A/cm^2 with a current break time of ~ 2 ns. The diodes can be connected in-series without any voltage dividing. The frequency limit, mainly determined by thermal processes, lies in the kilohertz range for power systems. The appearance of such devices was a breakthrough in the nanosecond technology. Rather rapidly was developed the basic circuitry of DSRD-based equipment, the demand for which was mainly determined by large-scale work in the field of super-wide-band radars, performed in the USSR and US. The highest-power DSRD generator created to date has a pulse power of 20 MW, pulse voltage of 80 kV, rise time of less than 1 ns, and operation frequency of 1 kHz.

In 1992–1994, the group of Academician G. A. Mesyats found another mechanism of nanosecond high-density current break in power diode recovery. It was found that at an extremely high reverse current density (1–50 kA/cm^2) in $p^+p'nn^+$ structure breaks abruptly (in 5–10 ns), being transferred to the load [10]. A calculation [11] demonstrated that the current break is due to an increase in the diode resistance in the initial stage of recovery, the increase being related not to the recovery of the pn junction, but to the formation in the p′-layer of a narrow region free of electron-hole plasma and having, therefore, a rather high resistance ($\sim 1\,\Omega$). In a stack of hundreds of in-series connected devices their total resistance becomes much higher than that of the load, with the result that the current is transferred therein. This process was named the SOS process.

SOS generators consist of a thyristor pulse generator, several magnetic pulse compression elements, and a sharpener comprising SOS diodes. The maximum pulse power obtained by now is 6 GW with a frequency limit of hundreds of hertz for power systems. Systems of this kind are a new breakthrough in the field of nanosecond technology. The most bulky and energy-consuming part of the systems are magnetic compressors of current pulses, since the pulse duration is commonly 0.3–0.6 μs for the forward current and 40–150 ns for the reverse current at an initial pulse duration in the thyristor pulse generator in the range 10–100 μs.

In 1996, one more mechanism of nanosecond diode recovery was proposed at the Ioffe Institute [12], in which the current density, similarly to the SOS process, lies in the kiloampere range and the working pulse duration can be increased approximately by an order of magnitude; this markedly simplifies the pulse compression systems.

In the new process, a markedly nonuniform plasma distribution is created in the stage of forward current in the n-base of the diode by reducing the injection coefficient of the n^+n junction. In this case, with reverse current flowing, the recovery starts at the n^+n junction, and the plasma region boundary moves toward the $p^+p'n$ junction. The junction is recovered and current breaks when the plasma front approaches the junction boundary. The SCR spreads, similarly to the case of DSRD, at a saturated velocity and, therefore, the current breaks very fast—in 1–2 ns and less. Such an "inverse" recovery process allows one to combine a high density of reverse current, large amount of charge accumulated in a diode, and very fast break of the current at a relatively long forward and reverse pulses. This means that in designing generators based on *inverse recovery diodes* (IRD) one can completely or partly reject magnetic compressors and build the systems based exclusively on semiconductor switches, which will dramatically reduce their size and increase efficiency.

3.3 Subnano- and picosecond range

Subnano- and picosecond commutation of high power by semiconductor devices became possible after the effect of wave breakdown in a silicon pn junction was discovered at the Ioffe Institute in studying the avalanche breakdown in silicon $p^+p'nn^+$ diodes at high ($\geq 10^{12}$ V/s) rate of voltage rise. The experimental arrangement is shown schematically in Fig. 7(a). To a reverse-biased ($U_r = 1$ kV) silicon diode with a static breakdown voltage of ~ 1.7 kV is applied a voltage pulse with $dU/dt \geq 10^{12}$ V/s. As a result, a decaying displacement current flows through the device for ~ 2 ns, and the voltage becomes as high as ~ 3 kV, i.e., nearly twice the static breakdown voltage. Then voltage across the device abruptly, in 70 ps, falls to ~ 100 V, and the current equally abruptly rises to $j \geq 0.6\,\text{kA/cm}^2$. This means that the n-base of the diode is filled with plasma having an average concentration of $\sim 10^{15}\,\text{cm}^{-3}$ in a time by one and a half order of magnitude shorter than the time of flight at saturated velocity! Further experiments enabled construction of a physical model of the process (Fig. 7(b)). At $t = t_0$ an SCR and a neutral region W_n with an electron concentration of $\sim 10^{14}\,\text{cm}^{-3}$ exist in the base of a reverse-biased diode. When the voltage rises rapidly, a displacement current flows through the SCR, with mainly conductance current flowing through the neutral region. During the rise time (~ 2 ns), thermal generation has no time to create in the SCR the number of carriers sufficient to initiate avalanche breakdown and, therefore, a superhigh voltage region Δ, in which the field strength exceeds the critical value, is formed at the pn junction. At the same time, the current flowing through W_n creates there a field

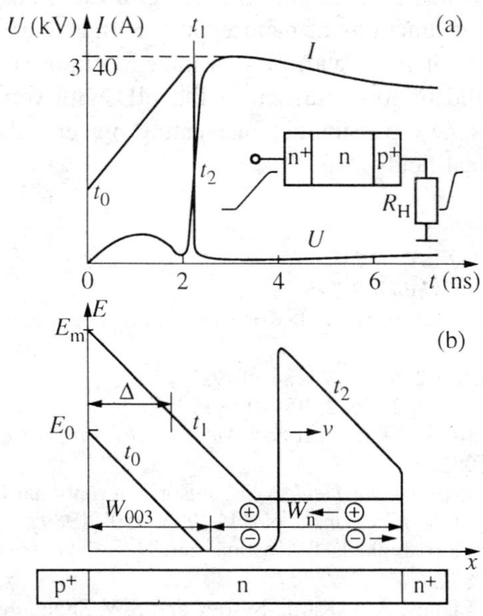

Fig 7. (a) Subnanosecond switching of diode. (b) Formation of impact-ionization wave.

sufficient for weak ionization by the majority carriers—electrons. The resulting holes are carried by the field to the superhigh voltage region in a time ~ 2 ns and initiate there a very intensive process of impact ionization forming high-density plasma. The field falls drastically in the plasma, but grows in the adjacent regions where holes initiating the breakdown already exist. In such a way fast impact-ionization wave travelling to meet the flow of holes is formed. The motion velocity of its front V_f is determined by the rate of ionization development in the region Δ and may exceed V_s by orders of magnitude. When the wave front reaches the n^+-contact, the n-base is completely filled with a highly conducting plasma and the device switches into the conducting state.

Devices whose operation is based on this process were named *avalanche pulse sharpeners* (AS). It was found that after some selection procedure these devices can be connected in-series to operate at rather high voltages of 15–20 kV. Moreover, the scatter of turn-on delay time in the devices (jitter) is very low (tens of picoseconds), which is highly important for quite a number of applications.

Further investigations have shown that impact-ionization waves can be used for ultrafast switching of not only $p^+p'nn^+$ diodes, but also $n^+p'nn^+$ transistors and $n^+p'np^+$ thyristor structures. Generally speaking, the possibility of using the impact-ionization wave of the TRAPATT type for ultrafast switching of a thyristor was discussed as far back as 1976 [14]. This possibility became reality after the phenomenon of wave breakdown in a diode was discovered. The first experimental results were reported in [15], but only in 1996 it became possible to overcome technical difficulties and obtain sufficiently reliable switching devices [16]. This device was named *fast ionization dynistor* (FID). The FID is switched by a short overvoltage pulse with a rise time of ~ 1 ns, formed by the technique described above. Geometrical parameters of layers constituting the $n^+p'nn^+$ structure must be chosen in such a way that carriers injected from the n^+ and p^+-emitters reach the collector junction plane only after the full pulse voltage is set across the junction. Under these conditions carriers initiate an impact-ionization wave whose travel is accompanied by area-uniform filling of the n-base with plasma, i.e., by switching of the device in fractions of nanosecond. Further current flow occurs by the conventional thyristor mechanism with plasma injection into base layers from n^+- and p^+-emitters. The fast and uniform turn-on of the FID with very low switching losses opens up fundamentally new possibilities of creating power pulse generators for quite a number of technological fields.

References

[1] I. V. Grekhov, *Electrotekhnika* **8** 2–5 (1986).

[2] Zh. I. Alferov, G. A. Ashkinazy, V. I. Korol'kov, and V. E. Chelnokov, *Electrotekhnika* **3** 210–213 (1979).

[3] L. Lorenz, *EPE-Journal* **2** No. 2, 77–84 (1992).

[4] D. W. Tsang, *EPE-Journal* **2** No. 2, 95–100 (1992).

[5] J. Tihanoyi, *Power MOSFET, Semiconductor Devices for Power Conditioning*, Plenum Press, p. 307–329, 1982.

[6] J. Baliga, *Power Semiconductor Devices*, Thomson International Publishing, USA, 1997.

[7] I. V. Grekhov, *Solid-State Electron.*, **2** No. 11, 923–930 (1989).

[8] I. V. Grekhov, A. V. Gorbatyuk, L. S. Kostina, and N. S. Jakovtchuk, *Solid-State Electron.* **26** No. 11, 1132 (1983).

[9] I. V. Grekhov, V. M. Efanov, A. F. Kardo-Sysoev and S. V. Shenderey, *Sov. Tech. Phys. Lett.* **9** 188 (1983).

[10] S. K. Lyubutin, S. N. Rukin, and S. P. Timoshenko, IX Symposium on High Power Electronics, Book of abstracts, p. 220–221, 1992.

[11] S. A. Darznek, A. Yu. Kotov, G. A. Mesyats, and S. N. Rukin, *Dok. Akad. Nauk* **334** No. 3, 304–306 (1994).

[12] I. V. Grekhov, *Megagauss-7*, Arzamas-16, Book of abstracts, p. 47, 1996.

[13] I. V. Grekhov and A. F. Kardo-Sysoev, *Sov. Tech. Phys. Lett.* **5** No. 15, 395 (1979).

[14] I. V. Grekhov, M. E. Levinstein, and V. G. Sergeev, *Sov. Phys. Semicond.* **10** No. 2, 345–349 (1976).

[15] V. M. Tuchkevitch and I. V. Grekhov, *New Principles of High Power Switching with Semiconductor Devices*, Leningrad: Nauka, 1988.

[16] V. M. Efanov, A. F. Kardo-Sysoev, I. G. Tchashnikov, and P. M. Yarin, *Proc. 11 Pulsed Power Conf.*, Baltimore, p. 22–25, 1997.

New physical phenomena in ferroelectric and magnetic crystals

S. B. Vakhrushev, V. V. Lemanov, R. V. Pisarev and V. A. Sanina

1 Introduction

Ferroelectricity and magnetism are two of the oldest scientific disciplines which have been developed at the Physico-Technical Institute since the late 1920s. Here we highlight only four topics in the field. Relaxor ferroelectrics, which were discovered at the Institute, and incipient ferroelectrics (or quantum paraelectrics) now attract great attention of researchers all over the world. In the present short review, results of microscopic studies of relaxors and new data on incipient ferroelectrics with impurities are discussed. In the field of magnetism, some new magnetooptical effects and a novel approach to magnetic phase transitions in crystals with two interacting subsystems are presented. Having only a limited room here, we cannot but paint these four topics only with a very broad brush.

2 Relaxor ferroelectrics

As early as in the fifties, in the course of a systematic study of complex perovskite-like compounds, Smolensky and Agranovskaya [1] discovered a new class of ferroelectrics—ferroelectrics with diffuse phase transitions (relaxor ferroelectrics). The most distinctive feature of these compounds is a broad maximum in the temperature dependence of the dielectric susceptibility $\varepsilon(T)$. Since the first observation of the relaxor phenomena, their properties have been studied widely, and the Ioffe Physico-Technical Institute was always one of the leading centers of such researches. It were Smolensky and Isupov who proposed the first semi-microscopic model of the relaxors, Popov was the first to observe an unusual frequency dispersion, and Bokov discovered the field-induced transition from the relaxor to the ferroelectric state. A large group of scientists from Smolensky's laboratory, including Isupov, Krainik, Siny, Yushin, and others, have performed an enormous number of studies of relaxors' properties. Many of the measurements became possible only due to successes in single crystal growth achieved by Mylnikova.

However, until recently, most studies were devoted to macroscopic (dielectric, elastic, etc.) properties of the relaxors. In the eighties, systematic studies of the microscopic features of the relaxors started. Several crystals, such as cubic perovskites $Na_{1/2}Bi_{1/2}TiO_3$ (NBT) and $PbMg_{1/3}Nb_{2/3}O_3$ (PMN), uniaxial tungsten bronze compound $Sr_{1-x}Ba_xNb_2O_6$ (SBN), etc. have been studied using the neutron and X-ray (both conventional and synchrotron radiation) scattering. It was found that the critical scattering is concentrated in a narrow quasielastic central peak and the q-dependence of this critical scattering intensity is well described by the Lorentzian, i.e., the static correlation function is an Ornstein–Zernike one. The temperature dependence of the ferroelectric fluctuations' inverse correlation length $\kappa = 1/r_c$ was determined [2, 5] (for the PMN it is presented in Fig. 1). At temperatures much higher than the average Curie temperature T_C^{av}, κ^2 follows the linear law $\kappa^2 \sim (T - T_C)$ in agreement with the mean field approximation. When approaching the transition region, the slope of the $\kappa^2(T)$ dependence starts to decrease, and at temperatures below T_C^{av}, κ^2 becomes practically temperature-independent. The observed behavior is absolutely different from that expected for the "normal" ferroelectric transition, but looks rather similar to that for spin glasses, or magnets with transition destroyed by random fields. So the conclusion can be made that the nature of the low-temperature state in relaxors is glass-like and the possible role of the frozen random electric fields can be revealed [2]. The temperature dependence of the peak intensity $I_0(T)$, proportional to the static susceptibility χ_0, was also analyzed. At high temperatures, it follows a Curie–Weiss law (unlike the dynamic susceptibility obtained from dielectric measurements). On further cooling, the $I_0(T)$ dependence levels-off, but is still scaled with κ^2 so that $I_0\kappa^2$ product is constant, down to some freezing temperature T_g (in the case of PMN, $T_g \approx 230$ K). Below T_g, κ^2 is temperature independent, while I_0 continues to increase, and $I_0\kappa^2$ is linearly increasing with temperature decreasing below T_g (Fig. 1(a)). The observed deviation of $I_0\kappa^2$ indicates that some changes of the scattering law take place, or a new scattering mechanism appears. To detect any difference from the Ornstein–Zernike correlation function, we have performed special high-resolution ($\approx 10^{-4}$ Å^{-1}) synchrotron radiation scattering measurements [3]. We have found that the $I(q)$ dependence measured with such a resolution was not Lorentzian-shaped, but at $q > k$ followed the power law $q^{-\alpha}$ with a power exponent $\alpha > 2$. Such a dependence could be attributed to the scattering on fractals, and, for the case of volume or mass fractals, α is equal to the fractal dimension D. In Fig. 1(b), the $\alpha(T)$ dependence is presented. One can see that at about 230 K a sharp change of the exponent from $\alpha = 2.3$ to $\alpha = 2.7$ takes place. This can be interpreted as a consequence of a change in the spatial organization of the ferroelectric fluctuations. Such effects have been observed in spin glasses and attributed to scattering on configuration fluctuations (different from the thermodynamic ones).

Additional corroboration of the appearance at T_g of some new objects different from the usual thermodynamic fluctuations was obtained from the dielectric spectra. Colla *et al.* [4] presented the results of measurements of dielectric losses in PMN at ultralow frequencies. The temperature dependence of the average relaxation time $\tau(T)$ was obtained, and it was demonstrated that, similarly to κ, τ becomes temperature-independent below T_g, attaining a value of about 10^{-5} s. Careful analysis of the shape of the spectra demonstrates that above T_g it is well described in terms of β-relaxation, whereas at lower temperature an appreciable deviation from it is observed, increasing as temperature is lowered. This deviation is observable only at low frequencies, and could be considered as an evidence of a new mechanism of losses related to the configuration fluctuations mentioned above.

The presented results suggest that a well defined temperature T_g, at which cooperative freezing takes place, exists in PMN and other relaxors. To study the characteristics of the low temperature phase formed as a result of this freezing, we have performed measurements with applied electric fields. We have studied the temperature dependences of the intensities of both Bragg and diffuse scattering and demonstrated that the

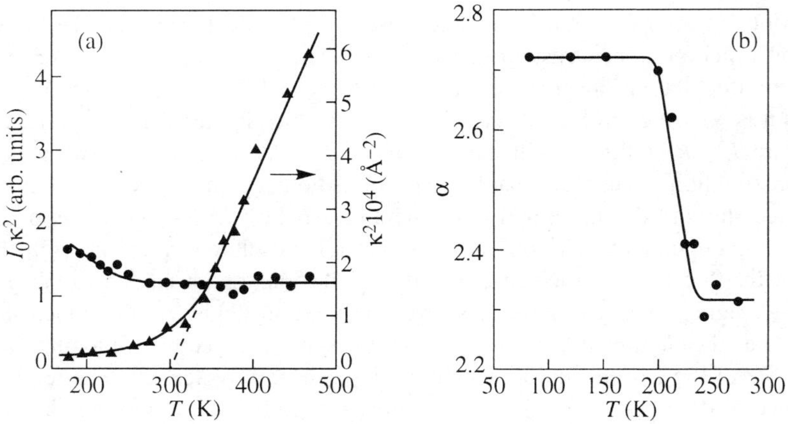

Fig 1. (a) Temperature dependences of the parameters, describing the neutron critical scattering measurements in PMN (I_0 is scattering intensity, extrapolated to $q = 0$, κ^2 is the square of the inverse correlation length). (b) Temperature dependence of the parameter fractal dimension α.

dependences measured in field-cooled (FC) and zero-field-cooled (ZFC) regimes coincide at temperatures above T_g and split at lower temperatures. The splitting indicates that below T_g the properties of PMN are history dependent and the low temperature phase is a nonergodic one. Additional evidence of the nonergodicity and the glass-like nature of the low-temperature state in relaxors is the logarithmic relaxation observed in PMN, first in neutron scattering measurements [5] and then in measurements of dielectric response in ZFC regime [6]. High-resolution single crystal X-ray measurements performed in collaboration with Ecole Centrale de Paris [7] confirmed the dipole-glass concept and gave no evidence in favor of the domain evolution interpretation proposed in [8].

Probably the least studied problem of relaxor materials is related to the lattice dynamics and especially to the critical dynamics. We have performed neutron inelastic scattering measurements and revealed the strongly damped low-lying excitation observable in a narrow region near the Brillouin zone center. It was impossible to study the dispersion of this excitation directly. We used the mode coupling approach to elucidate the parameters of this excitation from the analysis of distortions of the TA phonon lineshapes. The observed excitation can be considered as a quasioptic mode with strongly frequency-dependent damping. Substantial softening of this mode upon cooling was found at temperatures above 650 K. Below 650 K, the mode softening ceases and a strong central peak appears. This crossover of the critical dynamics can be attributed to some kind of phase transition to a superparaelectric state. This conclusion is supported by the appearance of a strongly q-dependent ($\gamma_{TA} \sim q^4$) damping of the TA phonons.

Summarizing the results presented above, we conclude that the freezing in relaxors has a cooperative nature with well-defined freezing temperature, which results in the formation of a macroscopically homogeneous glass-like state. No evidence of the existence of ferroelectric nanodomains was found. At very high temperatures (of about $3T_g$), the relevant phonon-like critical excitation exists. The observed crossover results in that the narrow central peak dominates in the spectrum of critical excitation, which explains the appearance of the frequency dispersion of the susceptibility and related properties.

3 Incipient ferroelectrics with impurities

In a broad family of ferroelectrics and related materials there is a number of crystals which would like to undergo a ferroelectric phase transition but cannot do this. Strontium titanate, $SrTiO_3$, potassium tantalate, $KTaO_3$, or PbTe are among such crystals. These crystals are baptized incipient ferroelectrics or quantum paraelectrics. They have a polar soft mode but exhibit no ferroelectric phase transition because of quantum fluctuations [9], similarly to quantum liquids ^{4}He and ^{3}He which do not transform into the solid state under normal conditions. In a sense, the quantum paraelectrics may be considered as marginal systems which are at the limit of stability of a paraelectric phase stabilized by quantum fluctuations. A small external perturbation may disturb the paraelectric phase and induce a transition into a ferroelectric phase. Indeed, both $SrTiO_3$ and $KTaO_3$ undergo a ferroelectric phase transition under uniaxial strain [10, 11]. For example, $SrTiO_3$ exhibits a ferroelectric phase transition under the action of only 1 kbar, which corresponds to an elastic strain of the order of 10^{-4}. The ferroelectric state can also be induced by electric fields [12–14]. An electric field of about 1 kV/cm is enough to induce the ferroelectric phase transition in $SrTiO_3$ at low temperatures. One observes similar effects in the case of incipient ferroelectrics with impurities, which may become real ferroelectrics with impurity concentration sometimes as low as 10^{-3}. Thus, one can see that small external perturbations of about 10^{-5} (electric fields as compared to atomic fields), 10^{-4} (elastic strains) and 10^{-3} (impurities) can indeed disturb the stability of quantum paraelectrics and force them to become ferroelectrics.

Here, we shall discuss ferroelectric phase transitions induced by impurities in $SrTiO_3$. Until very recently, only the effect of Ca^{2+} impurity had been studied in $SrTiO_3$ [15]. As for $KTaO_3$, there has been a long story of studying the effect of Li^+ (KLT) and Nb^{5+} (KTN) impurity ions on ferroelectric phase transitions [15–18]. The general conclusions of all these studies are as follows. A small amount of impurities induces a ferroelectric phase transition; the transition temperature varies as $T_C = A(x - x_c)^{1/2}$ where x_c is the critical impurity concentration (a quantum limit); at $x < x_c$, a polar glass-like behavior is possible; the mechanism of the induced ferroelectric phase transition is due to the interaction of dipole moments of impurity ions in off-center positions, as exemplified best of all by the case of small Li^+ ions in $KTaO_3$ substituted for K^+ host ions. To understand better the mechanisms of phase transitions induced in the incipient ferroelectrics by impurities, we studied a number of divalent cation impurities in $SrTiO_3$: Ca^{2+}, Cd^{2+}, Ba^{2+}, and Pb^{2+}.

All these impurities appeared to induce ferroelectric phase transitions [15, 19–21], and at low concentration x, the transition temperature follows the above-mentioned relation $T_C = A(x - x_c)^{1/2}$ where the critical concentration x_c is almost the same for Ca^{2+}, Cd^{2+} and Pb^{2+} and equals about 0.002 (Fig. 2). In the case of Ba^{2+}, $x_c = 0.035$, but one observes a polar glass-like behavior with nonzero mean-square value of electric polarization $\langle P^2 \rangle$ [22] at small concentrations, between 0.035 and about 0.002. So one may say that the critical concentration for all the impurities under study is almost the same ($x_c = 0.002$), even though individual characteristics of these ions are rather different as shown in Table 1. It can be seen that the radius of the Ca^{2+} and Cd^{2+} ions is less than that of the Sr^{2+} host ion, and the opposite is true for the Ba^{2+} and Pb^{2+} ions; the polarizability of the Ca^{2+} ion is less than, but comparable to the polarizability of the Sr^{2+}, ion but is an order of magnitude smaller than that of the Pb^{2+} ion. One can

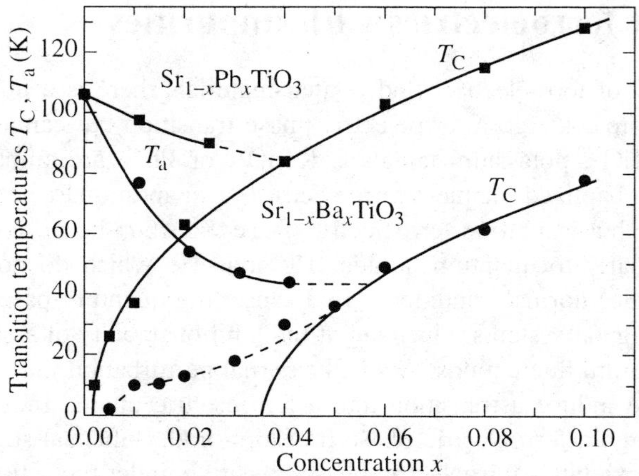

Fig 2. Phase diagram of $Sr_{1-x}Ba_xTiO_3$ and $Sr_{1-x}Pb_xTiO_3$ systems at low x values.

Table 1. Ionic radius, electronic structure and polarizability.

Ion	R (Å)	R/R (Sr^{2+})	nl^r	χ (10^{-24} cm^3)
Ca^{2+}	0.99	0.88	$3p^6$	0.5
Sr^{2+}	1.13	1.0	$4p^6$	1.0
Cd^{2+}	0.97	0.86	$4d^{10}$	1.8
Ba^{2+}	1.35	1.2	$5p^6$	1.7
Pb^{2+}	1.20	1.06	$6s^2$	4.9

conclude that the dominant role in the onset of ferroelectricity in the $SrTiO_3$ incipient ferroelectric is played not by characteristics of the impurity ions themselves and not by their off-center positions, if any, but by the properties of the $SrTiO_3$ host lattice and, first of all, by its extremely high polarizability. This means that the impurities act as a trigger of the induced ferroelectric phase transitions.

It is well known that in $SrTiO_3$ there is a structural (improper ferroelastic) phase transition at $T_a = 105$ K from a cubic (m3m) to a tetragonal (4/mmm) phase. The dependence of the transition temperature T_a on the impurity concentration x is shown in Fig. 2 for Ba^{2+} and Pb^{2+} impurities. It can be seen that the structural phase transition lines converge with the lines of the ferroelectric phase transition at $x \simeq 0.05$. We have an interesting situation here, when ferroelectric and structural instabilities may be made arbitrary close [23].

There is a special type of solid solutions when heterovalent ions are substituted for host ions in A and B positions simultaneously. An example is the $SrTiO_3$–$KTaO_3$ system [24]. Here, both end members are incipient ferroelectrics, so a question arises whether the ferroelectricity is possible in this system. It turned out that at low temperatures, a transition into a polar state is observed with a significant frequency dispersion of the dielectric constant. An example of the temperature dependence of the dielectric constant in 0.85 $SrTiO_3$–0.15 $KTaO_3$ solid solution is shown in Fig. 3. The frequency dispersion of the dielectric constant as a function temperature obeys a Vogel–Fulcher relation with $T_F = 26$ K. So, at this temperature there is a transition into a polar glass-like state.

In conclusion, all the impurities under study induce transitions into a ferroelectric or glass-like state with low critical concentration (about 10^{-3}), and these transitions are

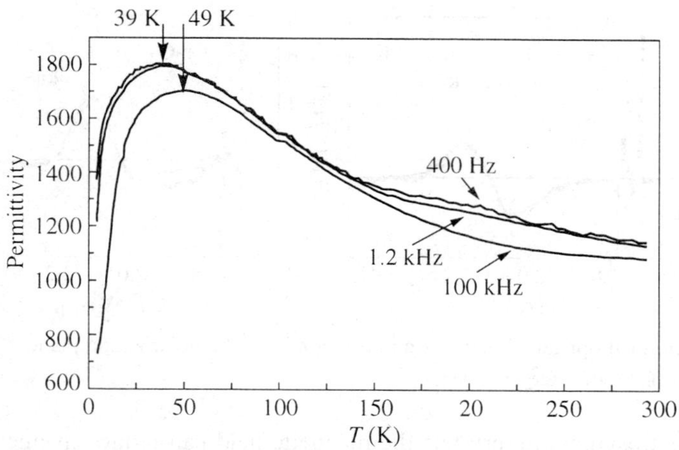

Fig 3. Temperature dependence of permittivity in $0.85SrTiO_3$–$0.15KTaO_3$ solid solution.

determined, first of all, by the very high polarizability of the host lattice. However, more research is to be done to establish mechanisms governing ferroelectric phase transitions in incipient ferroelectrics with impurities.

4 Magnetoelectric and nonlinear optical spectroscopy of magnetic crystals

4.1 Magnetoelectric spectroscopy

Linear and non-linear optical effects are closely related to breaking of different symmetry operations in space and time [25]. For example, when the *space-inversion* is broken, such effects as optical activity, linear electrooptical effect, second harmonic generation, etc., become allowed. Breaking the *time-reversal* leads to *nonreciprocal* magneto-optical effects such as Faraday rotation, Kerr effect, and others. All the above-mentioned effects have been studied in a large number of materials, and underlying microscopic mechanisms are quite well understood. In contrast, much less attention has been paid to optical effects which arise owing to a *simultaneous* breaking of the space inversion and time reversal. In such crystals a new kind of optical susceptibility becomes allowed:

$$\alpha_{ij}^{(nk)} \sim \mathrm{Re}\left(\mathrm{d}_{kn}^{i}\mathrm{m}_{nk}^{j}\right),$$

where α_{ij} is the magnetoelectric tensor, d_{kn}^{i} and m_{nk}^{j} are the matrix elements of the electric-dipole and magnetic-dipole moments, respectively. In optics, the tensor $\alpha_{ij}(\omega)$ gives rise to nonreciprocal effects not only in ferromagnets, but also in *antiferromagnets*. Some of effects of this kind have been observed recently [26–28]. Figure 4 shows the ellipticity ε and optical rotation Θ in reflection in antiferromagnetic Cr_2O_3 below the Neel temperature $T_N = 307$ K. This is the first observation of nonreciprocal effects in the absence of magnetization or magnetic field. The magnetoelectric contribution to the circular dichroism of third order in the magnetization was observed in reflection of light from lithium ferrite $LiFe_5O_8$ [28].

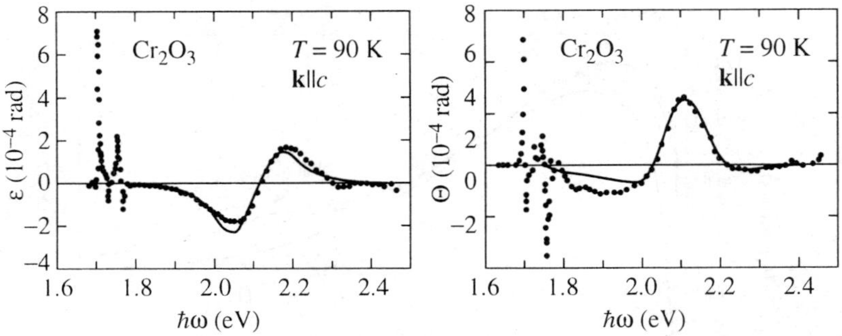

Fig 4. Nonreciprocal optical ellipticity ε and rotation Θ of the polarization plane of light reflected from the basal plane of a Cr_2O_3 crystal.

In non-centrosymmetric crystals the magnetic field can induce magneto-spatial dispersion effects, which are bilinear in the wave vector of light **k** and the magnetic field **B** [29]. These effects can be described by a contribution to the optical dielectric tensor ε_{ij} as

$$\Delta\varepsilon_{ij}(\omega, \mathbf{k}, \mathbf{B}) = \gamma_{ijkl} B_k k_l,$$

where $\hat{\gamma}$ is an axial fourth-rank tensor. Being due to a higher-order perturbation of the electronic structure, kB effects are smaller than the linear magneto-optical effects. However, an enhancement of kB effects might have been expected in magnetic semiconductors. A systematic study of the anisotropy and dispersion of kB-type birefringence was undertaken in cubic magnetic semiconductors $Cd_{1-x}Mn_xTe$, in which the tensor $\hat{\gamma}$ has two independent components A and g. Figure 5(a) shows the dispersion of normalized components of the A/x and g/x and Fig. 5(b)—the normalized Faraday rotation Θ_{FR}/x and quadratic Voigt birefringence Φ_{VB}/x^2. It provides an unique opportunity to compare the absolute values and the dispersion of the three different magneto-optical effects in four $Cd_{1-x}Mn_xTe$ samples. The Faraday rotation is three orders of magnitude larger than the kB effect. Near the band gap the kB and B^2 effects are comparable in magnitude at a field of about 1 T.

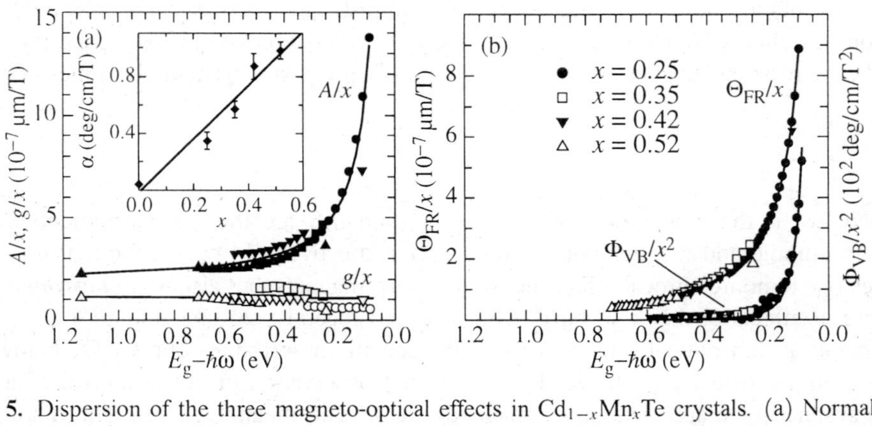

Fig 5. Dispersion of the three magneto-optical effects in $Cd_{1-x}Mn_xTe$ crystals. (a) Normalized components of symmetric A/x and antisymmetric g/x contributions to the kB effect. (b) Faraday rotation Θ_{FR}/x and Voigt birefrengence Φ_{VB}/x^2.

4.2 Nonlinear optical spectroscopy

Nonlinear optical susceptibilities, e.g., the susceptibility $\chi^{(2)}_{ijk}$ responsible for the second harmonic generation (SHG), are related to the microscopic structure of the medium. In the electric-dipole approximation, the optical nonlinearity is closely related to the distribution of positive and negative charges. The electric-dipole tensor $\chi^{(2)}_{ijk}$ vanishes in centrosymmetric media and this explains why piezoelectric and ferroelectric crystals have been most attractive for experimental studies and applications for more than three decades. The most magnetically ordered crystals are centrosymmetric and, therefore, the SHG is not allowed by the electric-dipole mechanism. Nevertheless, during last years a magnetization-induced SHG was successfully observed in ferromagnetic and antiferromagnetic crystals. Two cases have to be considered. The first arises in magnetics with *non-centrosymmetric crystal structure*, in which the nonlinear polarization $P_i(2\omega)$ can be written as

$$P_i(2\omega) = \chi^{(2)}_{ijk}(-2\omega, \omega, \omega)E_j(\omega)E_k(\omega) + \chi^{(3)}_{ijkl}(-2\omega, \omega, \omega, 0)E_j(\omega)E_k(\omega)M(0),$$

where $E_i(\omega)$ and $E_j(\omega)$ are the fundamental optical fields and $M(0)$ is the magnetization. Both contributions are of the electric-dipole type. The validity of this equation was verified in thin films of magnetic garnets on substrates with orientations (001), (110), (111), and (210) [30]. The possibility was demonstrated of separating unambiguously the crystallographic and magnetic contributions with the help of the rotational anisotropy of SHG.

A new kind of electric-dipole nonlinearity may exist in crystals with centrosymmetric structure owing to a *non-centrosymmetric spin order.* This mechanism was observed in antiferromagnetic Cr_2O_3 [31]. The spin structure below the Neel temperature $T_N =$ 307 K can be described by the magnetic group $\underline{3}$ m, in which the polar c-tensor of rank three is allowed. A very striking result was that the SHG spectra taken with the right and left circularly polarized light were identical for some transitions, but different for others. Such an unusual behavior was explained by interference between non-linear waves by the electric-dipole mechanism of c-type and magnetic-dipole mechanism of i-type. The interference between the two sources leads to a difference between the SHG spectra with opposite helicity, which allows distinguishing between the 180° antiferromagnetic domains [32–34]. Such domains are indistinguishable in linear optics, but with SHG they are observed in Cr_2O_3 and in a ferroelectric antiferromagnet, $YMnO_3$, with high contrast (Fig. 6).

(a) (b)

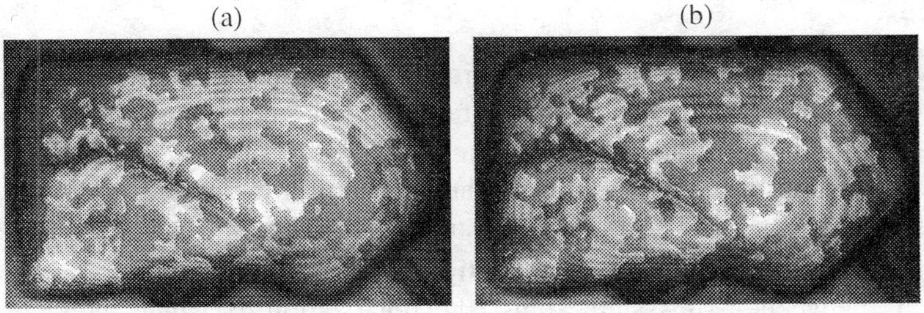

Fig 6. 180° antiferromagnetic domains in $YMnO_3$ as observed at the second harmonic 2ω for circularly polarized light with opposite helicity.

5 Phase transitions in magnetic crystals with interacting subsystems

Quite a number of magnetic crystals have several different magnetic ions in the unit cell, which can form two or more interacting magnetic subsystems with proper order parameters. A mutual influence of the subsystem states takes place, depending on the correlations of inter- and intra-subsystem interactions. Obviously, phase transitions in such crystals are also essentially sensitive to the value and symmetry of subsystem interactions.

If a magnetic crystal contains only one type of magnetic ions, some other, not necessarily magnetic, subsystems exist in the crystal and can interact with the spin subsystem. Orbital moments or local displacements of the magnetic ions in the lattice are examples of such subsystems. In these cases, the spin-orbital or magnetoelastic interactions take place in the crystal, respectively. The situation is completely identical with two interacting magnetic subsystems, when the order parameters are pseudospins and the corresponding interactions are identical with exchange spin interaction. A second order structural phase transition (an order-disorder type) is the most typical example of such a situation.

Our goal was to study phase transitions in crystals with two interacting subsystems and to analyze their peculiarities connected with the self-consistent mutual influence of subsystem states for a number of typical cases.

We now consider the simplest case. Let us assume that a unit cell contains two types (A and B) of magnetic ions with S^A and S^B spins, respectively. The exchange Hamiltonian in this case can be written as

$$H = \sum_{ij} J_{ij}^{AA} S_i^A S_j^A + \sum_{ij} J_{ij}^{BB} S_i^B S_j^B + \sum_{ij} J_{ij}^{AB} S_i^A S_j^B. \tag{1}$$

Here the values of J_{ij}^{AA}, J_{ij}^{BB}, and J_{ij}^{AB} are nonzero for the nearest neighbor ions i and j, they characterize the exchange interactions inside and between the A and B subsystems, respectively. Let us assume the following relation the exchange constants, between which is realized in magnetic crystals rather often: $J_{ij}^{AA} \gg J_{ij}^{AB} \gg J_{ij}^{BB}$. The effective Hamiltonian of the B-subsystem in the second-order perturbation theory with respect to the small parameter $J_{ij}^{AB}/J_{ij}^{AA} \ll 1$ can be written as

$$H_{\text{eff}}^{B} = \sum_{ij} J_{ij}^{AB} \langle S_i^A \rangle S_j^B - \sum_{ij} \left[\left(J_{ij}^{AB} \right)^2 / J_{ij}^{AA} \right] K_{ij}^{AA} S_i^B S_j^B + \sum_{ij} J_{ij}^{BB} S_i^B S_j^B. \tag{2}$$

Here $\langle S_i^A \rangle$ is the spin averaged over configurations of the A-subsystem. This value differs from zero at temperatures below the temperature of long-range ordering in A-subsystem, and it is zero in the paramagnetic state. The value $K_{ij}^{AA} = \langle S_i^A S_j^A \rangle - \langle S_i^A \rangle \langle S_j^A \rangle$ is the spin correlation function of the A-subsystem. The first sum in the effective Hamiltonian (2) describes the spin magnetization of the B-subsystem by the mean field $H_{mf}^{B} = J_{ij}^{AB} \langle S_i^A \rangle$. The second sum in (2) describes the effective exchange interaction of S^B spins through correlations between A-subsystem spins (the polarization exchange interaction). The polarization exchange is effective when the direct exchange interaction of S^B spins is weaker than the polarization interaction ($J_{ij}^{BB} \ll V_{ij}^{BB} = -[(J_{ij}^{AB})^2/J_{ij}^{AA}] K_{ij}^{AA}$), and the role of the mean field H^B_{mf} is considerably suppressed (for symmetry reasons or in the critical region near the phase transition in the A-subsystem).

The effective exchange interaction between impurity atoms in interstitial positions in metamagnets, which is caused by spin correlations in the host lattice, was first proposed by Ivanov and Shender [35]. The polarization f-d-f exchange for rare-earth magnetics and the Hamiltonian (2) were written in explicit form in [36] to explain the magnetic phase transition in $EuCrO_3$, induced by optical pumping. It should be emphasized that in contrast to the spin magnetization (spin orientation) effect of the mean field H^B_{mf}, the polarization exchange V^{BB}_{ij} can ensure the ordering of S^B spins with a proper ordering temperature. Such a proper ordering of optically excited magnetic Eu^{3+} ($^7F_1$) ions in $EuCrO_3$ crystal was reported in [36, 37].

Unlike the mean field H^B_{mf} which is zero in disordered state, the polarization exchange interaction is nonzero at any temperatures. However, the value of V^{BB}_{ij} interaction depends on temperature because the correlation function K^{AA}_{ij} is temperature dependent.

It is important to note that if the exchange constants J^{AA}_{ij}, J^{BB}_{ij}, and J^{AB}_{ij} are nonzero only for the nearest neighbors, the effective polarization exchange interaction has a long-range character. The V^{BB}_{ij} interaction length is determined by the correlation radius of the S^A spin fluctuations, and the V^{BB}_{ij} value is enhanced by the long-range interaction.

When the spin fluctuations in the A-subsystem have large correlation radii, the effect of the polarization interaction V^{BB}_{ij} is of particular importance. Such situations are realized for the critical region near a second order phase transition in the A-subsystem, or when the A-subsystem is of a low-dimensional type.

Because of the polarization interaction V^{BB}_{ij}, the magnetic phase transition in the main A-subsystem may be essentially dependent on the presence of the B-subsystem, in spite of the weakness of the proper interaction J^{BB}_{ij}. Let us consider the most important effects of polarization interactions on the phase transitions

1. The proper long-range ordering state, with the ordering temperature and symmetry, may arise in the B-subsystem owing to the polarization exchange interaction. Such a phase transition has been studied earlier [36, 37]. A similar type of structural phase transition was investigated in [38].

2. The polarization interaction V^{BB}_{ij} and the direct short-range exchange interaction J^{BB}_{ij} may have different signs. As a result, the frustrations may arise in the B-subsystem. In the critical region near the second-order phase transition point in the A-subsystem, the value of the mean field H^B_{mf} decreases, but the correlation radius of the critical fluctuations increases and, accordingly, the role of the polarization interaction V^{BB}_{ij} increases, too. For this reason, a random field or glass states in the B-subsystem may appear owing to the V^{BB}_{ij} interaction.

On the other hand, the presence of a random field in the B-subsystem will affect the character of the phase transition in the A-subsystem. In agreement with the Imre and Ma theorem [39], owing to Ising random field, the long-range ordering state becomes unstable to create domains of the same ordering type but of finite size. As a result, a diffuse-type phase transition in the A-subsystem may occur.

3. When the A-subsystem has the antiferromagnetic ordering, the situation in the critical region near the second-order phase transition is similar to the described in item 2.

In such a case, in the critical region in the A-subsystem, with the correlation radius of spin fluctuations $\xi \gg a$ (a is the lattice constant), the polarization interaction V_{ij}^{BB} becomes long-range and changes the sign at each lattice site. In this situation, we believe that the presence of a frustrating and long-range interaction can give rise to a glass state in the B-subsystem, despite the absence of a frozen random disorder.

The situation considered in item 3 is possible both for 3D and for 2D antiferromagnets. For the 3D case, the state with large correlation radius $\xi \gg a$ is realized only in the nearest vicinity of the critical point (Neel temperature). In the case of quasi-2D Heisenberg antiferromagnets of R_2CuO_4 type the condition $\xi \gg a$ is realized over a very wide temperature range above the quasi-2D Neel temperature [40]. So, such crystals are very convenient model objects for experimental investigations of the effect of polarization interaction on the critical phenomena.

The situation described in item 3, has been found in Eu_2CuO_4 antiferromagnet. In this case, the quasi-2D Heisenberg antiferromagnetic spin subsystem of the Cu^{2+} ions (in CuO_2 planes) plays the part of the A-subsystem, and the orbital subsystem of the Cu^{2+} ions—of the B-subsystem.

In Eu_2CuO_4, correlations of spin, orbital, and structural states have been observed, connected with the polarization orbital–orbital interaction through 2D Heisenberg antiferromagnetic spin fluctuations with very large correlation radius $\xi \gg a$.

The Ising orbital glass state exists in the Eu_2CuO_4 crystal over a wide temperature range, including $T \gg T_N$ [32, 33]. The inverse effect of the orbital Ising random field on the 2D spin subsystem determines its static and dynamic magnetic properties. Let us enumerate these properties [32, 33]:

—a jump of the static magnetic susceptibility at a temperature near T_N and near the temperature of orbital glass state onset;

—the existence of uniform, well-defined spin-wave excitations over a wide temperature range, including $T \gg T_N$;

—the appearance of a 2D random weak ferromagnetism;

—a change in the character of the quasi-2D phase transition, it is now associated with the appearance of domains having antiferromagnetically correlated spins, but restricted to a finite size in the temperature range at which the orbital glass state exists [42].

The specific properties of the Eu_2CuO_4 crystal are an example of the polarization interaction in low-dimensional systems.

6 Conclusion

This short review article reflects some of our recent interests in the field of ferroelectricity and magnetism: relaxor ferroelectrics, studied using microscopic techniques such as neutron and X-ray scattering; new ferroelectric systems on the base of $SrTiO_3$ quantum paraelectric; new magnetoelectric and nonlinear-optical effects in magnetic crystals with interacting subsystems and a novel approach to their theoretical treatment.

References

[1] *Ferroelectrics and Related Materials*, ed. G. A. Smolensky, Gordon and Breach, New York, 1981.

[2] S. Vakhrushev, A. Nabereznov, N. Okuneva, and B. Toperverg, *Ferroelectrics* **90** 173 (1989).

[3] S. Vakhrushev, A. Nabereznov, S. K. Sinha, Y. P. Feng, and T. Egami, *J. Phys. Chem. Solids* **57** 1517 (1996).

[4] E. V. Colla, E. Yu. Koroleva, N. M. Okuneva, and S. B. Vakhrushev, *J. Phys.: Condens. Matter* **4** 3671 (1992).
[5] S. Vakhrushev, B. Kvyatkovsky, A. Nabereznov, N. Okuneva, and B. Toperverg, *Physica B* **156-157** 90 (1989).
[6] E. V. Colla, E. Yu. Koroleva, N. M. Okuneva, and S. B. Vakhrushev, *Phys. Rev. Lett.* **74** 168 (1995).
[7] S. Vakhrushev, J.-M. Kiat, and B. Dkhil, *Solid. State Commun.* **103** 477 (1997).
[8] V. Westphal, W. Kleemann, and M. D. Glinchuk, *Phys. Rev. Lett.* **68** 847 (1992).
[9] K. A. Müller and H. Burkard, *Phys. Rev. B* **19** 3593 (1979).
[10] W. J. Burke and R. J. Pressley, *Solid State Commun.* **9** 191 (1971).
[11] E. Fuji and E. Sakuda, *Izv. AN SSSR* **41** 493 (1977).
[12] E. Hegenbarth, *Phys. Status Solidi* **6** 333 (1964).
[13] M. A. Saifi and L. E. Cross, *Phys. Rev. B* **2** 677 (1970).
[14] J. Hemberger, P. Lunkenheimer, R. Viana, R. Böhmer, and A. Loidl, *Phys. Rev. B* **52** 13159 (1995).
[15] J. G. Bednorz and K. A. Müller, *Phys. Rev. Lett.* **52** 2289 (1984).
[16] U. T. Höchli, K. Knorr, and A. Loidl, *Adv. Phys.* **39** 405 (1990).
[17] W. Kleemann, *J. Mod. Phys. B* **7** 2469 (1993).
[18] B. E. Vugmeister and M. D. Glinchuk, *Rev. Mod. Phys.* **62** 993 (1990).
[19] V. V. Lemanov, E. P. Smirnova, P. P. Syrnikov, and E. A. Tarakanov, *Phys. Rev.* **B54** 3151 (1996).
[20] V. V. Lemanov, E. P. Smirnova, and E. A. Tarakanov, *Phys. Solid State* **39** 628 (1997).
[21] V. V. Lemanov, *Phys. Solid State* **39** 1468 (1997).
[22] M. E. Guzhva, W. Kleemann, V. V. Lemanov, and P. A. Markovin, *Phys. Solid State* **39** 618 (1997).
[23] E. V. Balashova and A. K. Tagantsev, *Phys. Rev. B* **48** 9979 (1993).
[24] V. V. Lemanov, V. A. Trepakov, P. P. Syrnikov, M. Savinov, and L. Jastrabik, *Phys. Solid State* **39** 1642 (1997).
[25] R. V. Pisarev, *Ferroelectrics*, **162** 191 (1994); **183** 39 (1996).
[26] B. B. Krichevtsov, V. V. Pavlov, R. V. Pisarev, and V. N. Gridnev, *J. Phys. Condens. Matter* **5** 8233 (1993).
[27] B. B. Krichevtsov, V. V. Pavlov, R. V. Pisarev, and V. N. Gridnev *Phys. Rev. Lett.* **76** 4628 (1996); *Zh. Eksp. Teor. Fiz.* **110** 1505 (1996).
[28] V. N. Gridnev, B. B. Krichevtsov, V. V. Pavlov, and R. V. Pisarev *Pis'ma Zh. Eksp. Teor. Fiz.* **65** 65 (1997).
[29] B. B. Krichevtsov, R. V. Pisarev, A. A. Rzhevsky, V. N. Gridnev, and H.-J. Weber, *Phys. Rev. B* (May 1998); *Zh. Exp. Teor. Fiz.* (1998) (to be published).
[30] V. V. Pavlov, R. V. Pisarev, A. Kirilyuk, and Th. Rasing, *Phys. Rev. Lett.* **78** 2004 (1997).
[31] M. Fiebig, D. Fröhlich, B. B. Krichevtsov, and R. V. Pisarev, *Phys. Rev. Lett.* **73** 2127 (1994).
[32] M. Fiebig, D. Fröhlich, G. Sluyterman v.L., and R. V. Pisarev, *Appl. Phys. Lett.* **66** 2906 (1995).
[33] R. V. Pisarev, M. Fiebig, and D. Fröhlich, *Ferroelectrics* **204** 1 (1997).
[34] M. Fiebig, D. Fröhlich, St. Leute, and R. V. Pisarev, *Appl. Phys. B* **66** 265 (1998).
[35] M. A. Ivanov and E. F. Shender, *Sov. Phys. JETP* **42** 179 (1975).
[36] E. I. Golovenchits, B. D. Laikhtman, and V. A. Sanina, *JETP Lett.* **31** 1 (1980).
[37] E. I. Golovenchits, V. A. Sanina, and T. A. Shaplygina, *Sov. Phys. JETP* **53** 992 (1981).
[38] E. I. Golovenchits, V. A. Sanina, and G. Smolensky, *JETP Lett.* **40** 857 (1984);
E. I. Golovenchits, V. A. Sanina, A. Levin, Ju. Shepelev, and Ju. Smolin, *Sov. Phys. Sol. State* **29** 2036 (1987).
[39] Y. Imre and S-ka. Ma, *Phys. Rev. Lett.* **35** 1399 (1975).
[40] S. Chakravarty, B. Halperin, and D. Nelson, *Phys. Rev. B* **39** 2344 (1989).
[41] A. V. Babinskii, S. L. Ginzburg, E. I. Golovenchits, and V. A. Sanina, *JETP Lett.* **57** 299 (1993).
[42] E. I. Golovenchits, V. A. Sanina, and A. V. Babinskii, *Phys. JETP* **83** 386 (1996).

Wide-bandgap semiconductors

Yu. A. Vodakov, E. N. Mokhov, Yu. V. Zhilyaev, and V. I. Sankin

1 Introduction

It is common practice to relate, rather arbitrarily, to wide-bandgap semiconductor materials semiconductors with a forbidden gap width close to, or exceeding 2.3 eV, which corresponds to light wavelengths in the "green" or shorter-wavelength spectral ranges. Therefore, wide-bandgap semiconducting materials are considered to be basic semiconductors for optoelectronic devices operating in the short-wave visible and UV spectral ranges. Without these materials it is impossible to form full-color images with LED matrices, increase the density of laser recording of information, develop green and blue traffic lights with unlimited operation life, create UV photodetectors "blind" to the visible, etc. Another promising way of using wide-bandgap semiconducting materials is their possible use in devices capable to operate at temperatures much exceeding room temperature (up to 800–1000 °C). Therefore, with wide-bandgap semiconducting materials is associated a new avenue of research in modern semiconductor electronics—devices intended for operation under severe conditions: at high temperatures, strong electric fields, radiation, etc. The necessary qualities of semiconductor devices are stability, reliability, and long operation life in electronic equipment. The corresponding requirements are to be imposed on semiconductor materials, too. For fabricating barrier structures, a semiconductor material must also possess both n- and p-type conductance with controllable value. Among the presently known wide-bandgap semiconducting materials, these requirements are met in full measure by SiC, GaP, and GaN.

In what follows we discuss just these three materials and, partly, GaAs as a material used for comparison. Works on GaP were initiated in the early 1960s by D. Nasledov and B. Tsarenkov at the Ioffe Institute, investigations of SiC, by A. Ioffe at that same time and A. Regel in the middle 1960s at the Institute of Semiconductors; investigations of GaN were commenced in the middle 1970s at Vodakov's laboratory of the Ioffe Institute.

2 Silicon carbide

2.1 Technology of single-crystal growth

Silicon carbide is the only chemical compound in group IV of the periodic table. It crystallizes in a great number of stable crystalline modifications named polytypes. Mainly owing to works of scientists from the Institute of Semiconductors and the Ioffe Institute [1, 2], it has been established that SiC polytypes can be considered a family of wide-bandgap semiconductors with unique chemical, thermal, and radiation stability. The physicochemical properties of silicon carbide and its polytypism have been reviewed in sufficient detail [2–4]. At 2850 °C and sufficiently high pressure silicon carbide completely decomposes, without melting, onto Si and C, and beginning with $\sim$ 1800 °C it sublimates, dissociating into Si, Si_2C, SiC_2, and C. For this reason, bulk crystals of both abrasive and semiconducting SiC are obtained from vapor phase by sublimation or resublimation. Beginning with the late 1950s, crystals of semiconducting SiC has been grown by the Lely method [5]. In the USSR this method was developed into an industrial technique at the State Institute of Rare Metals (Moscow) under supervision of M. Reifman and implemented in practice at Podol'sk Chemical Metallurgical Works [6]. On the basis of these crystals, production of ultrastable LEDs for defense industry was brought to commercial level by the Ministry of Electronic Industry with direct participation of staff members of the Ioffe Institute (Yu. Vodakov and co-workers) and Leningrad Electro-Technical Institute (G. Kholuyanov and E. Violin). First studies of the electronic properties of SiC at the Institute of Semiconductors and the Ioffe Institute were based on these crystals. Unfortunately, the Lely method proved unsuitable for mass production of SiC devices because of its poor controllability.

In 1970, Yu. Vodakov and E. Mokhov proposed and later patented in Germany, France, UK, and US a new technique for growth of single-crystal SiC, named the sublimation-sandwich method (SSM) [7]. SSM also came into use for growing other wide-bandgap semiconductors, e.g., GaN [8].

SSM proved to be extremely convenient for fundamental studies of the mechanism of SiC growth [9–12], polytypism [14–17], specific features of SiC containing impurities [13, 17] and native defects [15, 17–19].

The advantages of SSM in SiC technology became more apparent after Yu. Vodakov and co-workers of the Ioffe Institute proposed to used instead of graphite new construction materials—Ta and TaC [20–22].

The result of SSM studies performed at the Ioffe Institute were used in developing techniques for sublimation growth of large bulk SiC crystals [23]. Presently, SSM is used to grow crystals (up to 35 mm in diameter) of 4H-, 6H-, and 15R-SiC polytypes in quite a number of places, including the Ioffe Institute and its small enterprise [24].

The controlled growth of main SiC polytypes at the Ioffe Institute became possible as a result of investigations into the nature of polytypism, performed by its staff members. It has been shown that the SiC polytypes are metastable phases with different Si:C ratios in which the concentration of carbon vacancies (or that of excess Si) monotonically increases with decreasing percentage of polytype hexagonality [15, 17]. A relative excess of Si favors growth of cubic SiC, and that of carbon (or Si deficiency) results in preferential growth of hexagonal polytypes, and in the first place of 4H-polytype. By varying the deviation from the stoichiometry, one can control the growth of polytypes. For example, the promising 4H-polytype can be grown in vapor of Group-IVb isovalent impurities (Ge, Sn, Pb) [15, 25–27].

The type of the substrate surface in growth also influences the polytype transformation. Using the method of artificial surface singularization, which was proposed at the Ioffe Institute and allowed achieving a strictly singular type of surface [11, 16, 28], it was found that in the absence of impurities on a singular surface having no structural defects the probability of formation of polytype modifications is even very high. This allowed E. Mokhov to explain the appearance of a layer of a new polytype by spontaneous nucleation of a cluster-nucleus on the singular growth surface (having no microsteps) as a result of interaction between adatoms [17].

2.2 Electronic properties of main silicon carbide polytypes

At the Ioffe Institute, much attention has been paid to studies of the electronic properties of SiC. While the optical properties of SiC polytypes have been studied by a group of American scientists headed by W. Choyke [29], major contribution to investigation of fundamental semiconductor parameters of hexagonal SiC polytypes was made by staff members of the Ioffe Institute, and in the first place by G. Lomakina [1]. Activation energies for main dopants in p- and n-SiC, effective masses and their anisotropy, anisotropy of electrical conductance for ten polytypes [30], carrier mobility and its concentration and temperature dependences had been determined [31]; forbidden gap widths, exciton binding energies, and valence band splitting parameters for 4H- and 6H-SiC found from electroabsorption [32, 33]; and the position of the absolute minima of the conduction band in the main polytypes evaluated. These data are given in Table 1 for n- and p-type SiC, figures in brackets are present-day data. It can be seen that the main parameters of promising SiC polytypes, and in the first place those of 4H-SiC, had been determined sufficiently accurately by the middle 1970s.

Table 1. Properties of SiC polytypes

	3C-SiC	15R-SiC	6H-SiC	4H-SiC
Excitonic forbidden gap width at 0 K, E_{gx} (eV)	2.39	2.986	3.023	3.265
Binding energy of free exciton, ΔE_{ex} (eV)	0.04	0.04	0.078	0.025
Forbidden gap width at 77 K, E_g (eV)	2.43	3.02	3.09	3.29
Ionization energy for shallow donor (N):				
for E_{D1} (meV)	53 (54)	47 (34)	100 (85)	35 (45)
for E_{D2} (meV)	—	100 (111)	150 (130)	90 (100)
Anisotropy of electrical conductance, $\rho_{\parallel}/\rho_{\perp}$	1	1.5	3.7	0.9
Electron mobility at 300 K, $\mu_n(\perp C)$ (cm^2 V^{-1} s^{-1})	>1000 (890)	500	360 (500)	800
Energy of Al impurity level, E_{Al} (eV)	0.26–0.27 (0.25)	0.24	0.24–0.25	0.24
Energy of Ga impurity level, E_{Ga} (eV)	0.34	0.28–0.3	0.35	0.27
Energy of B impurity level:				
$E_{B,shallow}$ (eV)	(0.3)	0.39	0.39 (0.3)	0.35
$E_{B,deep}$ (eV)	0.73 (0.73)	0.6–0.7	0.6–0.7	<0.70 (0.65)
Hole mobility at 300 K, μ_p (cm^2 V^{-1} s^{-1})	60	60	60 (100)	80 (115)

Table 2. The avalanche breakdown field and temperature coefficient for SiC polytypes

Polytype	$E_{max}(\| C)$ $(10^{-6}\,V\,cm^{-1})$	K_T $(10^4\,K^{-1})$
4H-SiC	5.9	−4.6
6H-SiC	5.5	−3.25
10H-SiC	4.6	−1.4
15R-SiC	5.4	−1.93
21R-SiC	4.7	−1.2

G. Lomakina *et al.* were the first to study in detail tensoresistive parameters of p-SiC, which allowed developing for the first time high-temperature strain sensors on its base [34, 35].

For identifying more reliably and completely the nature and state of impurity centers in the SiC lattice, resonant methods have been used (ESR, NMR, Electron Nuclear Double Resonance). These works have been performed by P. Baranov and Yu. Vodakov with co-workers of the Ioffe Institute in cooperation with Kazan State University, Kiev Institute of Semiconductors, and Moscow State University. For the first time, SiC samples were preliminarily enriched with isotopes to ensure correct interpretation.

It was found that acceptor impurities in SiC (B, Al, Ga) have several chemically nonequivalent states. They mainly substitute silicon in SiC (A_{Si}). For boron this result was obtained for the first time at the Ioffe Institute [36] and is now commonly accepted (previously it was believed that boron substitutes C). Another state is an associate of impurity with a silicon vacancy ($A_{Si}-V_C$) [37]. More complicated is the behavior of Be: in addition to Be_{Si} and (Be-V_C) there exists an interstitial center Be_I^{+2}.

In SiC, Mo [38] is responsible, similarly to V [39], for deep levels lying near the midgap, being of interest for obtaining semi-insulating SiC.

SiC crystals doped with a rare earth Er were first grown by P. Baranov and co-workers [40]. These crystals can be used to fabricate LEDs with luminescence at 1.53 μm, which is important for fiber optics.

A donor impurity of phosphorus could be introduced directly during growth or by transmutation doping and identified by ESR for the first time by A. Zabrodskii and Yu. Vodakov with co-workers [41].

Studies of the impurity optical absorption, performed under supervision of G. Dubrovskii [42], allowed him to explain the observed specific features by the presence in hexagonal SiC polytypes of a one-dimensional natural superlattice (NSL) along the major symmetry axis C (first noticed by L. Esaki). The alternation of cubic and hexagonal planes along the C axis, typical of SiC polytypes, creates in them an individual one-dimensional superstructure and the corresponding miniband electronic spectrum. Even though this concept has not been analyzed in detail theoretically, experimental data suggest its plausibility. In addition to the behavior of impurity absorption bands and their characteristic polarization, this is indicated by the anisotropy of electrical conductance [30] and anomalous avalanche breakdown [43] first discovered in hexagonal SiC polytypes by staff members of the Ioffe Institute. The breakdown field along the NSL $E_{max}(\| C)$ axis was found to exceed substantially the theoretical value and to be higher than that along the direction perpendicular to the C axis (for instance,

Table 3. The comparative characteristics of wide-bandgap and classic semiconductor parameters

Parameter	6H-SiC	4H-SiC	GaN	GaAs	Si
Relative dielectric constant:					
ε_0	6.6	6.6	5.35		
ε_s	9.7	9.7	9.0	12.8	11.8
Heat conductivity, λ (W cm^{-1} K^{-1})	2–4	2–4	1.3	0.5	1.5
Breakdown field, E_{max} (10^{-5} V cm^{-1})	20–50	30–60	50	3	2
Maximum carrier velocity at 300 K V_{max} (10^{-7} cm s^{-1})	2	2	2.7	2	1
Intrinsic carrier concentration at 300 K, n_i (cm^{-3})	10^{-6}	10^{-7}	10^{-9}	10^{7}	2×10^{10}
Temperature at which $n_i = 10^{13}$ cm^{-3}, T_{max} (°C)	730	840	850	360	200
Electron mobility, $\mu_n(\parallel C)$ (cm^2 V^{-1} s^{-1})	100	800	500	8500	1300
Hole mobility, μ_p (cm^2 V^{-1} s^{-1})	100	115	150		840

Table 4. The comparative quality criteria of different semiconductors

Semiconductor	K_c (W Ω GHz^2)	K_T (W Ω GHz^2)	f_M (GHz)
Si	10^6	1.3×10^5	350
GaAs	2×10^6	10^5	130
α-SiC (GaN)	10^8	10^6	7×10^3

for 6H-polytype $E_{max}(\perp C) = 3.7\times10^{-6}$ V cm^{-1}). In the former case the breakdown field was characterized by an anomalous negative temperature coefficient K_T, while in the latter, by a positive coefficient commonly observed for the avalanche breakdown ($K_T = +1.5\times10^{-4}$ K^{-1} for 6H-polytype). The breakdown fields and their temperature coefficients depended on a specific polytype, which must be the case for a superlattice. This is can be seen from Table 2 for $(N_d - N_a) = 10^{18}$ cm^{-3}.

Further studies carried out at the Ioffe Institute under supervision of V. Sankin [43, 44] allowed these features to be explained by the Bragg reflection of electrons from the SL potential inside the first miniband on applying a strong electric field along the SL axis. In this case electrons moving along the major symmetry axis in hexagonal polytypes cannot acquire in strong electric fields an energy sufficient for avalanche energy multiplication, and the avalanche breakdown in this direction occurs at much higher voltages, whereas in the perpendicular direction there are no such limitations and the avalanche breakdown shows classical behavior. Studies of SiC polytypes in pre-breakdown electric fields $(0.1-4)\times10^6$ V cm^{-1} validated this mechanism and allowed revealing principal phases of the Vannier–Stark localization process developing in NSL and largely determining the behavior of SiC polytypes in strong electric fields parallel to the C axis [44].

All these investigations made it possible to compare parameters of wide-bandgap semiconductors and other promising semiconductor materials (Table 3), to determine whether or not they show promise for use in electronics, and to calculate the design of concrete devices. Table 4 presents the quality criteria with respect to electric K_c ($\sim V_H^2 E_M^2$) and thermal K_T ($\sim V_H \lambda T$) loads and cutoff frequency f_M.

3 Fabrication of SiC device structures

3.1 Epitaxial growth

Abroad, SiC device structures are fabricated by chemical vapor deposition (CVD) [4]. Unfortunately, this method does not allow performing controlled polytype transformations or achieving the maximum possible level of doping. For this reason special original methods have been developed at the Ioffe Institute for fabrication of device structures by growth from liquid phase or by sublimation.

Growth from liquid phase

Growth of epilayers from the liquid phase has long been known. Among the commonly used solvents are molten Si, Cr, Sc, Y, Dy, other rare earths, and Mn, but device structures are obtained from molten silicon. Since the special container graphite is a matter of serious difficulty, containerless growth techniques have been developed at the Ioffe Institute for preparing SiC epilayers.

Layers of p- and n-type SiC were obtained by the method of zone melting under a temperature gradient [46]. Since reproducible growth of the 4H-polytype is due to enrichment of the growing SiC surface with carbon [15, 25], Sc, Tb, and Dy melts enriched with C were used to grow 4H-polytype layers on substrates of other polytypes.

V. Dmitriev from Chelnokov's laboratory of the Ioffe Institute developed a method of containerless liquid phase epitaxy in which a Si melt maintained at 1450–1700 °C is suspended in an RF field [47]. An important advantage of this method is that it can produce high-quality multilayer structures with layers of n- and p-type. The dopants (N and Al) are introduced directly in the course of epitaxial growth.

Sublimation growth of doped SiC layers

To fabricate device structures by sublimation, the SSM was used at the Ioffe Institute [7]. This method allowed obtaining doped p- and n-type SiC layers of thickness in the range from several micrometers to several millimeters. To ensure controlled doping, detailed studies of SiC doping have been performed [13]. Data on the maximum concentrations have been obtained for more than 20 impurities in SiC; it was established that the concentration of an impurity introduced strongly depends on the crystallographic orientation of the substrate, temperature and rate of growth, vapor phase composition, and the extent of deviation from the stoichiometry. These effects were explained by the lack of equilibrium at the phase boundary, which is due to the low rate of relaxation processes occurring in the solid phase [17].

Only Al, Be, B, N, and Ge possess high solubility in SiC ($\geq 10^{20}\,\mathrm{cm}^{-3}$). The maximum level of SiC doping with acceptor impurities is achieved at high growth temperatures ($T_g > 2400\,°\mathrm{C}$) on (0001) Si substrates [13]. The maximum level of doping with Group-V impurities was, on the contrary, obtained at the lowest temperatures ($T_g \approx 1900\,°\mathrm{C}$) on (0001) C substrates.

The possibilities of the SSM allowed calculating the elementary capture coefficients K_i for the most important electrically active impurities—N, B, Al, Ga, and P, depending on the growth temperature, substrate orientation, and impurity concentration. With increasing temperature, K_i decreases for N and P and increases for B, Al, and Ga [13]. The strong dependence of the doping level on growth conditions can be explained by specific features of processes occurring in the chemisorbate layer. The adsorption of most impurities (except for those belonging to Group V) is stronger on the Si face owing to the high surface energy of this face. At the same time Group-V impurities are better adsorbed on the (0001) C face. The primary reason for the decrease of K_i in SiC with growing concentration of a donor impurity is its desorption in the form of N_2 or P_2 molecules [13]. With increasing partial pressure of N or P, their capture is reduced by several orders of magnitude. As a result, the concentration of Group-V impurities (especially P, As, Sb) in SiC turns out to be much lower than the expected solubility limit.

The knowledge of these laws allowed improving the technology of fabrication of bulk doped SiC crystals and device structures. It is with the use of the SSM that structurally perfect p-n junctions with uniform breakdown were first fabricated from SiC at Vodakov's laboratory of the Ioffe [48]. These junctions made it possible to commence pioneering studies of the behavior of charge carriers in SiC in strong electric fields.

4 Introduction of point defects into SiC

4.1 Impurity diffusion in SiC

The diffusion of the most important impurities in SiC (Li, Be, B, Al, Ga, P, O) has been studied in most detail by researches of the Ioffe Institute [17, 18, 49].

The mechanism of diffusion of most impurities in SiC is rather complicated, which is due to the existence of several impurity states with different diffusion mobilities. The diffusion of Li and B in p-SiC occurs via interstitials. Other impurities diffuse mainly via vacancies, but, as a rule, in complexes (except N and P).

According to the model proposed in [17, 50], a mobile state for acceptor impurities is a metastable associate $(A_C - V_C)$, where A_C is an impurity atom substituting C, and V_C is C vacancy. As traps for a fast diffusing state act structural defects, impurities, and even native B atoms (at $C_B > 10^{19}\ cm^{-3}$) [50] in the SiC lattice.

The presence of states with different mobilities accounts for the effect of a strong dependence of diffusion coefficients on the initial state of an impurity. The diffusivity of acceptor impurities strongly depends on the conduction type and doping level, owing to changes in the share of the diffusion-active state. For example, in a heavily doped n-SiC(N) the rate of acceptor diffusion falls, and in p-SiC(Al), on the contrary, rises dramatically [49].

For B, Al, and P impurities data were obtained on unstimulated truly solid-state diffusion initiated solely by thermal activation of substitution impurities occupying regular lattice sites [17]. In this case, the rate of migration is very low ($D < 10^{14}\ cm^2\,s^{-1}$ at $T = 2400\,°C$), being by 2–4 orders of magnitude lower than in the case of ordinary diffusion. These data allowed estimating the energy of V_{Si} migration in SiC at ~ 2 eV and that of its formation, at ~ 7 eV.

Thus, the works mainly performed at the Ioffe Institute have demonstrated, on the one hand, that, except for B and Be, the possibility of using diffusion of other impurities in SiC for fabricating device structures is very limited. On the other hand, the knowledge of the fundamental aspects of diffusion gave deeper insight into the behavior of point defects, including native ones, in wide-bandgap semiconductors with higher bonding energy (SiC, GaN).

4.2 Introduction of native defects

The width of the homogeneity region in SiC is still disputable. It is believed that SiC samples contain some excess of Si, with Si content varying between 10^{-5} and several percent, as reported in different works. Investigations performed by the Ioffe Institute staff members have shown that SiC crystals grown with excess of Si do contain such nonstoichiometric defects V_C which affect the electrical and optical properties of the material, but their concentration does not exceed $(3-5)\times10^{17}\,cm^{-3}$. At the same time, their major part in such crystals is in an electrically inactive state in the form of clusters [18].

Scientists from the Ioffe Institute (Vodakov's laboratory) have proposed a number of techniques for controlled deviation from stoichiometry, toward excess of either Si or C [51]. It is significant that the properties of crystals grown at temperatures below 2300 °C in an excess of Si markedly differ from those of standard samples grown by the Lely method. Analysis shows that SiC with an excess of Si contains nonequilibrium native defects in the form of associates and clusters. The presence of the latter accounts for the anomalously high thermal stability of nonequilibrium centers and for the enhanced microbrittleness of crystals [18, 51].

Thus, the investigations performed by the Ioffe Institute researches have established that the behavior of the ensemble of point defects in SiC is mainly governed by the presence of considerable concentrations of excess native defects and clusters of these. Controlling their state, one can obtain a stable SiC with novel properties. This fact has been used in developing new types of highly stable short-wave SiC LEDs and stabilizer diodes.

5 Ion implantation

To fabricate p-n structures by ion implantation, an understanding is necessary of the behavior of nonequilibrium native defects introduced in implantation. Staff members of a number of laboratories at the Ioffe Institute, Kurchatov Institute of Atomic Energy, and Moscow Engineer-Physical Institute performed a comprehensive study of SiC crystals, irradiated with high-energy particles (ions, fast electrons, reactor neutrons), in relaxation annealing. It was confirmed that the interaction of point defects in thermal treatment results in the formation of exceedingly stable clusters and associates of point defects existing up to temperatures of 2200–2500 °C. Even though large clusters do not presumably introduce levels into the forbidden gap, they promote, by generating simple native point defects in high-temperature annealing, stabilization or formation of electrically and optically active complexes composed of only native defects (D_1 center) or even more stable complexes with an impurity atom (D center). For example, acceptor impurities of Group IIIa form stable associates, which are deep acceptors, with carbon vacancies [52].

In view of the aforesaid, "hot" implantation must inhibit drastically the formation of clusters composed of nonequilibrium point defects. It is in this way that V. Gusev and K. Demakov [53] could obtain high-quality p-n junctions, suitable for fabricating various device structures, by ion implantation.

6 Silicon carbide devices

6.1 Optoelectronic devices—LEDs and UV photodetectors

The first photosensitive and light-emitting injection and prebreakdown SiC structures were discovered by O. Losev in crystals with spontaneously formed p-n junction [54]. However, means of controlled fabrication of effective electroluminescent SiC diodes were only found in the middle 1960s. For the first time, red, yellow, and green SiC LED of this kind were fabricated by Al and B diffusion at the Institute of Semiconductors by Yu. Blank and co-workers [55]. Further investigation and development resulted in the fabrication of stable single- and multiple-element yellow, green, and blue LEDs and also red and blue-violet single-element LEDs by the epitaxial-diffusion technology with the use of SSM [56–58]. To control the color of electroluminescence, various impurities and various SiC polytypes have been used [59]. The results of these studies performed at the Ioffe Institute, reviewed in [56, 59], were passed on to research institute *Sapfir*, where highly stable SiC LEDs were developed, and to the Design Office at the Ioffe Institute for production of multiple-element LED arrays [58]. Both these designs are presently used in important weaponry. SiC LEDs emit a stable light flux and have practically infinite operation life (they have been functioning in some armament systems for more than 20 years without a single failure). Also at the Ioffe Institute (Chelnokov's laboratory) were fabricated by liquid phase epitaxy SiC LEDs for blue, violet, and near ultraviolet spectral ranges [60, 61].

Studies of microplasma breakdown have resulted in the development of p-n structures with a uniform field of microplasmas ($\sim 10^7\,\mathrm{cm}^{-2}$) on a large surface area at Vodakov's laboratory of the Ioffe and extra stable (10^{-2}% per degree) prebreakdown single-and multiple-element SiC LEDs used as star simulators and reference emitters in optoelectronic equipment.

SiC is one of the most promising materials for UV photodetectors. UV photodetectors absolutely blind to visible light were developed at the Ioffe Institute on the basis of Schottky barriers and p-n structures (SSM) [62]. The photodetectors are sensitive in the spectral region 400–150 nm, with the photosensitivity at maximum (200–340 nm) equal to $0.15\,\mathrm{A\,W}^{-1}$, quantum efficiency as high as 0.9, and detection limit of $10^{-14}\,\mathrm{W\,Hz}^{-1/2}$, and admissible incident light power of up to $10^4\,\mathrm{W\,cm}^{-2}$. Also, avalanche photodiodes were developed with a photocurrent multiplication coefficient at 67-V bias equal to 25 and a threshold sensitivity of $5 \times 10^{-13}\,\mathrm{W\,Hz}^{-1/2}$.

6.2 Voltage stabilizers

Voltage stabilizers for voltages in the range 13–20 V were also developed on the basis of p-n structures with uniform microplasma breakdown [63].

6.3 Strain sensors and heat-sensing devices

These devices were developed at the Ioffe Institute (Vodakov's laboratory) [64] on the basis of an epitaxial p-type layer heavily doped with B and Al and an electrically insulating high-resistance p-n junction serving simultaneously for temperature measurement.

The total sensitivity of the sensor was 40 at 300 K and 20 at 1100 K. The operating life in air at 600 °C was longer than 2000 h. The upper limit of working temperatures (900 °C) was determined by the stability of contacts in air. The temperature coefficient of resistance of the heat sensor varied from 3.7 to 1% in the same interval of temperatures.

6.4 Devices of power and microwave electronics

The possibility of developing silicon carbide devices of this kind is evident from Tables 3 and 4. From the very beginning of work on SiC, this direction dominated in investigations of US scientists. In our country, first high-quality rectifying SiC p-n structures were fabricated in 1968 by Yu. Vodakov [56]. Further on, this direction of research was pursued at the Ioffe Institute by V. Chelnokov's co-workers. A rather comprehensive review of these works was given in 1995 by P. Ivanov and V. Chelnokov [65].

6.5 Rectifying diodes

It was shown for the first time in [66, 67] that, owing to the anomalously high electric strength of SiC, requirements to base modulation are much less stringent for SiC diodes than those for Si or GaAs. Diodes operating at reverse voltages of 300 V were studied at Chelnokov's laboratory at temperatures up to 500 °C and their working capacity and short switching time were demonstrated. By now, high-temperature (400–500 °C) SiC-based diodes have been developed with reverse voltages of about 1.5 kV and forward currents of several amperes. Prototype diodes with reverse voltages exceeding 4.5 kV and SiC-based Schottky diodes with reverse voltages of up to 2 kV and low forward voltage drops have been reported.

6.6 Transistors

The most advanced works on SiC are related to transistors of various types. By now, power levels of the order of 2 W at 1 GHz, power amplification of about 10 dB at 5 GHz and 2 dB at 10 GHz, and an efficiency of about 16% have been achieved with a 4H-SiC-based FET having a Schottky gate. The output power, reduced to 1 mm of gate (0.9 W), already exceeded that for GaAs transistors by a factor of 1.5. Scientists of the Ioffe Institute made a substantial theoretical contribution to this field of research [68–71]. For MOS transistors, the problems have been solved concerning the quality of the SiO_2-n-SiC interface [72, 73], which allowed fabricating high-quality MOS SiC transistors. By now, first memory cells showing promise for use in programmable read-only memory devices have been fabricated on the basis of transistor n-p-n structures.

Thus, quite a number of pioneering investigations, which demonstrated theoretically and experimentally the promise of SiC for the modern electronics, have been performed at the Ioffe Institute. Unfortunately, the technological equipment available now in Russia has as yet given no way of taking advantage in optimal manner of such a unique semiconductor as SiC.

7 High-temperature electronics based on wide-bandgap semiconductors—GaAs and GaP

The interest in $A_{III}B_V$ for use in electric conversion electronics had remained intense since the 1950s, and in 1967 Zh. Alferov proposed to use $A_{III}B_V$ (and in the first place

GaAs) in power semiconductor devices [74]; even earlier, in 1965, work in this direction was commenced using two technological methods: vapor-phase and liquid-phase epitaxy. The fundamental properties of GaAs and GaP indicated that these materials can be, in principle, used to fabricate fast power devices operating at higher temperatures than similar silicon-based devices do. However, by the beginning of works on $A_{III}B_{V}$-based power semiconductor devices, it was unclear whether or not this possibility can be realized (as indicated by the results of first works of foreign scientists in this direction [75, 76]). An opinion formed that GaAs cannot be used in power semiconductor devices because of the impossibility to obtain pure homogeneous crystals with long diffusion lengths of minority carriers. Thus, the problem of whether or not GaAs will be a material for high-temperature power devices reduced to the problem of obtaining a sufficiently pure material with long diffusion length of minority carriers to ensure the necessary high operation voltage, on the one hand, and sufficiently small forward voltage drop, on the other.

To solve the formulated problem the vapor phase epitaxy in an open chloride system was chosen as a basis. This process combines the possibility of obtaining in a single technological cycle multilayer homo- and heterostructures of large area ($\geq 120\,\text{cm}^2$), including n- and p-type layers of thickness in the range from 50 Å to 200 μm, simplicity, technological effectiveness, low toxicity of the starting components, and possibility of their exhaustive purification.

In the first place, processes of vapor-phase epitaxy have been studied, a method for analyzing in situ the vapor phase composition and determining the kinetic characteristics of reactions in chemical-gas-transport system by absorption spectroscopy developed, a model of vapor-phase epitaxy in this system created, conditions for epilayer growth optimization found, and, as a final result:

- a technique was created for growing pure and donor- or acceptor-doped ($10^{14}-10^{18}\,\text{cm}^{-3}$) GaAs epilayers of thickness in the range from fractions to hundreds of micrometer;
- the problem was solved of obtaining ultrapure gallium arsenide with a total concentration of shallow donor and acceptor centers of about $10^{12}\,\text{cm}^{-3}$, total concentration of deep recombination centers not higher than $10^{12}\,\text{cm}^{-3}$, and $\mu_{1.6\,\text{K}} = 2.5\times10^{6}\,\text{cm}^2\,\text{V}^{-1}\text{s}^{-1}$.

The problem of creating a high-quality p-n junction was solved by shifting it away from the metallurgical interface substrate-epilayer into the region of a high-quality epilayer—by bringing the p-n junction into coincidence with the metallurgical boundary of two epilayers with different types of doping.

This problem was solved using a multichamber reactor proposed and designed in 1966. This reactor allowed fabricating in a single process, without extracting substrates from the reactor, a wide variety of multilayer structures [75, 76] (first foreign publications on multichamber reactors appeared in 1977). The installation comprises several channels, each being an independent reactor; during growth of a multilayer structure the growth chamber with a substrate is connected, in less than 1–2 s, from one source to another to change the type of doping and/or solid solution composition without interrupting the growth, with high-temperature channel switch used for this purpose. Such a design ensures a fast change of the vapor phase composition and much reduces the so-called dead volume (volume of gaseous component mixture flowing over a substrate from the instant of switching to attainment of new stationary state in the vapor phase). To prevent deposition on reactor walls, the zone of mixing between the source and the substrate is maintained at a higher temperature than that of the source. Also, a design was used in which a substrate mounted on a special holder was transferred from one channel into another. It is in a reactor of this kind that GaP_xAs_{1-x}-GaAs-GaP_xAs_{1-x} heterostructures with emission line edge narrowing on increasing current, and also superlattices, power GaAs and GaP diodes, power diodes with Schottky barrier, bipolar transistors, and power and ultrafast thyristors have been fabricated.

To obtain n-type epilayers, S, Sn, and Te were used as dopants; Zn, Cd, and Mg gave a p-type conduction. For all types of devices, high turn-off voltage was provided for by an undoped layer.

The development of the technology of multilayer epitaxial structures in the system GaAs-GaP with $\geq 20\,\mathrm{cm}^2$ area, thickness in the range 0.1–200 μm, and $< 5\%$ nonuniformity of electrophysical properties over the thickness of an epilayer and the whole structure made possible fabrication of basic elements of high-temperature electronics from GaAs and GaP for temperatures of up to 350 and 500 °C, respectively.

The following devices, described in [76], have been developed:

Based on gallium arsenide

(i) Power p-n junctions with the following parameters: structure area $\sim 2.5\,\mathrm{cm}^2$, $U_{\mathrm{rcv}} =$ 2000 V, $I_{\mathrm{d}} = 200$ A, $U_{\mathrm{d}} = 1.4-1.86$ V at $I = 200$ A, $T_{\mathrm{p-n}} \approx 300\,^\circ\mathrm{C}$, $T_{\mathrm{case}} =$ $260-280\,^\circ\mathrm{C}$, reverse recovery time $t_{\mathrm{rr}} = 0.2\,\mu\mathrm{s}$;

(ii) Power Schottky diodes with the following parameters: $I_{\mathrm{d}} = 50$ A, $U_{\mathrm{rcv}} = 150$ V, $T_{\mathrm{w}} = 200\,^\circ\mathrm{C}$, τ_{rec} tens of nanoseconds;

(iii) Power bipolar GaAs transistors with the following parameters: $I_{\mathrm{c-c}} = 10$ A, $U_{\mathrm{c-c}}$ up to 600 V, β up to 48°, $T_{\mathrm{w}} = 300\,^\circ\mathrm{C}$;

(iv) Power GaAs thyristors with the following parameters: anode current $I_{\mathrm{A}} = 100$ A, static turn-on voltage $U_{\mathrm{t}} = 1000$ V, $U_{\mathrm{rcv}} = 1000$ V, $T_{\mathrm{w}} = 270\,^\circ\mathrm{C}$, τ_{rec} hundreds of nanoseconds, $\tau_{\mathrm{turn-on}}$ tens of nanoseconds;

(v) Ultrafast GaAs thyristors with a current rise time constant 100 ps, pulse current up to 150 A, switching voltage up to 1000 V, $U_{\mathrm{rcv}} \leq 1000$ V.

Based on gallium phosphide

(i) diodes with $I = 10$ A, $U_{\mathrm{rcv}} = 400$ V, $T = 500\,^\circ\mathrm{C}$, f_{cutoff} up to 50 MHz, and leakage currents remaining low under reverse bias at 400 °C for hundreds of hours;

(ii) discrete GaP FETs with a p-n gate and a Schottky barrier, differential slope 2–3 mA V^{-1}, $T = 300-450\,^\circ\mathrm{C}$;

(iii) GaP FET matrices with the same parameters as discrete devices;

(iv) dynistors with switching voltages up to 100 V and $T = 450\,^\circ\mathrm{C}$;

(v) stabilizer diodes with a stabilizing voltage in the range 10–100 V, $I = 200$ mA, $T = 450\,^\circ\mathrm{C}$.

8 Gallium nitride

Gallium nitride has recently not only found application as the best optoelectronic wide-bandgap semiconducting material for the short-wave visible UV spectral regions, but, as it follows from Tables 3 and 4, also proved to be actually close in its parameters to SiC, which makes possible its use in power and microwave electronics. However, the lack of technology for bulk GaP preparation and appropriate substrates for epitaxial growth hinders the development of the device direction. The effort has been concentrated on obtaining bulk GaP layers suitable as substrates for device structures.

For this purpose, two varieties of CVD growth have been developed at the Ioffe Institute: growth in the chloride system and sublimation deposition.

8.1 Growth in the chloride system

GaN samples were grown on Al_2O_3, $LiGaO_2$, $LiAlO_2$, Si, and SiO_2 substrates in the system Ga-HCl-NH_3-H_2. A substrate preparation technique was developed which allows separating GaN layers from substrates to give the so-called free-standing GaN—bulk samples of gallium nitride. The quality of GaN layers was monitored by X-ray diffractometry and from luminescence spectra. The halfwidth (FWHM) of rocking curves was 6–10 angular minutes. The photoluminescence spectrum at 77 K had a maximum at 3.462 eV, near the fundamental absorption edge. Emission at < 2.3 eV is either very weak, or absent at all.

The main results obtained in growth of bulk GaN layers by this method:

(i) A method has been developed for obtaining bulk GaN layers of area up to several cm^2 and up to 1 mm thick without mechanical substrate removal.

(ii) At high growth rates (50–100 $\mu m\,h^{-1}$) a crystal refaceting is observed: on $(10\bar{1}2)$ sapphire substrates grow (0001) GaN layers which can also be separated without substrate grinding-off.

(iii) A technique has been developed of obtaining easily detachable bulk GaN layers on amorphous substrates (SiO_2).

8.2 Sublimation deposition

Growth was performed in a reactor with RF heating to 1100–1250 °C in ammonia flow on SiC or sapphire substrates of area up to 4 cm^2. Polycrystalline GaN or metallic Ga was used as vapor source in sublimation.

The most important results of the investigations performed are the following:

(i) The possibility has been demonstrated of growing GaN layers by SSM with high deposition rates (up to 1 $mm\,h^{-1}$) [8, 51, 77, 78];

(ii) Bulk single-crystal 0.5–0.8 mm thick GaN layers have been grown without preliminary deposition of Al_2O_3 buffer layer onto the (0001)Si substrate;

(iii) GaN layers are characterized by high structural perfection, with FWHM of rocking curve of 5–10 angular minutes. The photoluminescence spectra of the best samples show an clearly pronounced isolated exciton line with a halfwidth up to 4 meV [77]. There is practically no visible luminescence. The sufficiently high quality of the layers grown is also indicated by the fine structure of ESR spectra, observed in samples containing Fe, Ni, Mn, and V impurities [77, 78], which allowed for the first time identification of these impurities.

Thus, the above-described techniques for growth of bulk GaN layers of acceptable crystal perfection can form a basis for development of a method of preparation of substrates for fabrication of device structures based on Group-III nitrides AlN-GaN-InN.

9 Conclusion

Wide-bandgap semiconducting materials, such as nitrides of Group-IIIa elements and silicon carbide, possess a number of unique properties making them highly promising for development of electronic devices. The wide bandgap of these materials allows their use in optoelectronics for full-color imaging. Devices based on wide-bandgap materials can operate at high temperatures (up to 1000°C), in strong electric fields, and under hard radiation. The breakdown fields for SiC and GaN are more than order of magnitude higher than those for Si or GaAs (Table 2). Among SiC polytypes, 4H-SiC is of primary interest for device engineering. All SiC polytypes show significant anisotropy of electrical conduction and anomalous avalanche breakdown due to the presence in them of a one-dimensional natural superlattice.

Original methods have been developed at the Ioffe Institute for growing single-crystals of SiC and GaN and epitaxial layers based on them. Among these, the most promising is the sublimation sandwich method which ensures controlled growth and doping of these materials with various dopants and produces the most important SiC polytypes at growth rates as high as 1–2 mm h^{-1}. The growth and doping in these wide-bandgap materials are essentially nonequilibrium, so that their properties strongly depend on growth conditions (temperature, stoichiometric composition of vapor phase, etc.). The properties of these materials are also affected by native defects introduced during growth and doping. This is favored by the extremely high thermal stability of nonequilibrium native defects which can persist in the form of associates or clusters to temperatures above 2000°C.

SiC has been used to fabricate single- and multicolor LEDs with emission in the yellow, green, and blue-violet spectral regions, UV photodetectors, voltage stabilizers, strain sensors and heat-sensing devices, and also diodes and transistors for power and microwave electronics. Multiple-layer epitaxial structures in the system GaAs-GaP and thick GaN layers have been grown by vapor phase epitaxy.

References

[1] G. A. Lomakina, *Silicon Carbide 1973*, South Carolina Univ. Press, p. 520, 1974.

[2] Yu. M. Tairov and Yu. A. Vodakov, *Topics in Appl. Phys.* vol. 17, Springer Verlag, p. 31, 1977.

[3] G. Pensl and W. J. Choyke, *Physica B* **185** 199 (1993).

[4] H. Morkoc, S. Strite, G. B. Gao, M. E. Lin, B. Sverdlov, and M. Burns, *J. Appl. Phys.* **76** 1363 (1994).

[5] J. A. Lely, *Ber. Deut. Ceram. Ges.* **32** 229 (1955).

[6] A. A. Glagovskii, *et al. Problems of Physics and Technology of Wide Bandgap Semiconductors*, Leningrad, p. 226, (in Russian) 1980.

[7] Yu. A. Vodakov and E. N. Mokhov, *USSR Patent* No. 403275, 1970; *UK Patent* No. 1458445, 1977; *Germany Patent* No. 2409005, 1977; *USA Patent* No. 414572, 1979; *France Patent* No. 2 264 589, 1975.

[8] Yu. A. Vodakov, E. N. Mokhov, and A. D. Roenkov, *USSR Patent* No. 1136501, 1983.

[9] E. N. Mokhov, M. G. Ramm, A. D. Roenkov, and Yu. A. Vodakov, *Mat. Sci. & Eng.* **B46** 317 (1997).

[10] Yu. A. Vodakov, E. N. Mokhov, M. G. Ramm, and A. D. Roenkov, *Krist. und Techn.* **14** 729 (1979).

[11] Yu. A. Vodakov, E. N. Mokhov, M. G. Ramm, and A. D. Roenkov, *Springer Proceed. in Phys.* vol. 56, p. 323, 1992.

[12] S. Yu. Karpov, Yu. N. Makarov, E. N. Mokhov, *et al., J. Cryst. Growth* **173** 408 (1997).

[13] Yu. A. Vodakov E. N. Mokhov, M. G. Ramm, and A. D. Roenkov, *Springer Proceed. in Phys.*, vol. 56, p. 329, 1992.

[14] Yu. A. Vodakov, E. N. Mokhov, A. D. Roenkov, and D. T. Saidbekov, *Phys. Status Solidi* **51a** 209 (1979).

[15] Yu. A. Vodakov, G. A. Lomakina, and E. N. Mokhov, *Sov. Phys. Solid State* **24** 780 (1982).
[16] G. V. Saparin, S. K. Obyden, P. V. Ivannikov, E. N. Mokhov, and A. D. Roenkov, *Scanning* **19** 269 (1997).
[17] E. N. Mokhov, *Thesis*, Ioffe Institute, 1998.
[18] Yu. A. Vodakov and E. N. Mokhov, *Inst. Phys. Conf. Ser.* No. 137, ch. 3, p. 197, 1994.
[19] A. I. Girka and E. N. Mokhov, *Phys. Solid State* **37** 1855 (1995).
[20] Yu. A. Vodakov,E. N. Mokhov, M. G. Ramm, and A. D. Roenkov, *USSR Patent* No. 882247, 1981.
[21] Yu. A. Vodakov, A. D. Roenkov, M. G. Ramm, E. N. Mokhov, and Yu. N. Makarov, *Phys. Status Solidi* **202** 177 (1997).
[22] E. N. Mokhov, M. G Ramm, A. D. Roenkov, and Yu. A. Vodakov, *J. Crystal Growth* **181** 254 (1997).
[23] Yu. M. Tairov and V. F. Tsvetkov, *J. Cryst. Growth* **52** 146 (1981).
[24] A. Yu. Maksimov, A. A. Maltsev, N. K. Yushin, and I. S. Barash, *Tech. Phys. Lett.* **21** 20 (1995).
[25] Yu. A. Vodakov, E. N. Mokhov, A. D. Roenkov, and M. M. Anikin, *Sov. Tech. Phys. Lett.* **5** 147 (1979).
[26] Yu. A. Vodakov and E. N. Mokhov, *USSR Patent* No. 913762, 1981.
[27] E. N. Mokhov, A. D. Roenkov, Yu. A. Vodakov, G. V. Saparin, and S. K. Obyden, *Inst. Phys. Conf. Ser.*, vol. 142, ch. 1, p. 245, 1996.
[28] Yu. A. Vodakov, E. N. Mokhov, M. G. Ramm, and A. D. Roenkov, *USSR Patent* No. 1398484, 1986.
[29] W. J. Choyke, *Silicon Carbide-1968*, Ed. H. K. Henish and R. Roy, Pergamon Press, p. 141, 1969.
[30] G. A. Lomakina and Yu. A. Vodakov, *Sov. Phys. Solid State* **15** 83 (1973).
[31] G. A. Lomakina, G. F. Holuyanov, R. G. Verenchikova, *et al.*, *Sov. Phys. Semicond.* **6** 1133 (1972).
[32] V. I. Sankin, *Sov. Phys. Solid State* **17** 1191 (1975).
[33] G. B. Dubrovskii and V. I. Sankin, *Sov. Phys. Solid State* **17** 1847 (1975).
[34] G. A. Lomakina, *Sov. Phys. Solid State* **17** 1808 (1975).
[35] G. A. Lomakina, Yu. A. Vodakov, and E. N. Mokhov, *Sov. Phys. Solid State* **19** 1643 (1977).
[36] A. G. Zubatov, I. M. Zaritskii, S. N. Lukin, E. N. Mokhov, and V. G. Stepanov, *Sov. Phys. Solid State* **27** 197 (1985).
[37] P. G. Baranov, I. V. Ilyin, and E. N. Mokhov, *Sol. State Commun.* **100** 371 (1996).
[38] K. F. Dombrowski, M. Kunzer, U. Kaufmann, J. Schneider, P. G. Baranov, and E. N. Mokhov, *Phys. Rev. B* **54** 7323 (1996).
[39] St. G. Muller, D. Hofmann, A. Winnacher, E. N. Mokhov, and Yu. A. Vodakov, *Inst. Phys. Conf. Ser.* vol. 142, ch. 2, p. 361, 1996.
[40] P. G. Baranov, I. V. Ilyn, and E. N. Mokhov, *Solid State Commun.* **103** 291 (1997).
[41] A. I. Veinger, A. G. Zabrodskii, G. A. Lomakina, and E. N. Mokhov, *Sov. Phys. Solid State* **28** 917 (1986).
[42] G. B. Dubrovskii, *Thesis*, Ioffe Institute, 1987.
[43] Yu. A. Vodakov, D. T. Litvin, V. I. Sankin, E. N. Mokhov, and A. D. Roenkov, *Sov. Phys. Solid State* **19** 502 (1985).
[44] V. I. Sankin and I. A. Stolichnov, *Phys. Solid State* **31** 489 (1997).
[45] H. Matsunami, *Physica B* **185** 65 (1993).
[46] D. T. Saidbekov, *Properties of Doped Semiconductors*, Moscow, p. 81–84 (in Russian) 1977.
[47] V. A. Dmitriev and A. E. Cherenkov, *J. Cryst. Growth* **128** 375 (1993);
E. N. Mokhov, B. S. Machmudov, M. M. Usmanova, and G. F. Uldashev, *Sov. Tech. Phys. Lett.* **8** 77 (1982).
[48] A. O. Konstantinov, D. H. Litvin, and V. I. Sankin, *Sov. Tech. Phys. Lett.* **7** 1335 (1981).
[49] Yu. A. Vodakov and E. N. Mokhov, *Silicon Carbide-1973*, South Carolina Univ. Press, p. 508, 1974.

[50] E. E. Goncharov, E. N. Mokhov, and G. G. Ryabova, *Properties of Doped Semiconductor Materials*, Moscow, p. 116, (in Russian) 1990.
[51] E. N. Mokhov and Yu. A. Vodakov, *Inst. Phys. Conf. Ser.* No. 155, Ch. 3, p. 177, 1997.
[52] P. G. Baranov, I. V. Ilyin and E. N. Mokhov, *Solid State Commun.* **100** 371 (1996).
[53] V. M. Gusev *et al.*, *Sov. Phys. Semicond.* **9** 820 (1975).
[54] A. G. Ostroumov and A. A. Rogachev, *Physics, Problems, History and Men*, Leningrad, p. 183–217 (in Russian) 1986.
[55] Yu. S. Blank, Yu. A. Vodakov, and A. A. Mostovskii, *Sov. Phys. Solid State* **5** 2228 (1963).
[56] Yu. A. Vodakov, *Thesis*, Ioffe Institute, 1968.
[57] G. F. Kholuyanov and Yu. A. Vodakov, *Silicon Carbide 1973*, South Carolina Univ. Press, p. 574, 1974.
[58] A. S. Barash, *et al.*, *Technique Connection Facilities*, vol. 5, p. 41, (in Russian) 1982.
[59] Yu. A. Vodakov, *et al.*, *Springer Proceed. in Phys.*, 1992, vol 56, p. 311, 1992.
[60] B. I. Vishnevskaya, *et al.*, *Zh. Tekh. Fiz.* **16** 56 (1990).
[61] V. A. Dmitriev, *et al.*, *Sov. Phys. Semicond.* **23** 23 (1989).
[62] R. G. Verenchikova, *et al.*, *Sov. Phys. Semicond.* **26** 565 (1992).
[63] S. A. Poltinnikov, Yu. A. Vodakov, and G. A. Lomakina, *Pis'ma Zh. Tekh. Fiz.* **12** 261 (1986).
[64] Yu. A. Vodakov, *et al.*, *Pribory i Sistemy Upravleniya* **1** 23 (1981).
[65] P. A. Ivanov and V. E. Chelnokov, *Semiconductors* **29** 1003 (1995).
[66] A. E. Otblesk and V. E. Chelnokov, *Probl. Phys. Techn. Wide-Bandgap Semicond.* Gatchina, p. 197, (in Russian) 1980.
[67] V. E. Chelnokov, *et al.*, *Proc. 6th ISPSD* (Davos, Switzerland) p. 253, 1994.
[68] P. A. Ivanov and B. V. Tsarenkov, *Sov. Phys. Semicond.* **25** 1152 (1991).
[69] M. M. Anikin, *et al.*, *Sov. Tech. Phys. Lett.* **15** 636 (1989).
[70] P. A. Ivanov, *Semiconductors* **28** 662 (1994).
[71] V. A. Dmitriev, *et al.*, *Sov. Tech. Phys. Lett.* **17** 1115 (1991).
[72] P. A. Ivanov, *et al.*, *Semiconductors* **27** 1146 (1993).
[73] P. A. Ivanov, *et al.*, *Semiconductors* **28** 668 (1994).
[74] Zh. I. Alferov, *Sov. Phys. Semicond.* **1** 358 (1967).
[75] Yu. V. Zhilyaev, *Thesis*, Ioffe Institute, 1971.
[76] Yu. V. Zhilyaev, *Thesis*, Ioffe Institute, 1991.
[77] C. Wetzel, D. Volm, B. K. Meyer, K. Pressel, S. Nilson, E. N. Mokhov, and P. G. Baranov, *Appl. Phys. Lett.* **65** 1033 (1994).
[78] Yu. A. Vodakov, E. N. Mokhov, A. D. Roenkov, M. E. Boiko, and P. G. Baranov, *J. Cryst. Growth* **183** 10 (1998).

Physical kinetics of deformation and fracture of solids

V. S. Kuksenko, A. M. Leksovskij, S. P. Nikanorov, A. B. Sinani, A. I. Slutsker, and B. I. Smirnov

Introduction

Physics of strength, including problems of plasticity, has been a traditional subject of research at the Physico-Technical Institute since its foundation in 1918. Many well-known scientists have made significant contributions into this area of physics. Among them are A. F. Ioffe, A. P. Alexandrov, S. N. Zhurkov, N. N. Davidenkov, Ya. I. Frenkel, P. P. Kobeko, A. V. Stepanov, F. F. Vitman, V. A. Stepanov, N. A. Zlatin, A. N. Orlov, and others.

The classical works of A. F. Ioffe on rock salt (1920s) and of A. P. Alexandrov and S. N. Zhurkov on quartz and glasses (1930s) revealed an extremely important role of defects (structural heterogeneity) in the strength of materials and demonstrated that it is possible to reach the theoretical strength.

However, the structural heterogeneity which gives rise to high local overstresses proved to be not a single important factor in the mechanical fracture of solids.

It was shown that dynamic heterogeneity—local fluctuations of the vibrational energy of atoms—causes atomic-molecular regroupings up to scissions of interatomic bonds. The ideas developed by Ya. I. Frenkel "triggered" the investigations of the kinetics of fracture started by S. N. Zhurkov in the early 1950s.

Thus research into the physics of fracture of solids at the Institute relies on the concept of the structural-dynamic heterogeneity of solids as a crucial combined factor in the phenomenon of mechanical fracture.

The research efforts of six laboratories of the Institute, which form the group of laboratories for strength of solids, are aimed at studying most important aspects of physics of fracture and plastic deformation in the framework of the general concept of the structural-dynamic heterogeneity.

The investigations carried out by these laboratories combine the phenomenological methods and the use of a wide set of modern physical spectroscopic, resonance, diffraction, emission, microscopic, and other techniques. They are aimed at obtaining detailed information on the processes proceeding in stressed solids at the subatomic, atomic-molecular, superatomic, and macroscopic levels.

A wide range of solids with different compositions and structures are being studied. They are crystals with ionic and covalent bonds, metals, alloys, polymers, glasses, ceramics, and composites.

This review briefly describes the major results obtained in the investigations carried out at the Institute.

1 Phenomenological studies

As early as in the 1930s, A. V. Stepanov put forward the hypothesis that there is no purely brittle fracture of crystals. Fracture of crystals is always associated with plastic deformation [1]. These ideas, which were confirmed later, showed that plastic deformation should necessarily be included into the consideration when developing the physics of strength.

1.1 Plastic deformation

The kinetic nature of plastic deformation was quite evident. Indeed, under the action of an applied constant load plastic deformation (ε) increased with time (t). Such a characteristic as plastic deformation rate (instantaneous or mean) $\dot{\varepsilon} = \mathrm{d}\varepsilon/\mathrm{d}t$ followed from this fact in the most natural way.

The studies of the $\dot{\varepsilon}(\sigma, T)$ dependences (σ is the stress and T is the temperature) were started at the Institute in the 1930s. For many metals, polymers, and other materials, a fairly general expression

$$\dot{\varepsilon}(\sigma, T) = \dot{\varepsilon}_0 \exp\left(-\frac{Q_0 - q\sigma}{kT}\right) \tag{1}$$

was obtained for a wide range of mean temperatures.

The fact that the temperature dependence (1) is in the Boltzmann form indicates that plastic deformation is controlled by elementary thermal fluctuation acts. The magnitudes of the initial barrier Q_0 (reduced by the applied stress σ) ranged from 1 to 3 eV and $\dot{\varepsilon}_0 = 10^{12}\,\mathrm{s}^{-1}$.

A significant innovation was the use of laser interferometry to measure the creep rate, due to which it became possible to determine experimentally the instantaneous rate of plastic deformation with the accuracy of up to 1% for the deformation increase of 0.15 μm [2]. This technique has revealed a nonmonotonous stepwise behavior of deformation.

1.2 Fracture

Nothing had been said about the fracture kinetics before S. N. Zhurkov started his systematic studies. It had been generally believed that fracture of a solid is the critical event which occurs when a definite level of stress, the "strength limit," is reached. S. N. Zhurkov showed first of all that breakdown of a solid can take place under stresses much lower than the "strength limit", however, a certain time is needed for the breakdown to occur. The lower the applied load, the longer the required time. So the notion of "lifetime" (τ), the time from the moment of load application to the eventual breakdown of a solid, was introduced into the physics of fracture [3]. The finite value of lifetime τ implied that in a stressed solid the process of preparation of the solid for the eventual event—breakdown of a solid into parts—proceeds. Thus fracture of solids proved to be a kinetic phenomenon (a process) rather than a critical act. Lifetime τ is the integral characteristic of the fracture process and is nearly proportional to the inverse mean rate of this process.

The magnitude of τ was found to depend drastically on the applied stress (σ) and temperature (T). It was shown that for a wide range of solids with diverse compositions and structures, in the region of not too low and not too high stresses, the $\tau(\sigma, T)$ dependence is described by Zhurkov's formula [4, 5]

$$\tau = \tau_0 \exp\left(\frac{U_0 - \gamma\sigma}{kT}\right) \tag{2}$$

where $\tau_0 \approx 10^{-12}$ s for any solid, U_0 is close to the dissociation energy of interatomic bonds in a solid (which is determined by the chemical composition of the solid) and is independent of the structural state of the solid, and γ depends on the structural state and characterizes the level of local overstresses.

The Boltzmann form of dependence (2) and magnitudes of U_0 and τ_0 suggest that preparation of a solid for breakdown is controlled by the elementary acts of the type of scission of interatomic bonds due to thermal fluctuations.

It was found that at low temperatures (below the Debye temperature) a number of objects exhibit an athermal fracture kinetics, which led to the hypothesis that quantum fluctuations of atom coordinates (tunnel transitions) participate in the elementary acts of the process governing fracture at low temperatures [6].

Thus the studies of the phenomenological aspects of fracture have led to the important conclusion that fracture is a kinetic phenomenon rather than a critical event, and the underlying mechanism for this process is the fluctuational dynamics.

Speaking about the relation between plastic deformation and fracture, the similarity between the kinetic expressions (1) and (2), from the point of view of both the form and meaning of the parameters entering into them (Q_0 and U_0; $\dot{\varepsilon}_0$ and τ_0), should be noted. This suggests that the elementary acts of these two processes are closely connected.

1.3 Electronic aspects of the physics of deformation and fracture

It is quite evident that the electronic subsystem participates in determining the mechanical properties of bodies because all interatomic interactions occur through the electronic shells of atoms. Nevertheless, a detailed study of the electronic processes proceeding in stressed solids are needed to gain a better insight into the physics of plastic deformation and fracture. The phenomenological studies carried out at the Institute give valuable information on this subject as well.

A. V. Stepanov observed the appearance of electric charges on the surface of ionic crystals during plastic deformation [7]. Later it was found that application of a static electric field changes the plastic deformation rate of many crystals and also polymers [8, 9].

The plastic deformation rate turned out to increase in HTSC ceramics at transition into the superconducting state. The effect of mechanical stresses upon the temperature of transition to superconductivity was also revealed [10].

New data were also obtained for fracture. For instance, application of a static electric field was found to sharply decrease the lifetime of polymers [6].

Of particular importance was the comparison of the kinetic behaviors of two types of fracture, i.e., the mechanical fracture (disintegration of a body under the action of a mechanical load) and electrical one (breakdown of an insulator in an electric field). It has been found for polymers that the electrical fracture exhibits the kinetic character similar to that of the mechanical one and the barriers for the elementary acts of both processes and also the temperatures of transition from the classical to quantum dependences nearly coincide [6]. Since the electrical fracture is a purely electronic process (including formation of free electrons and their accumulation in traps up to critical concentrations), the coincidence between the barriers points to the important role played by electronic processes in the preparation of objects for the mechanical fracture. A decrease in the electrical lifetime at polymer loading has also been demonstrated.

For ferroelectrics, the effect of electric field on the strength and also the influence of mechanical loading on the breakdown electric field have been established [11].

Thus the obtained information clearly shows that further experimental and theoretical investigations of the behavior of the electronic subsystem during loading of solids and also of the participation of electronic processes in plastic deformation and fracture are highly important for a deeper understanding of this subject.

2 Microscopic studies

2.1 Plastic deformation

Of fundamental significance for the physics of plasticity of crystals was the hypothesis, and its subsequent experimental confirmation, of the existence of specific linear defects in the crystalline lattice, i.e., dislocations [12]. The nucleation, motion, multiplication, and interaction of dislocations—the problems of evolution of the dislocation structure—became the subject of research in many laboratories all over the world. At the Ioffe Institute, a number of important aspects of this problem have been solved and being intensely studied at present.

A new method which provided plastic deformation of crystals solely along one system of planes was suggested [13]. Owing to this, the dislocation phenomena could be studied in the simplest conditions of single slippage. The experiments using this technique revealed that an increase in the dislocation density results from multiplication of mobile screw dislocations through the mechanism of double transverse slippage, and parameters of this slippage were quantitatively determined for the first time [13].

It was shown that, at not too high temperatures, the prevailing elements of dislocation structure of deformed crystals are dislocation dipoles [13].

An important detalization of the kinetics of dislocation motion was carried out. Using the deformation curves and dynamic techniques, the thermally activated and athermal components in the mechanism of dislocation motion were separated [14]. For instance, it has been found that the motion of dislocations through local short-range barriers produced by point defects occurs due to thermal fluctuations, while the barriers resulting from the far-range stress fields produced by the dislocation themselves, their ensembles, and inclusions of the particles of another phase are overcome only due to mechanical forces, i.e., athermally [13, 15].

The *in situ* investigations of plastic shifts at helium temperatures (1–4 K) revealed the state of the crystalline lattice with the strength approaching the theoretical one and also anomalies in the temperature dependences of deformation due to local heatings in the slippage bands [16].

A new phenomenon, the dislocation-dynamic diffusion, was discovered [17]. It consists in the intense motion of atoms or molecules along nuclei of incipient or moving dislocations. This phenomenon is observed up to liquid helium temperatures. The dislocation-dynamic diffusion is a new, earlier unknown, mechanism through which the environment (gaseous or liquid) affects the mechanical properties of solids. On the basis of the results obtained in the studies carried out at the Ioffe, a new original technique for introducing impurities into the materials and also an efficient method for reducing the energy spent on dispersion of different materials in specially chosen media (which can find a wide application) have been suggested.

The photon and electron emission was observed at plastic deformation of ionic crystals [18]. The emission arises at crossings of dislocations, which points to the appearance of high local stresses leading to formation of electron and hole centers. Emission was also observed at plastic deformations of polymers, which is also the evidence of high local stresses at chain molecules.

The effect of an electric field on the dislocation structure of ionic crystals in the absence of mechanical loading has been revealed. An increase in the dislocation density due to the field-induced displacement of dislocation dipoles has been detected [19].

Of fundamental significance for the physics of plasticity is the theoretical development of the problems of evolution of dislocation ensembles. A synergetic theory of formation of dislocation structures and localization of slippage at plastic deformation has been developed [20]. Such important phenomena as formation of spatially ordered systems of slippage lines, appearance of cell-like and misoriented dislocation structures, and thermal instability of plastic deformation at low temperatures have been explained.

In addition to the achievements described above, new results on plastic phenomena have been obtained. They include migration of block boundaries at elevated temperatures [21], similarity between temperature dependencies of micro- and macrocreep [15], the structure of dislocation nuclei [17], and anisotropy of elastic properties of crystals which determines both the theoretical strength and dislocation microplasticity [14].

Dislocations were shown to play diverse important roles in fracture of crystalline bodies. For instance, they can produce local stresses, thereby leading to nucleation and growth of cracks. At the same time, moving dislocations can cause slowing down and arrest of the crack growth by relaxing stresses.

Thus the dislocation processes can lead both to strengthening and loss of strength by a stressed body. Therefore, it is evident that the effect of dislocations on fracture of solids should necessarily be taken into account in the investigation of fracture of solids.

2.2 Fracture

Fracture at static loading (under constant stress)

The data on the microscopic processes proceeding during preparation of a stressed solid for breakdown were obtained at the Ioffe Institute for the first time. The specific feature of these investigations were the in situ experiments, i.e., measurements immediately on stressed samples.

The infrared spectroscopic technique allowed detection of high local overstresses (exceeding the mean stresses by tens of times) on backbone interatomic bonds resulting from the structural heterogeneity of oriented polymers (bodies with the one-dimensional structure) [4, 5].

In the electron paramagnetic resonance and infrared spectroscopic experiments, scissions of overstressed bonds—the primary acts of fracture—were observed in the same polymers. The kinetics of scission accumulation pointing to the thermal fluctuations mechanism of bond scissions was traced [4, 5].

The X-ray, light, and ultracold neutron diffraction techniques proved to be highly efficient in detecting finest submicroscopic cracks (with the sizes of tens, hundreds of Angstroms), i.e., primary fracture nucleation sites, in stressed polymers and metals. The kinetics of accumulation and enlargement of submicrocracks indicating that the thermal fluctuation mechanism acts at this stage of fracture as well was traced [4, 5, 22].

Using the electron microscopic technique, direct observations of individual submicrocracks in crystals, metals, polymers, and model composite materials were performed [5, 23]. Tunnel electron microscopy detected formation of incipient crack-like defects at the surfaces of metals [24]. From the results obtained in the studies of stressed solids, the specific features of nucleation and evolution of finest cracks were deduced. They are as follows:

(i) crack nucleation in heterogeneous plastic solids occurs in a stepwise fashion within the limits of structural elements at the sites of nonuniform development of plastic deformation;

(ii) immediately after nucleation the crack rapidly grows, but then it growth rate slows down because of plastic relaxation of local stresses;

(iii) at the stage of rapid crack growth, the field of elastic stresses surrounding the crack is able to promote nucleation of another crack or accelerate the growth of the existing neighboring crack;

(iv) in all the cases, crack accumulation, formations of clusters of cracks, and enlargement of cracks take place in a stressed heterogeneous plastic body.

This is in fact the kinetics of microfracture in which the role of plastic dislocation processes is evident. In the most clear form the effect of these processes manifests itself, for instance, in decorating the dislocations by impurity atoms which drastically decreases the mobility of dislocations. In this case, the lifetime (the time to breakdown) of a solid was found to increase sharply, which means that the fracture rate becomes much lower [25]. This implies that the lifetime of a loaded polycrystalline solid is controlled first of all by the expectation for the acts of nucleation of primary cracks, when motion of dislocations produces local overstresses necessary for crack nucleation.

It is apparent that an extremely important problem of the kinetics of macrofracture of a solid is formation of the fracture nucleation site which leads to formation of the fault whose growth results in the eventual breakdown of a solid.

Along with the direct microscopic observations of clusters of primary cracks and appearance of larger and larger cracks, valuable information on the fracture nucleation sites can be obtained by the acoustic emission technique involving detection of acoustic pulses resulting from nucleation and rapid growth of cracks. This technique is able to detect the appearance and growth of cracks with the sizes from microns and their coalescence into larger cracks. This process was found to be very complicated. In a solid having considerable microplasticity, formation of a cluster is accompanied by unloading of the relevant portion of a solid, which decreases the immediate danger of this cluster as a fracture nucleation site [26–28]. In a solid with a lower plasticity, this cluster, by preserving the field of increased stress, can act as an active fracture nucleation site [29].

Thus the scientists of the Ioffe Institute were the first to carry out experimentally the detalization of the kinetics of fracture, from the submicroscopic level (nucleation of primary cracks) to macrofracture.

Fracture at high-speed (impact) loading

Information on nucleation and growth of submicrocracks and their coalescence into a fault leading to the eventual breakdown of a body was also obtained in the experiments using laser interferometry and light scattering and involving the shock wave loading of transparent objects (the time interval from 10 ns to 100 μs) [30, 31].

The shock-wave loading of metals revealed a new phenomena—microcumulation [32]. Under the action of the shock wave, the collapse of defects of the type of micropores or microcracks occurs. The temperature and pressure in the region of defect increase by orders of magnitude and the microcumulative jet with the velocity of substance outflow reaching the velocity of sound in solids is formed. The discovered phenomenon explains a number of effects observed on exposure of a solid to a shock wave.

Another type of a high-speed loading is penetration of a projectile (with the speed of 2 km s^{-1}) and cumulative jets (with the speed of 4–5 km s^{-1}). The experiments on ceramics and glasses using the pulsed X-ray photographing revealed the dependence of the response of a medium on the ratio between the speed of penetration and crack growth velocity [33, 34]. If the speed of penetration is lower than the crack growth velocity, the kinetics of fracture of the medium determines its effective resistance at the path along which the projectile moves. If the speed of penetration is higher than the crack growth velocity, resistance to penetration drastically increases because the projectile is ahead of the fracture wave and interacts with the undamaged hard material whose displacement is determined by the kinetics of deformation.

3 Fracture prediction. Strengthening

The kinetic nature of fracture creates the physical basis for prediction of macrofracture—forecasting the moment of breakdown of a stressed body. The equation for lifetime (2) itself allows such a prediction, but with the logarithmic accuracy because of statistical scatter of τ.

The possibility of a more precise prediction follows from the revealed laws of evolution of cracks in a stressed body. For heterogeneous bodies (the major part of bodies have just this type of structure), the stage of accumulation of primary cracks up to a certain critical mean concentration followed by the stage of coalescence and enlargement of cracks has been revealed [29, 34]. As the studies showed, the duration of the first stage (accumulation of the primary cracks) τ_1 constitutes the major part of the lifetime τ, though at the second stage the system of cracks goes through a number of enlargement steps up to formation of a fault and breakdown of a body. The enlargement stages proceed faster and their net time τ_2 constitutes a small fraction of τ.

On the basis of these data, a two-stage model of the fracture kinetics was suggested. In the model, the total time to breakdown (lifetime) $\tau = \tau_1 + \tau_2$ [35].

The main goal is to identify the transition from the first stage to the second one. After it the breakdown of the body is to occur fairly soon. This problem is being successfully solved by using the acoustic emission technique in laboratory conditions and practical situations.

It has been shown that the first stage—the stage of chaotic crack formation—is characterized by the Poisson flow of acoustic signals with nearly equal amplitudes. When crack enlargement starts, their amplitudes vary, and what is most important, the flow of signals markedly differs from the Poisson one [29]. Thus for a stressed solid brought to breakdown, τ, τ_1, and τ_2 can be found.

So detection and analysis of the spectra of acoustic emission signals are most suitable tools for estimating the time to macrofracture τ_2.

At present the approach to prediction of macroscopic fracture developed by researchers of the Ioffe is used for the technical diagnosis of large-size constructions, in the mines where the danger of rock bursts exists, and prediction of earthquakes.

For the high-speed penetration, using the established laws for penetration with the speed of 2 km s^{-1}, the prediction of the penetration depth for the speed of 15–20 km s^{-1} (which can hardly be achieved in laboratory conditions) is made [36].

The studies of the phenomenology and microscopics of the fracture kinetics have made it possible to suggest new methods for strengthening of materials and prolonging their service time.

For instance, the well-known method for strengthening of polymers by orientational drawing involves two kinetics: strengthening of the molecular structure of the polymer and scissions of chain molecules. The optimal temperature and velocity regimes for the orientational drawing have been found. As a result, the tensile strengths of polymers of up to 10 GPa were obtained [37, 38]. This is the highest strength obtained in the world and is lower than the theoretical strength only by a factor of 2–3.

Achievements have also been made in prolonging the service life of metals. It is evident that since the fracture kinetics of metals involves accumulation of microcracks during the time the metal is subjected to loading, elimination of cracks must result in a longer lifetime. Indeed, by properly choosing the temperature-baric conditions of treatment of the metal which has "exhausted" a part of its lifetime, it is possible to "heal" microcracks through the dislocation mechanism and thus restore the initial lifetime. Using multiple successive operations, a many-fold increase in the lifetime has been achieved [39].

4 Conclusions

Owing to the investigations of the physics of strength and plasticity at the Physico-Technical Institute, many key problems in this field have been solved. They are the possibility to reach the theoretical strength, kinetic nature of response of solids to loading, micromechanisms of plastic deformation and fracture, ratio between brittleness and plasticity, technical problems of increasing the strength of materials, prediction of fracture, and others.

At the same time, the problems which require further thorough studies in this direction have become clear. They are as follows:

(i) detalization of the elementary acts of deformation and fracture, including electronic phenomena;

(ii) elucidation of the participation and role of quantum effects in these processes;

(iii) a deeper understanding of the relation between the processes proceeding in stressed solids at different structural levels, from the subatomic (electron) level to the macroscopic one;

(iv) elucidation of the mechanisms through which different physical factors (electric and magnetic fields, irradiation, superconducting transitions, and others) affect the plasticity and strength of solids and also of the mechanisms of the influence of mechanical loading on physical properties of solids;

(v) looking for wider application areas of the laboratory results.

The progress in solving these problems is of great significance for the development of the physics of strength of solids and, on a greater scale, of the technology of the 21st century.

References

[1] A. V. Stepanov, *Foundations of Real Crystal Strength*, Nauka, Moscow 1974 (in Russian).

[2] V. A. Stepanov, N. N. Peschanskaya, and V. V. Shpeizman, *Strength and Relaxation Phenomena in Solids*, Nauka, Leningrad 1984 (in Russian).

[3] S. N. Zhurkov, *Int. J. of Fracture Mech.* **1** No. 4, 311 (1965).

[4] V. R. Regel', A. I. Slutsker, and E. E. Tomashevsky, *Kinetic Nature of Solids Strength*, Nauka, Moscow 1974 (in Russian).

[5] Zhurkov Anniversary Issue, *Int. J. of Fracture* **11** No. 5, (1975).

[6] V. R. Regel, and A. I. Slutsker, *Trans. of St. Petersburg Academy of Sci. for Strength Problems*, **1** 226 (1997).

[7] A. V. Stepanov, *Z. f. Phys.*, No. 2, 537 (1932).

[8] A. N. Kulichenko and B. I. Smirnov, *Sov. Phys. Solid State* **23** No. 4, 595 (1981).

[9] V. I. Nikolaev and B. I. Smirnov, *Sov. Phys. Solid State* **27** No. 11, 2028 (1985).

[10] B. I. Smirnov, T. S. Orlova, and V. V. Shpeizman, *J. of Mech. Behav. of Materials* **3** No. 4, 245 (1992).

[11] L. V. Zhoga and V. V. Shpeizman, *Sov. Phys. Solid State* **34** No. 8, 1382 (1992).

[12] A. H. Cotrell, *Dislocation and Plastic Flow of Crystals* Clarendon Press, Oxford 1953.

[13] B. I. Smirnov, *Dislocation Structure and Hardening of Crystals*, Nauka, Leningrad 1981 (in Russian).

[14] S. P. Nikanorov and B. K. Kardashev, *Elasticity and Dislocation Anelasticity of Crystals*, Nauka, Moscow 1985 (in Russian).

[15] B. K. Kardashev, S. B. Kustov, A. B. Lebedev, and S. P. Nikanorov, *J. Alloys and Compounds* No. 211/212, 83 (1994).

[16] O. V. Klyavin, *Plasticity Physics of Crystals at Helium Temperatures*, Nauka, Moscow 1987 (in Russian).

[17] O. V. Klyavin, N. P. Likhodedov, and A. N. Orlov, *Int. Rev. J. Progr. Surf.* **33** 259 (1990).

[18] V. A. Zakrevskii and A. V. Shuldiner, *Phil. Mag. B.* **71** No. 2, 127 (1995).

[19] S. V. Krishtopov, A. N. Kulichenko, and B. I. Smirnov, *Sov. Phys. Solid State* **32** No. 8, 1377 (1990).

[20] G. A. Malygin, *Phys. Solid State* **37** No. 1, 1 (1995).

[21] B. I. Smirnov, R. S. Chudnova, and V. V. Shpeizman, *Poverkhnost': Fizika, Khimia, Mekhanika* No. 6, 143 (1982) (in Russian).

[22] V. S. Kuksenko and V. P. Tamuzh, *Fracture Micromechanics of Polymer Materials*, Martinus Nijhof Publishers 1981.

[23] A. M. Leksovskij, B. L. Baskin, A. J. Gorenberg, G. H. Usmanov, and V. R. Regel, *Sov. Phys. Solid State* **25** No. 4, 1096 (1983).

[24] V. I. Vettegren', V. N. Svetlov, and S. Sh. Rakhimov, *Phys. Solid State* **38** No. 2, 323 (1996).

[25] G. A. Malygin and G. V. Vladimirova, *Problems of Strength and Plasticity of Solids. On the 100 Birthday Anniversary of N. N. Davidenkov*, p. 96, Nauka, Moscow 1979 (in Russian).

[26] A. M. Leksovskij, A. P. Tishkin, and A. Abdumanonov, *Proc. IAES-13*, p. 18, Nara, Japan 1996.

[27] A. M. Leksovskij, *Trans. of Inst. of Fund. Technolog. Research of Polish AS* **36** 235 (1995).

[28] A. P. Tishkin, A. Abdumanonov, and A. M. Leksovskij, *Techn. Phys. Lett.* **21** No. 8, 857 (1995).

[29] V. Kuksenko, N. Tomilin, E. Damaskinskaya, and D. Lockner, *Pageoph.* **146** No.2, 253 (1996).

[30] N. A. Zlatin, G. S. Pugachev, and S. A. Leont'ev, *Sov. Phys. Techn. Phys.* **26** No. 7, 1507 (1981).

[31] E. N. Bellendir, G. S. Pugachev, and A. B. Sinani, *Proc. of the 6th USSR Conf. on Fracture Physics* p. 256, Kiev 1989 (in Russian).

[32] E. L. Zil'berbrand, G. S. Pugachev, and A. B. Sinani, *Techn. Phys. Lett.* **20** No. 3, 236 (1994).

[33] A. S. Vlasov, Yu. A. Emel'yanov, E. L. Zil'berbrand, A. A. Kozhushko, A. I. Kozachuk, G. S. Pugachev, and A. B. Sinani, *Techn. Phys. Lett.* **23** No. 2, 117 (1997).

[34] A. A. Kozhushko, I. I. Rykova, and A. B. Sinani, *J. de Physique IV*, Colloque c 3, supp. 1 au *Journ. de Phys. III* **1** 117 (1991).

[35] S. N. Zhurkov, V. S. Kuksenko, and V. A. Petrov, *Theor. and Appl. Fracture Mechanics* **1** 271 (1984).

[36] N. A. Zlatin, *Ballistic Installations and their Use in Experimental Research*, p. 344, Nauka, Moscow 1974 (in Russian).

[37] A. V. Savitsky, I. A. Gorshkova, I. L. Frolova, and G. N. Shmikk, *Polym. Bull.* No. 12, 195 (1984).

[38] V. A. Marikhin, L. P. Myasnikova, D. Zenke, R. Hirte, and P. Waigel, *Polym. Bull.* No. 14, 287 (1984).

[39] V. Sklenicka, V. I. Betekhtin, and K. Kucharova, *Script. Metallurg.* **25** 2159 (1991).

Optical data processing and holography

Yu. N. Denisyuk, S. B. Gurevich, and V. B. Konstantinov

1 Introduction

Holography—method of recording and reproducing wave fronts by registration of the interference patterns between object and referent beams—was proposed by D. Gabor in 1948 [1], but only in the early 1960s, when the first lasers appeared, it became a functional instrument. Academician B. P. Konstantinov initiated projects on optical data storage at the Ioffe Institute. The investigations started at laboratories headed by A. N. Zaidel, S. B. Gurevich, A. P. Komar and at specially formed laboratory which was later united with the laboratory of technical TV-systems and changed its title to Laboratory of Optoelectronics and Holography. Since 1988 it is has been headed by Yu. N. Denisyuk. Since the early 1970s some experiments have been performed in the laboratories headed by Yu. A. Dunaev, B. P. Zakharchenya, M. P. Petrov, and B. T. Kolomietz.

In the first stages of development of laser holography many investigations were concerned with applications of static holograms. The resolution of holographic methods was determined [2] and, for the first time, the method of holographic resolvometry for investigating the resolution of high-resolution photographic materials was proposed [3]. The holography was also applied for the determination of electron concentration in pulse laser-induced plasma. The course of the process was investigated with the help of cinema-holography methods proposed by Yu. I. Ostrovsky [4]—one of the pioneers of practical application of holographic interferometry methods. The informational aspects of holography were studied experimentally and theoretically by S. B. Gurevich and his colleagues [5]. The possibility of making holographic cinematography and television were studied, also hologram transmission by communication channels was analyzed. It was shown that a transmission through binary channel allows one to reproduce half-tone information [6].

Just in the 1960–70s it became the practical fields of applications of static holograms—holographic optical elements, holographic interferometry, obtaining the 3D images, etc. Later the radiation sources were improved, and the first semiconductor lasers were created at the Ioffe Institute. Many works were aimed at creating new recording materials and devices for systems of optical data processing and hologram recording.

2 Investigations and development of new recording materials and devices

An intensive search for nonsilver recording media and reusable systems had resulted in the 1960s in the creation of semiconductor light sensitive materials [5, 7]. Chalcogenide glassy semiconductors (CGS) change their transparency under illumination. This is due to changes in the absorption band, taking place as a result of photostructural transformations in CGS. Also the refractive index changes. The information stored in CGS can be erased by heating to the saftening temperature of a material. It is also possible to erase the recording with the help of a high-power laser. As_2Se_3 and As_2S_3 proved to be the most sensitive. The resolution of CGS films, obtained by thermal evaporation, was about 9000 mm^{-1}. Changes in the CGS structure may lead to changes in the band gap and move the fundamental absorption edge in a wide range of wavelengths. The refractive index changes also under illumination with light. High values of Δn, obtained on this materials provide a high efficiency of read out. The read out by inactive light does not destroy the recordings, hence the material for this light has only phase component. This fact results in a high diffraction efficiency—40%. Pulsed reversible recording is also possible and can be used in systems of point by point optical memory.

In the mid-1970s, some works on creating a recording material based on type I phase transition were performed in the laboratory of B. P. Zakharchenya [8]. It soon became clear that VO_2 is the optimal for most applications. It has a transition temperature (340 K) and the maximum variation of permittivity, which is more than 0.5 for the refractive index in the visible range. Such high changes in refraction allow one to produce an interference reflection structure with a thin (50 nm) modulating layer of VO_2, which is a bistable optical element. It can be switched by heating (the light pulse is absorbed and the material is heated). One can use this structure as a reversible registration medium for holography (diffraction efficiency—10^{-2}, sensitivity—3×10^{-4} J cm^{-2}) both in real-time mode and for archive recording. The experiments showed that this structure can be successfully used in different holographic applications, including picosecond holographic interferometry of electron–hole plasma in semiconductors in the IR-range.

Nonlinear heat reflection response is so short because of the small thickness of VO_2 film, that it can ensure passive mode self-synchronization in a solid state laser.

3 Holographic interferometry and non-destructive testing

Many technical and scientific problems in hydro- and aerodynamics, materials science, etc. can be solved with the help of methods of optical data processing and holography. These methods allow one to measure the pressure and temperature, the kinetics of their changes and obtain spatial distribution with a very high accuracy. All these methods are non-destructive, they do not influence processes that are investigated in the experiments. Holographic interferometry allows one to compare the state of the objects at different instants of time. This is impossible in conventional interference measurements.

In the 1960–70s, special modernization of devices for analyzing aeroballistic traces (for investigations of processes of space-craft landing on the Earth and other planets) was taking place. An aeroballistic trace is a close volume with supervised compound of gas, into which a model of spacecraft or aircraft is shooted. High-speed photography, spectrography, and different radiomethods furnish information about movement of models and processes in gases near the models, but they are too expensive and consume a lot of energy. In the laboratory headed by Yu. A. Dunaev these experiments were performed using methods of holographic registration [9]. Investigations were carried out in two directions. The first—with single-exposure holograms fixing all the

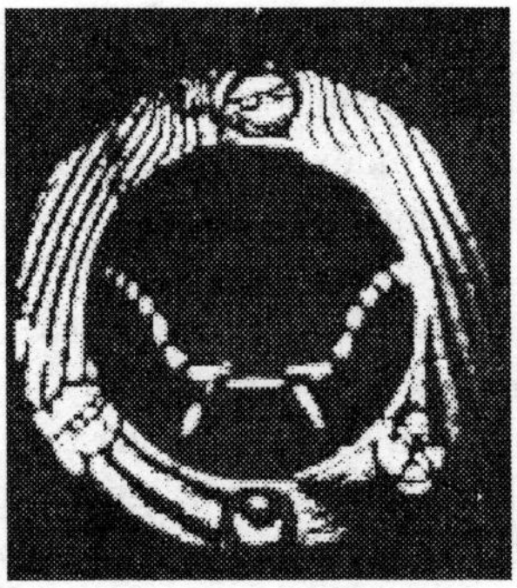

Fig 1. Interferogram of AHV legs at cycling.

needed information. The hologram was reconstructed by shadow methods of optical filtration or interferometry for image visualization, which allows processing the obtained information at all times with the help of He-Ne DC lasers. The second—multipath holographic interferometry using nonlinear properties of photographic emulsion. Both these methods increase 10–30 times the sensitivity and provide a possibilities of investigating supersonic gas flows with low density.

In the early 1980s, a compact holographic device for space research was created [10]. The methods and equipment developed gave rise to scientific application of holography. Holographic interferometry by a portable holographic camera has been used in the following fields: materials science (crystal growth, protein separation by electrophoresis under zero gravity in a space station), measurement of the compressibility of a liquid at low pressure. It was shown that real-time holographic interferometry gives a unique possibility to estimate qualitatively and quantitatively the heat transfer processes at the interface between gas and liquid. Compression heats a gas more then a liquid, and the temperature difference between them can be calculated by recording the change in the number of interference fringes immediately after compression. The distortion of the fringes in the boundary layer actually gives a curve of temperature variations in the layer.

Holographic interferometry is also effective for non-destructive testing. Nowadays many people suffer from different cardiovascular diseases. For many of them the only chance to stay alive is surgical implantation of artificial heart valves (AHV). They can live long if AHV functions properly. AHV must open and close up to 10^9 times during at least 20 years. Thus, the quality testing during manufacture is the main guarantee of the reliability of valves. Quantitative and qualitative testing of the mechanical properties of AHV and of hydrodynamical parameters can be effectively done by holography. The interferogram in Fig. 1 demonstrates AHV legs deformation after cycling.

On the basic of a holographic camera for space materials science research a TV holographic device for outside spacecraft window surface was designed. Figure 2 shows a micrometeor trace image from "Mir" space station.

The possibility of generating and observating a nonlinear solitary strain wave (soliton) in a solid waveguide is of special interest in connection with the undeniable potential application in solid state physics and mechanics. The theory of longitudinal nonlinear strain wave propagation was developed by A. M. Samsonov *et al* [11]. Analysis of the results obtained produced correct estimations of the initial conditions necessary to

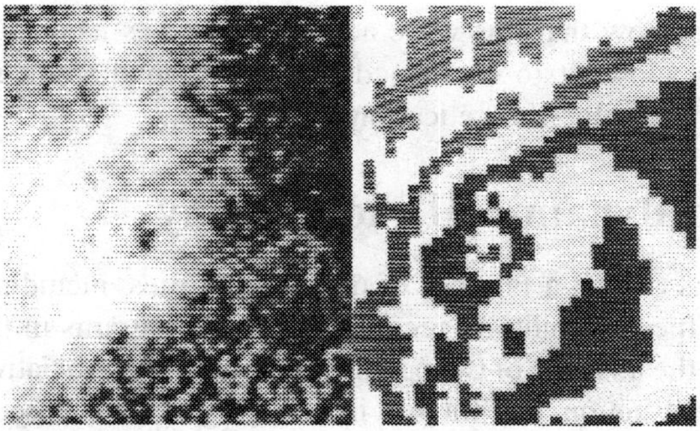

Fig 2. Image of a micrometeor trace on the "Mir" window.

generate a solitary strain wave and enabled first experiments in soliton generation [12]. However, there was no possibility to observe the soliton evolution in motion, and, therefore, no opportunity, either, to confirm or to disprove the main features of strain soliton, namely, the wave shape conservation and the right proportion of an amplitude and wave length, when the soliton is moving along a rod. Moreover, the measured wave length was essentially smaller than the predicted one. For these reasons, the theory was improved and experimental approach refined. The experiments were performed with the help of pulsed laser holographic interferometry. The apparatus consisted of a device producing a wave in a solid, a synchronizer, and a holographic interferometer. Two synchronized pulses from Q-switched ruby lasers were used. One of them was used to generate the initial shock wave and the other for recording the wave pattern inside the rod. Solitons were formed in a transparent solid rod by an initial shock wave induced in a liquid surrounding the rod by laser evaporation of a metallic target placed near the end of the rod. Observations were made in the perpendicular direction. A pulse from the second laser synchronized with a certain stage of strain wave propagation was used for holographic recording of the wave pattern. The polystirol rod was probed by a plane light beam (40 mm in diameter) at different regions. The set-up allowed changing precisely the delay time between the pulses from the laser inducing the shock wave and from the recording one. A wave pattern in a waveguide and surrounding water is shown in Fig. 3. Thus, using the holographic interferometry, the strain soliton in a nonlinearly elastic solid waveguide was recorded together with its evolution. The data

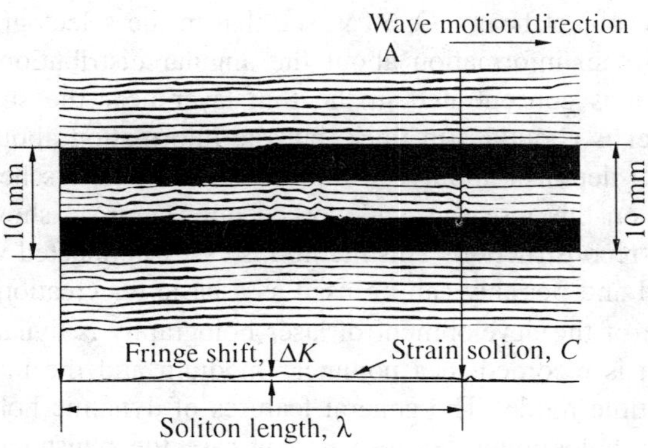

Fig 3. Wave pattern in waveguide.

obtained allowed to show that this wave actually preserves its shape and amplitude-to-length ratio, which allows it to be named the longitudinal strain soliton. The results obtained agree well with the theoretical predictions.

4 Static and dynamic 3D holography

Deep 3D holograms exhibit a number of useful properties, including high spectral and angular selectivity, the possibility of recording many holograms in the same region of a recording material, the absence of twin image, possibility of associative readout and using white light for reconstruction, etc. Interest in 3D holograms increased now owing to their promising application as computer interconnectors. In the known intellectual systems such as human brain, electronic computers or the in optical computer, which is now under development, linear relations are used. The holography proposes an alternative solution of the problem—transmission of signals supported by superimposed light waves interconnection between the last provided by superimposed gratings. Unfortunately, application of deep holograms is difficult because of the necessity for producing thick-layered light-sensitive media. Yu. N. Denisyuk proposed a method of so-called pseudo-deep holograms [13–16]. Although they have the same angular and spectral selectivity as deep 3D holograms, pseudo-deep holograms do not require application of a thick light-sensitive material. Pseudo-deep hologram recording and reconstruction is relatively simple. A light-sensitive layer, on which a fan of beams propagating from a line object is recorded, is inclined at a small angle relative to the plane in which these beams propagate. In the reconstruction, a special slit allows passage of only those beams which are situated in the object plane and cuts off all the others.

The method of recording selectograms without referent beam was developed by Yu. N. Denisyuk and his research group. It allows one to obtain 3D holographic images of objects illuminated with natural light sources. This method is based on a phenomenon similar to that used in Lippman's photography. Comparison between these two methods shows that a selectogram similarly to Lippman color photograph, is a recording of a pattern of standing waves, formed as a result of interference of the object wave and a "twin" wave—the wave split from the object wave with an optical element. These methods are different because the way of spreading of twin wave is different. In Lippman's photography this wave is formed by a mirror and is opposed the object wave. In the selectogram method the "twin" wave is formed by a diffraction grating and is aligned with the object beam. One can see that in the selectogram as in Lippman's photography, the main information about the angular distribution of the intensity of registered radiation is concentrated in the thin layer near the surface of the optical element. This layer is about 1 μm thick, but the information about fine details of the image extends to a depth of about 1–2 millimeters. The new scheme of projection of 3D images based on this research was worked out. Figure 4 shows the initial object and its white-light reconstruction. This method is easily connected with a TV 3D image projection method and possibly can be used as a basic for creation of a 3D display.

The next stage of the development of laser holography is dynamical holography in which a hologram is recorded in a nonlinear medium and the interference pattern is registered in real time mode. The general features of dynamic holography differ from those of the static holography, because in this case the registered radiation starts to interact with the hologram. The dynamic holography allows solving some problems

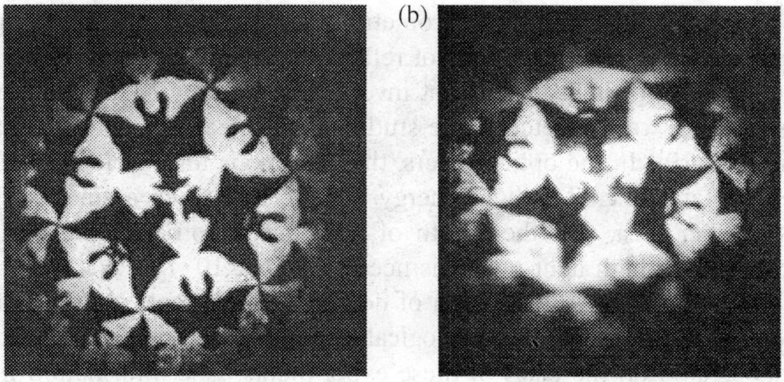

Fig 4. (a) Object and (b) its reconstruction by a selectogram.

of optical information processing in real time, such as compensation of the wave front distortion, caused by the influence of optical nonlinearity of the media through which the light is passing. In M. P. Petrov's laboratory, investigations of photorefractive crystals ($LiNbO_3$, $Bi_{12}SiO_{20}$, $Bi_{12}TiO_{20}$, $KNbO_3$, etc.) [17] were performed. They may be of interest for holographic memory systems, holographic interferometry, optical data processing methods, and investigations of nonlinear phenomena in optical fibers. Spatial-time modulator was developed (image converter), where the recording of the image is created with the help of lateral electrolytic effect geometry and an external electric field aligned the light beams. The refractive index is modulated by the lateral component of an internal electric field created by the spatial charge inside the sample. This configuration provides a diffractional efficiency by an order of magnitude greater than that in a similar US device PROM.[1] It also has low noise rate. On the basis of this analyzer, a spectrum analyzer for processing multichannel acoustic signals in real time (2D Fourier processor) has been created.

Photorefractive crystals are electrolytic media in which the refractive index changes under the action of an external electric field. This fact can help one to control the Bragg diffraction in 3D holograms. One can obtain dielectric multiplexing, while recording many holograms. This means that the recording is controlled not by changing the angle of reference beam (angle multiplexing) or the wavelength of recording beams, but with an external electric field. This control principle is interesting for holographic memory tasks, optical fiber communications (when switching the spectral channels), for semiconductor and variable mode lasers.

If, in the process of hologram recording in photorefractive crystals, unstable conditions take place (changes in electric fields, position and contrast of the interference pattern) waves of solid charge occur, and, hence, electric field and refractive index waves appear. This is attributed to the photorefractive effect, so this waves were named photorefractive waves. They exhibit an extraordinary dispersion principle—their frequency is inverse to the wave vector. The time of relaxation of these waves depends on the carrier drift length and, when this length is large enough, the relaxation time may much exceed the period of oscillations. In this case it is very simple to ensure the resonance conditions for excitation of photorefractive waves, so that their amplitude will also grow and so will do the diffraction efficiency of moving holographic gratings. These investigations can be successfully applied. They allow creating more optimized

[1] Pockels Readout Optical Modulator.

systems of adaptive holographic interferometry, and also can be used for registration of faint (about 1 Å) vibration amplitudes of reflecting surfaces.

The phenomenon of light wave front inversion caused by Mandelshtamm–Brilluen simulated scattering was detected while studying nonlinear effects in optical fibers for the first time [18, 19]. In the optical fibers, the conditions for development of nonlinear effects are very favorable. The light energy is concentrated in a small section limited to the central fiber area, and the length of light interaction with the object is about 1–100 meters. So no high light power is needed (about 10–100 W).

Such a simple process of generation of nonlinear electrooptical effects allows fabricating optical elements that can do all logical operations. The model of circular optical memory was created on the basis of these experiments. The information from this device is represented by a train of light pulses that are moving for any long time you need inside the light circle. To avoid distortions and losses, these pulses must be amplified, with their form and relative time position restored. This problem was solved with the use of Raman scattering in the fiber. Such a system is the first in the world.

References

[1] D. Gabor, *Proc. Roy. Soc.* **197** 454 (1949).
[2] B. P. Konstantinov, A. N. Zaidel, V. B. Konstantinov, and Yu. I. Ostrovsky, *Zh. Tekh. Fiz.* **36** 1718 (1966).
[3] A. N. Zaidel, V. B. Konstantinov, and Yu. I. Ostrovsky, *J. Sci. Appl. Photo and Cinematogr.* **11** 381 (1966).
[4] A. N. Zaidel, G. V. Ostrovskaya, Yu. I. Ostrovsky, and T. Ya. Chelidze, *Zh. Tekh. Fiz.* **36** 2208 (1966).
[5] S. B. Gurevich, V. B. Konstantinov, V. I. Sokolov, and D. F. Chernykh, *Transmission and Processing of the Information with the Help of Holographic Methods* ed. S. B. Gurevich, Sov. Radio: Moskva, (in Russian) 1978.
[6] B. P. Konstantinov, S. B. Gurevich, and G. A. Gavrilov, *Zh. Tekh. Fiz.* **39** 347 (1969).
[7] S. B. Gurevich, B. T. Kolomiets, V. B. Konstantinov, V. M. Lyubin, and I. A. Maurer, *Pis'ma Zh. Tekh. Fiz.* **1** 1076 (1975).
[8] A. A. Bugaev, B. P. Zakharchenya, and F. A. Chudnovski, *Phase Transition Metal-semiconductor and its Applications*, Nauka: Leningrad, (in Russian) 1979.
[9] A. K. Beketova, A. F. Belozerov, and A. N. Berezkin, *Holographic Interferometry of Phase Objects*, Nauka: Leningrad, p. 232, (in Russian) 1979.
[10] S. B. Gurevich and V. B. Konstantinov, *Opt. J.* **10** 3 (1996).
[11] G. V. Ostrovskaya, *Opt. J.* **64** 39 (1967).
[12] Yu. N. Denisyuk, *Zh. Tekh. Fiz.* **60** 59 (1990).
[13] Yu. N. Denisyuk, *Dok. Akad. Nauk SSSR* **144** 1275 (1962).
[14] Yu. N. Denisyuk, *Opt. Spectrosc.* **15** 522 (1963).
[15] Yu. N. Denisyuk, *Opt. Spectrosc.* **18** 276 (1965).
[16] Yu. N. Denisyuk, *Opt. J.* **64** 5 (1997).
[17] M. P. Petrov, S. I. Stepanov, and A. V. Khomenko, *Photorefractive Crystals in Coherent Optics*, Nauka: St. Petersburg (inRussian) 1992.
[18] E. A. Kuzin, M. P. Petrov and B. E. Davidenko, *Pis'ma Zh. Tekh. Fiz.* **10** 833 (1984).
[19] M. P. Petrov, V. I. Byelotizky, E. A. Kuzin, and V. V. Spyrin, *Quantum Electronics* **22** 1245 (1995).

Physics of atomic collisions and its applications

V. V. Afrosimov, Yu. S. Gordeev, and M. P. Petrov

1 Introduction

The physics of atomic collisions deals with processes occurring when atomic particles—atoms, molecules or their ions—interact between themselves or with electrons and photons. Data on these processes give the main body of information on the atomic and molecular structure, interatomic, and intermolecular potentials and mechanisms of particle interactions. Moreover, characteristics of elementary processes occurring in collisions between individual particles determine to a considerable extent the parameters of low density matter, e.g., gases and plasmas. For this reason, data on collisions are widely used in analyzing phenomena in plasma physics, quantum electronics, astrophysics, and gas dynamics, and developing analytical methods, technologies, and devices which use particle beams and radiation generated in collisions.

2 Elementary processes

The physics of atomic collisions became an independent area of research at the threshold of the 1940–1950s. Until then special problems related to processes involving atomic particles had been solved within the frame of those scientific and technology problems where they had arisen. However, at that moment it was well realized that the number and complexity of the atomic collision processes are so great that essentially new experimental and theoretical methods are needed to elucidate their mechanism. Shortly thereafter original theoretical models appeared in physics of atomic collisions and a rich arsenal of experimental methods was developed based on mass-spectrometry, electron spectroscopy, and spectroscopy of electromagnetic radiation in a wide range of wavelengths and on an analysis of the kinematics of interacting atomic particles. Means of distinguishing and studying an elementary act of collision of two individual particles under well-known initial conditions were developed. One of the leading schools in the physics of atomic collisions was founded at the Physico-Technical Institute by V. M. Dukel'skii, N. V. Fedorenko, and O. B. Firsov.

Until the early 1950s views dominated that atomic collisions can be described in terms of the impact mechanism whereby an impinging particle directly acts on a single electron of an atom and contributes an amount of energy necessary to a transition of the electron to a higher bound level (excitation) or into the continuum (ionization). From simple physical considerations it follows that the transition probability has its maximum

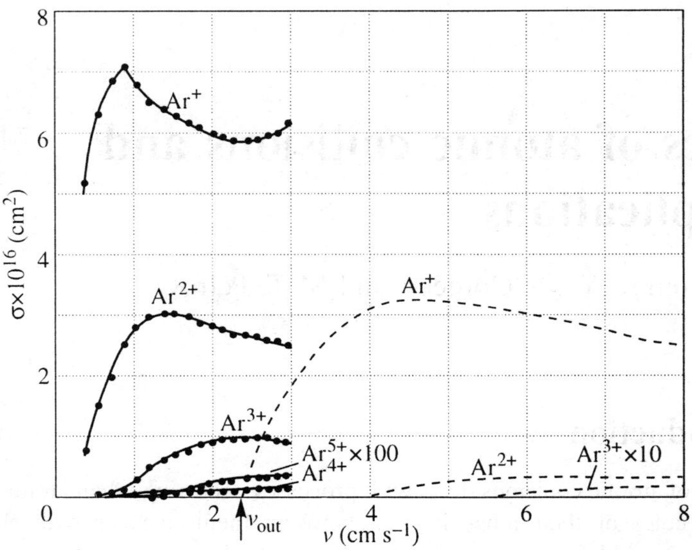

Fig 1. Cross section for formation of ions of various charge in collisions of electron (dotted curves) and He ion (full curves) with Ar atoms. The arrow marks a typical velocity of an electron of the outer shell of an atom, $\nu_{out} = e^2/h$.

at a relative velocity of colliding particles approximately equal to the orbital velocity ν of the electron.

These views were seriously questioned in the 1950s. Experiments performed at the Institute showed that ionization in ion–atom collision occurs at velocities much lower than those in electron impact ionization (Fig. 1).

A decisive step on the way to an essential change in the views of the day was accomplished in the early 1960s. New experimental methods using the coincidence technique to analyze and record both atomic particles participating in an individual collision act were developed at the Institute. This enabled for the first time studying inelastic processes and, in addition, recording simultaneously the trajectories of the colliding particles. In the first experiments of this cycle [2] it was revealed that 2p vacancies arise in the inner L-shells with about 100% probability when the colliding Ar ions and atoms approach one another to distances approximately equal to the dimensions of these L-shells.

It was revealed that at energies as low as 15 keV the cross sections of formation of 2p vacancies reach their maximum values equal to the geometric dimensions of L-shells. This result was unexpected and contested as absolutely improbable at that time. Indeed, the binding energy of 2p electrons in Ar atom is about 250 eV and it was expected that the cross section of 2p vacancy formation would be at its maximum at the energy of impinging ions of about 15 MeV being by approximately ten orders of magnitude lower than the observed experimental cross section at an energy of 15 keV. Thus, the experiments showed that ion-atom collisions are absolutely inelastic, whereas after the knowledge of the day those should be absolutely elastic. Studies [1, 2] performed at the Ioffe gave rise to the use of a known but undeveloped approach to inelastic processes between atomic particles. Accordingly, an atomic collision is treated as creation of a quasi-molecule, i.e., a system of two particles with rapidly changing internuclear distance.

This quasi-molecule arises during the approach of two particles and disintegrates when these particles fly apart. The quasi-molecule lifetime is determined by the time of flight of this particle along a distance of the order of its dimensions, being 10^{-15}–10^{-16} s at velocities $\nu \approx 10^7$–10^8 cm s^{-1}. When existing, this quasi-molecule has a system of levels which must undergo an exceedingly strong reconstruction: from that comprising levels of two isolated particles at large distances to that with levels of a united atom with a nuclear charge equal to the sum of charges of the interacting particles at a distance shorter than the dimensions of inner electron shells. A certain shift of the levels is unavoidable, since the inner shells of the atoms are occupied and during close encounter of the particles, each shell experiences a double "overpopulation." As a result, quite a number of levels occupied by electrons in a quasi-molecule may be shifted towards the continuum boundary. When the particles fly apart, the level system must return to that of the isolated particles. However, as a result of transitions of electrons from the shifted levels to empty levels, excitation and ionization may occur both in the quasi-molecule itself and during decay of autoionization states after separation of the particles. In the subsequent works, it was asserted that it is this system dynamics that accounts for the large cross sections of the inelastic processes and, also, the participation of many electrons in the elementary processes occurring during collisions.

Based on the results obtained at the Ioffe Institute, Fano and Lichten [3] made further progress in analyzing the quasimolecule level features in detail. They could simplify essentially the problem and showed that the transitions of inner electrons in the complex many-electron systems can be described on the basis of an analysis of an extremely simple system—a single electron in the field of two Coulomb centers in a wide range of internuclear distances. This made possible using two important features of such a system: the energy levels of the system can be calculated accurately and may cross regardless of their symmetry. Figure 2 shows a system of levels of individual electrons calculated for an Ar–Ar quasi-molecule at various internuclear distances by analogy with the system of H_2^+ levels. On the basis of this model, origin of the experimentally discovered 2p vacancies at slow collisions of Ar ions and atoms is explained by a strong shift of one of the levels formed from the 2p level of Ar atom, namely, 4fσ, crossing in the process a number of empty levels.

The experimental Auger-spectroscopy of the quasi-molecule advanced and originally realized at the Ioffe [4] was of importance for the development of the quasi-molecular model. For quasimolecular spectroscopy Auger electrons emitted in autoionization transitions in an excited quasi-molecule at certain values of internuclear distance were analyzed. For this purpose, spectra of Auger electrons were measured via their coincidences with ions escaping at a fixed angle, i.e., at a prescribed minimal distance of close approach between the nuclei of the colliding particles. In such a way, the energies and widths of the levels of the quasi-molecule were determined. Thus, towards the early 1980s, two main models describing the atomic collisions—the model of impact interaction at higher velocities and the quasi-molecular model for velocities lower than characteristic velocities of the electrons in an atom—were formulated. In their terms, the ionization was explained either as a direct expulsion of an electron by impinging particles or through excitation of an autoionization state decaying in a quasi-molecule during the collision or in isolated atomic particles after their separation. This, however, gave no answer to an essential question—how the observed ionization can be explained in those cases when the collision velocity is low and the system has no autoionization

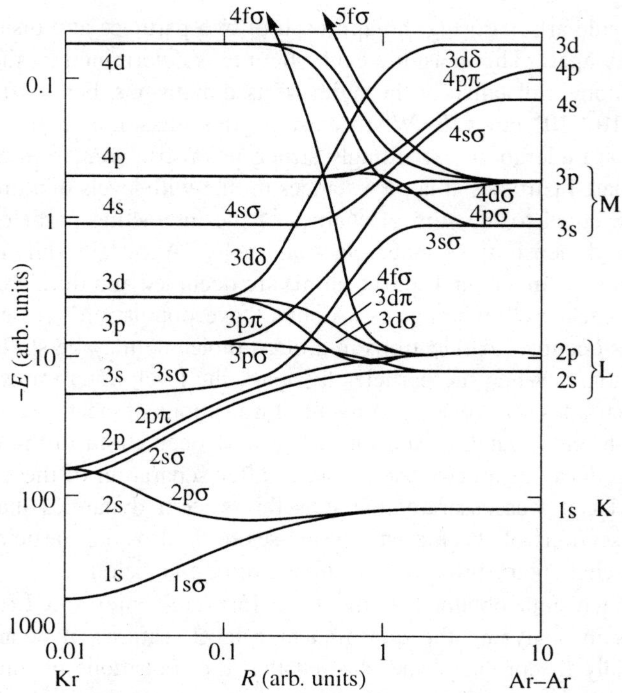

Fig 2. Dependence of energy levels of individual electrons (molecular orbitals) in a system formed by colliding particles—Ar–Ar quasimolecule—on the internuclear distance R. Threshold values of levels: at $R \to \infty$—"separated Ar atoms", $Z_{SA} = 18$; at $R \to 0$—"united Kr atom", $Z_{UA} = 18 + 18 = 36$.

states (e.g., in slow collision of a proton with a hydrogen atom).

The research made in collaboration between the Ioffe and FOM-Institute for Atomic and Molecular Physics (Amsterdam) in the 1980s [5] showed that another important ionization mechanism—direct ionization due to exit of a quasi-molecular level into continuum—does really exist. The direct ionization may be most successfully described by analyzing the quasi-molecule energy surfaces in the complex plane of internuclear distances [6]. An example of such a surface is presented in Fig. 3 where one can see that motion over this surface permits a system to be transferred from one bound state to another and to escape into continuum.

The development of the direct ionization model accomplished within the framework of a joint program of the Ioffe and the Oak-Ridge National Laboratory (USA) [7] produced a solution to the ancient problem of accounting for electron transitions at any velocity. The developed method is based on expanding the wave function of the system in a set of eigenfunctions of the Sturm–Liouville problem, rather than by the traditional atomic or molecular formalism. This technique opens up possibilities of tackling with a more general task—the dynamic many body problem of quantum mechanics, which allows it to be used not only in atomic collisions but, also, in the cases of electron impact and photoionization. With the use of the original experimental methods developed at the Ioffe, a number of pioneer works on determination of cross sections of elementary processes have been accomplished to make up a database for adjacent fields of science

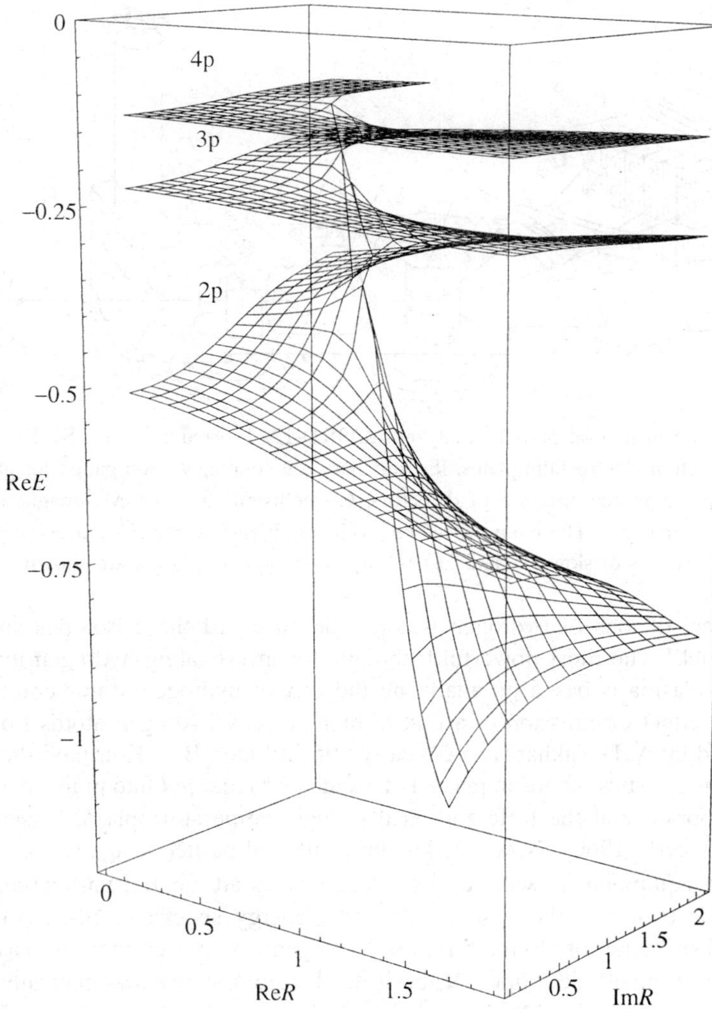

Fig 3. Real part of potential energy of H^+–H system against complex internuclear distance.

and formed a basis for some applications. This is exemplified below by two applications which have been mainly developed at the Ioffe and became independent areas of research.

3 High temperature plasma diagnostics using atomic collisions

3.1 Diagnostics of hydrogen ions

It is well known that studies of high temperature plasma are, in the first place, related to works on controlled fusion (CF). Any successful development of these works is impossible without effective methods of plasma diagnostics to determine its parameters: temperature, density, composition, confinement time, energy, etc. Of particular interest

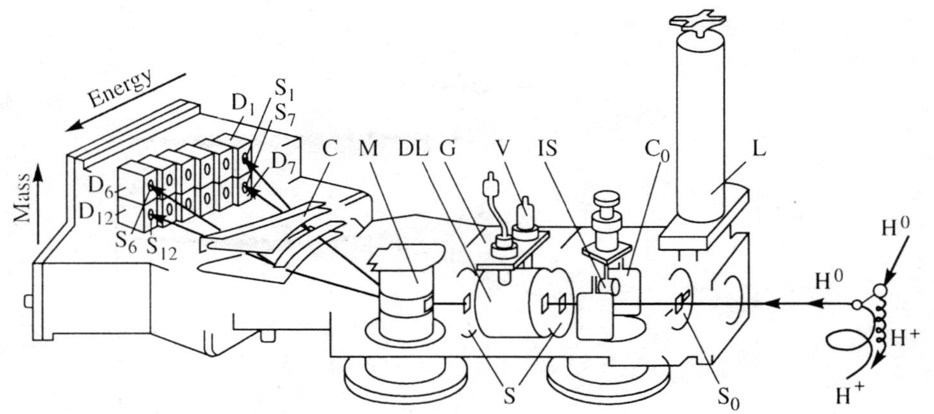

Fig 4. Schematic of neutral particle analyzer ACORD-12. L–vacuum valve, S_0, S—collimating slits, C_0—deflection electrostatic plates, IS—auxiliary ion source, V—ion gauge for gas pressure control, S_1–S_{12}—entrance aperture of detectors, G—collision chamber, M—magnetic analyzer, C—electrostatic analyzer. The main processes producing high-velocity H^0 hydrogen atoms and a scheme of trajectories of slow H^+ and fast H^+ hydrogen ions in plasma are shown.

are parameters of ions of hydrogen isotopes because just these isotopes are a "thermonuclear fuel." The most powerful technique for investigating hydrogen ions in high temperature plasma is based on analyzing the flux of hydrogen atoms emitted by the plasma. The effect of emission of a flux of high-velocity hydrogen atoms from plasma was predicted by A. D. Sakharov in the early 1950s. Later, B. P. Konstantinov proposed to use it for diagnostics of ions in plasma. This idea was first put into practice in N. V. Fedorenko's laboratory at the Ioffe and used in high temperature plasma installations in the late 1950–early 1960s. Now it is known as neutral particle diagnostics.

The main equipment in service of this diagnostics are neutral particle analyzers—devices aimed to detect the absolute intensity, energy spectrum, mass composition, and spatial distribution of atomic fluxes of hydrogen isotopes emitted by plasma. Such devices were originally developed at the Ioffe. Figure 4 shows schematically a neutral particle analyzer of this kind [8], intended to analyze simultaneously fluxes of hydrogen and deuterium atoms in the energy range 100–50 000 eV. High-velocity atoms produced in plasma by charge exchange or recombination and entering the analyzer are ionized by collision with gas in a stripping cell G and analyzed in parallel magnetic and electric fields created by magnet M and capacitor C. The energy spectrum of hydrogen atoms is detected by a lower row of channeltrons used as detectors D_7–D_{12} of secondary ions, whereas the upper row D_1–D_6 deals with those of deuterium. With three rows of detectors, a simultaneous analysis can be performed of all three isotopes—hydrogen, deuterium, and tritium.

The most important result obtained by neutral particle diagnostics in the 1970s was that the neoclassical nature of the ion energy balance in most tokamaks and stellarators was established [9, 10].

In recent years, another important problem of atomic analyzer application has arisen as to how the density ratio of different hydrogen isotopes in a plasma is to be measured. The hydrogen isotope ratio in the plasma should be controlled to optimize the operation of a thermonuclear reactor. Measuring the ratio of deuterium and tritium fluxes with an atomic analyzer can provide the ratio of deuterium and tritium densities in plasma core. At present a system for controlling the isotopic composition of plasma for the project of the international fusion reactor ITER [11] is being developed on the basis of this diagnostics.

3.2 Diagnostics of products of fusion reactions—α-particles—in thermonuclear plasma

Investigation of the behavior of high energy α-particles in thermonuclear deuterium–tritium plasma and development of diagnostic methods for this purpose is of top priority for fusion reactor design. The α-particles are produced by the reaction

$$D^+ + T^+ = {}^4He^{2+}\,(3.5\,\mathrm{MeV}) + n\,(14\,\mathrm{MeV}) \qquad (1)$$

having a slowing down time of order of 1 s in case of classical slowing down. During this time interval they are assumed to transfer a larger portion of their energy to the plasma, providing for a self-sustaining thermonuclear burning. However, α-particles may be lost from the plasma owing to instabilities of various kinds, arising in the process of slowing down. In this case, the thermonuclear burning becomes less effective, demanding an extra supply of heating power into the plasma, and the reactor efficiency decreases progressively.

To control the behavior of α-particles, their energy distribution should be measured. For this purpose, a high-energy neutral-particle analyzer GEMMA-2 [12], which can be used for investigating neutral α-particles in the 0.5–4.0 MeV range, has been developed at the Ioffe. An essential feature of this analyzer is its ability to operate under conditions of a high neutron background inherent in deuterium–tritium plasma installations. GEMMA-2 analyzers were supplied from the Ioffe Institute to the three largest world's tokamaks—JET (England–EURATOM), TFTR (USA), and JT-60U (Japan)—and used there in intensive investigations of α-particles and other ions of MeV energies in plasma.

On TFTR tokamak a GEMMA-2 analyzer was used in combination with injection of Li pellets producing a cloud of helium-like ions neutralizing α-particles. The clouds moves across the plasma, which enables measuring the spatial distribution of the α-particle density. In Fig. 5 typical radial distributions of α-particles, arising in D–T fusion reactions, are presented for two given energies before and after crash occurring in the process of so-called "sawtooth" oscillations intrinsic to tokamaks [14].

One can see that the pre-crash distribution is of a classical type, but after the crash α-particles are redistributed along the major radius of the torus. It is of importance to control this phenomenon and to take it into consideration when running a thermonuclear reactor, since the outflow of α-particles from the plasma center, resulting from the sawtooth oscillations, can essentially upset the thermonuclear burning.

The experiments on measuring the isotopic ratio in plasma and studies of thermonuclear α-particles, carried out recently with the use of neutral particle diagnostics, have placed this diagnostics among the most advanced techniques for controlling future thermonuclear reactors and, in particular, the international thermonuclear reactor ITER which is under design now.

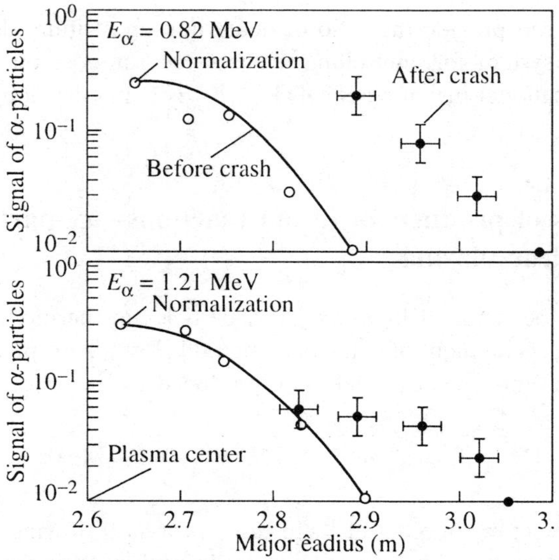

Fig 5. Radial distributions of α-particles in TFTR tokamak: before sawtooth crash (○), after crash (•), calculation assuming classical slowing down without any losses and without sawtooth oscillations (full curves). Calculated and experimental data are normalized at $R = 2.64$ m.

4 Atomic collisions in solids

4.1 Diagnostics of element composition and crystal structure

Investigations of atomic collisions in solids are stimulated, first of all, by the need for developing diagnostic methods for solids and surfaces. The possibility of widely using data on atomic collisions in studying the parameters of solid state samples arises since particle beams are convenient tools for exciting the states inherent to a large ensemble of particles, which constitutes part of a sample. It is no less important that a number of effects arising in a sample are determined by pair collisions of impinging particles with those of a target. The reason for this analogy is that a number of inelastic processes occur in close encounter of these particles when other particles in solid target are far away and their influence is small or at all negligible. So, the major problems consisted in studying atomic collision mechanisms in solids and creating new methods of diagnostics.

One of examples is the development of a method for quantitative nonstandard analysis of the elemental composition of solids. The presently widely used method of secondary ion mass spectrometry (SIMS), using sample sputtering with a beam of fast particles, exhibits a record-breaking sensitivity to impurities (as high as 10^{-9}) and enables layer-by-layer analysis of a sample with high depth resolution, but is not quantitative because of a strong dependence of secondary ion yield on sample composition and surface conditions. The fraction of ions in a total flow of sputtered particles of the same element may vary several orders of magnitude. Yet in stationary sputtering the formation rate of neutral particles is practically the same for different elements—sample components. Therefore, a method of secondary neutrals mass spectroscopy (SNMS) was developed in which the sputtered neutral particles are ionized by electron impact before analysis [15]. The feasibility of a quantitative analysis in the mass spectrometry of

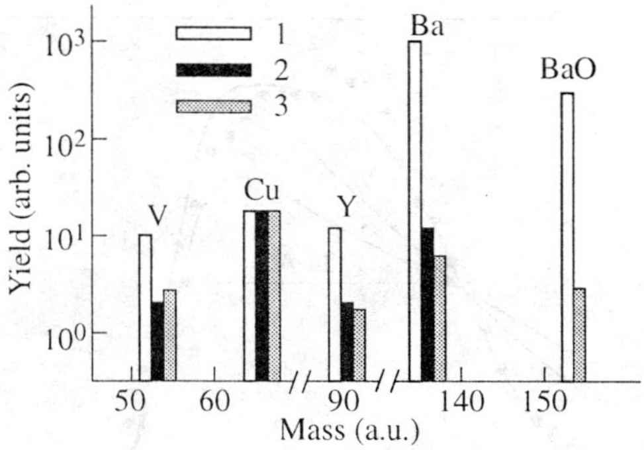

Fig 6. Data of HTSC-$Y_{0.5}V_{0.5}Ba_2Cu_3O_7$ ceramics composition, obtained by analyzing particle flux sputtered by Ar^+ ion beam. 1—stoichiometric composition, 2—by ion flux, 3—by atom flux.

neutral particles is ensured by separation of the emission and ionization of the sputtered particles. Contrary to the SIMS in which ionization is a result of a complicated interaction between a departing particle and the surface, the efficiency of neutral particle ionization is determined by the conditions preset by an experimenter.

The capabilities of secondary neutrals mass spectrometry are illustrated in Fig. 6 which shows mass spectra of atoms and ions sputtered in ion bombardment of HTSC-ceramics. One can see that yield of atoms reflects the composition adequately, while discrepancies in ion yield are as high as two orders of magnitude.

Another example of application of atomic collisions is the diagnostics based on investigating the scattering of the impinging ions themselves experiencing collisions in the subsurface layer of a sample. It is well known that the composition and perfection of the crystal structure of micrometer-thick layers of solids have been studied with advantage by the Rutherford backscattering method based on probing a sample with ions of MeV energies with the backscattered ions registered and their energy analyzed. This method provides a depth resolution of 10–20 nm.

However, the development of microelectronics, involving transition to submicron films and multilayered structures, demands a much higher depth resolution. This problem was solved by using medium-energy ions (100–200 keV). In this case, the energies of backscattered ions were analyzed with electrostatic analyzers which give much higher resolution in energy with the high solid angle of the instrument preserved. With He^+ ions of 100-keV energy, the depth resolution is as high as 0.5 nm for the subsurface layer [16].

One more important advantage of the method of medium-energy ion scattering is that the large backscattering cross sections ($\sigma \sim E^{-2}$) enable considerable reduction of the radiation loads on the samples under investigation.

In Fig. 7, the Ba peak in the energy spectrum of 100 keV He^+ ions scattered by a $YBa_2Cu_3O_7$ film about two lattice constant thick (2.5 nm) illustrates the high depth resolution. Experimental points are compared with the model (full curves) obtained under assumption of layer-by-layer growth (curve 1) and pyramidal island growth (curves 2 and 3). With the total amount of material deposited onto a substrate preserved, comparison of experimental data with model calculations shows with certainty that in this case there is a layer-by-layer film growth at early stages.

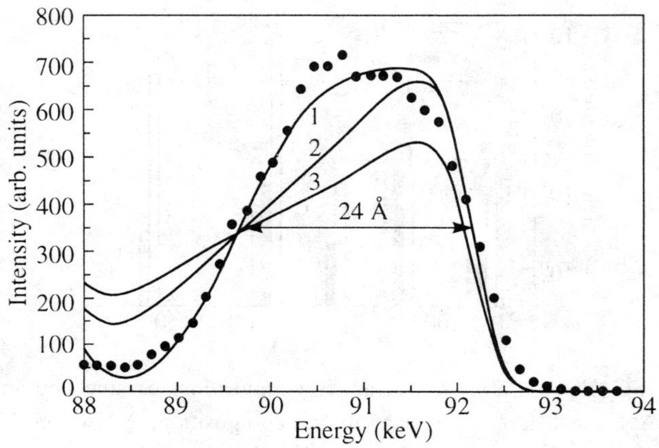

Fig 7. Part of energy spectrum of He^+ ions with initial energy $E = 100$ keV backscattered from a 2.5-nm-thick $YBa_2Cu_3O_7$ film: experiment (•), model calculation assuming layer-by-layer growth (curve 1), model calculations for various forms of island growth for three-layer pyramids (curve 2), and for the 4-layer pyramids (curve 3).

In this method, the characteristic X-radiation emitted when vacancies created by the probing ions in the inner shells of atoms of a sample are filled, is recorded along with the backscattered ions. Recording the characteristic radiation enables investigation of light impurities in heavy matrices, which is commonly a complicated problem for the Rutherford backscattering method.

The crystal lattice perfection is generally studied in the canalization regime when a sample is oriented so that the ion beam direction coincide with a given low-index crystallographic axis. In this case, the flux of backscattered ions and quanta of the characteristic radiation decreases markedly (30–50-fold for "ideal" single crystals). Measuring the fluxes of ions and quanta as a function of the angle between the incident ion beam and a low-index crystallographic axis $\chi(\psi)$ gives two parameters: full width of a curve at half maximum $2\psi_{1/2}$ and the minimum yield χ_0, i.e., fluxes of ions and quanta at $\psi = 0$. Comparing the measured widths of the angular dependence $2\psi_{1/2}$ and the minimum yield χ_0 with analogous values for an ideal crystal one can determine the misorientation angle of the crystallites and the fraction of polycrystals in a sample [17].

Thus, a comparison of the angular functions of the characteristic radiation yield for various elements gives a possibility to evaluate the imperfection of the corresponding sublattices and to reveal significant differences in chemical composition between polycrystalline impurities and highly oriented phase. In our experiments the angular functions of the yield for the oxygen K-lines were measured for the first time, which opened up prospects analyzing the oxygen sublattice imperfection affecting significantly HTSC film properties.

It should be noted that all these characteristics are obtained in a single experiment without sample destruction (for HTSC films, the temperature of the superconducting transition and the critical current are retained).

An important problem is the formation of charged states of the particles as a result of ion collisions with surface.

A knowledge of mechanisms responsible for particle charge state formation as a result of ion collisions with surface is of importance in such fields as thermonuclear plasma interaction with materials of the first wall and preparation and analysis of thin films. Electron exchange between an atomic particle and the solid surface is the main process producing a particle charge in the energy range from several eV to several keV. In spite of a great number of experimental and theoretical works performed in this field, there is as yet no generally accepted opinion about the mechanisms responsible for particle charge state formation.

To study the formation of charged states of particles ejected under ion bombardment of solids, a method has been used [15] which enabled, for the first time, measuring the energy distributions of neutral atoms (ionized by electron impact before analysis) and positive and negative ions simultaneously. Experimentally, the velocity dependences of ion and atom formation probabilities have been investigated for 12 metals with different ionization potentials and electron affinity energies. The results obtained [18] enabled evaluation of the validity of various theoretical approximations, since the time of interaction in a "particle-surface" system and, consequently, the particle velocity is a crucial factor in charge formation, involved in all the theoretical models. It was found that the probability of positive ion formation P^+ is described adequately by an exponential dependence on velocity ν: $P^+ = \exp(-\nu_0/\nu)$. Such a dependence is predicted by the model of "electron tunneling" developed in the late 1970s. This model considers the secondary ion formation to result from a single-electron resonance transition between the level of a particle and the conductivity band of a metal. At the same time, the results obtained with negative ions contradict drastically the predictions of the model of single-electron transition; hence, they indicate a many-electron character of electron interchange. It has been shown that the main mechanism of secondary negative ion formation is a two-electron transition of Auger type in which two electrons of a metal occupy, accordingly, the ionization level and the level of electron affinity of a flying-away particle.

5 Analysis of electron structure of clusters

In recent years, an ample program of investigations of the electron structure of nanometer-scale objects—thin films, interfaces, clusters, and fullerenes—has been accomplished at the Ioffe Institute. One of the research areas was the investigation of the evolution of the electron structure in going from atoms to clusters of various sizes and further to solids. This problem is of fundamental physical interest and also of importance for development of novel types of nanostructures and evaluation of their properties.

The electronic structure of clusters has been investigated by analyzing the energy spectra of photoelectrons and Auger electrons and, also, the spectra of energy losses in electron collisions with complexes containing various number of atoms [19]. In Fig. 8, changes in the binding energy of the core level 3d and $M_{4.5}$ VV Auger electrons are shown as functions of the number of atoms in silver clusters N_A on the inert graphite surface. It turns out that already at the early stage of cluster formation, with $N_A = 2$–4, there occur significant changes in the core level energy, Auger electron energy being almost constant. It might be that this effect is related to a promotion of the quasimolecular orbitals 5g6 and 6h6 in combining of atoms into clusters. In further growth of cluster dimensions, the most important are changes in the binding energy of valence electrons. Investigations of the plasmon excitation in clusters [20] have shown

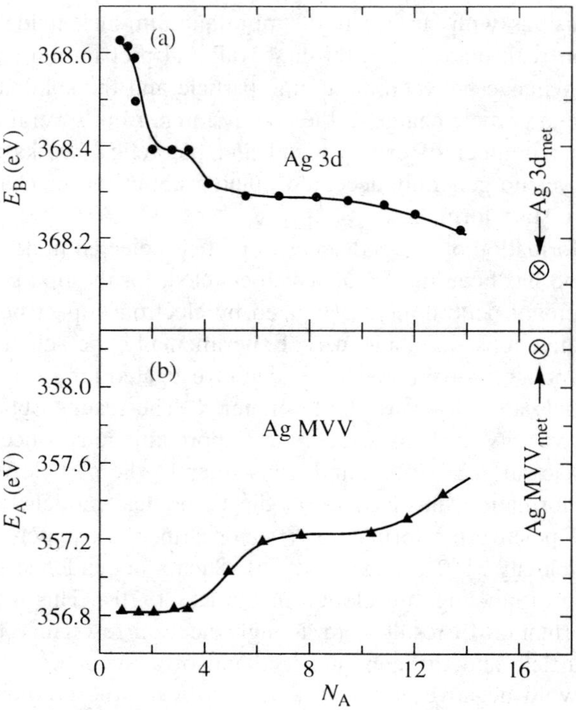

Fig 8. (a) Binding energies of bulk Ag 3d level and (b) kinetic energy of Auger Ag MVV electrons versus the number of atoms in silver clusters. Separate points denote the respective energies for bulk silver metal.

that this process is not of threshold character, with plasmon excitation observed even in clusters of 2–4 atoms. On the whole, silver cluster characteristics are similar to those of silver metal when the number of atoms in a cluster is as high as $N_A = 20$.

Similar investigations of As clusters in GaAs bulk and Cu clusters in the bulk of a Cu/SO_2 composite film have shown that in all cases under investigation the electron properties of clusters and the corresponding metals are much the same at cluster dimensions of the order of 10 Å. It is these small clusters that are of interest for a number of applications and, in the first place, for preparation of new materials and structures for nanoelectronics.

References

[1] N. V. Fedorenko and V. V. Afrosimov, *Zh. Tekh. Fiz.* **26** 1941 (1956).

[2] V. V. Afrosimov, Y. S. Gordeev, M. N. Panov, and N. V. Fedorenko, *Sov. Phys. Tech. Phys.* **9** 1248 (1965);
ibid. 1256; *ibid.* 1265;
V. V. Afrosimov, Yu. S. Gordeev, A. M. Polyansky, and A. P. Shergin, *Sov. Phys. JETP* **30** 441 (1970).

[3] U. Fano and W. Lichten, *Phys. Rev. Lett.* **14** 627 (1965);
W. Lichten, *Phys. Rev.* **164** 131 (1967).

[4] V. V. Afrosimov, G. G. Meskhi, N. N. Tsarev, and A. P. Shergin, *Sov. Phys. JETP* **57** 434 (1983).

[5] P. H. Woerlee, Yu. S. Gordeev, H. de Waard, and F. W. Saris, *J. Phys. B* **14** 527 (1981).
[6] S. Yu. Ovchinnikov and E. A. Solov'ev, *Comments on Atomic and Molecular Physics* **XXII** 69 (1988).
[7] S. Yu. Ovchinnikov and J. H. Macek, *Phys. Rev. A* **55** 3605 (1997).
[8] A. B. Izvozchikov, M. P. Petrov, S. Ya. Petrov, F. V. Chernyshev, and I. V. Shustov, *Sov. Phys. Tech. Phys.* **37** 201 (1992).
[9] V. V. Afrosimov and M. P. Petrov, *Plasma Diagnostics*, ed. M. I. Pergamet, Moscow, Energoizdat, vol. 4(1), p. 167, (in Russian) 1981.
[10] V. V. Afrosimov and M. P. Petrov, *Plasma Diagnostics*, Moscow, Energoatomizdat, vol. 5, p. 135 (in Russian) 1986.
[11] A. I. Kislyakov, V. I. Afanassiev, A. V. Khudoleev, S. S. Kozlovskij, and M. P. Petrov, *Diagnostics for ITER* Workshop in Varenna (Italy) 1997.
[12] A. I. Kislyakov, A. V. Khudoleev, S. S. Kozlovskij, and M. P. Petrov, *Fusion Eng. Des.* **34–35** 107 (1997).
[13] S. S. Medley, D. K. Mansfield, A. L. Roquemore, R. K. Fisher, H. H. Duuong, J. M. McChesney, P. B. Parks, M. P. Petrov, A. V. Khudoleev, and N. N. Gorolenkov, *Rev. Sci. Instrum.* **67** 3122 (1996).
[14] M. P. Petrov, N. N. Gorelenkov, R. V. Budny, Z. Chang, D. S. Darrow, H. H. Duong, R. K. Fisher, R. J. Fonk, H. W. Herrmann, G. R. MkKee, *et al. Fusion Energy 1996, Proc. IAEA Fusion Energy Conf.* (IAEA Vienna) vol. 1, p. 261, 1997.
[15] B. N. Makarenko, A. B. Popov, A. A. Shaporenko, and A. P. Shergin, *Pis'ma Zh. Tekh. Fiz.* **14** 609 (1988).
[16] V. V. Afrosimov, G. O. Dzyuba, R. N. Il'in, M. N. Panov, V. I. Sakharov, and I. T. Serenkov, *Preprint PhTI: Leningrad*, No. 1395, p. 20, 1989.
[17] V. V. Afrosimov, G. O. Dzyuba, R. N. Il'in, M. N. Panov, V. I. Sakharov, and I. T. Serenkov, *Phys. Tech. Phys.* **41** 76 (1996).
[18] A. P. Shergin, and Yu. S. Gordeev, *Proc. ESCAMPIG-92*, St. Petersburg, p. 37, 1992.
[19] Yu. S. Gordeev, M. V. Gomoyunova, A. K. Grogorjev, V. M. Mikoushkin, I. I. Pronin, S. E. Sysoev, V. V. Shnitov, and N. S. Faradzhev, *Phys. Solid State* **36** 1298 (1994).
[20] Yu. S. Gordeev, M. V. Gomoyunova, V. M. Mikoushkin, I. I. Pronin, and S. E. Sysoev, *Phys. Solid State* **36** 973 (1994).

Study of ohmic H-mode in the TUMAN-3 tokamak

V. E. Golant and S. V. Lebedev

1 Introduction

Substantial energy loss from plasma is an essential obstacle to accomplishing controlled fusion in tokamaks. The reason for fast heat loss is a high level of turbulence due to growth of various microinstabilities. Until recently, no effective methods of instability stabilization were found, with the plasma confinement being much worse than that predicted by the theory of transport in magnetically confined plasma. Another difficulty from the reactor point of view is the growth of turbulence with increasing power input. This results in an unfavorable dependence of the energy confinement time τ_E on heating power P. According to the present views $\tau_E \propto P^{-\alpha}$, where $\alpha = 0.5–0.7$. Such a dependence prevents raising the energy storaged in plasma by increasing the heating power. Hence, with no effective method of turbulence suppression available, the only possibility to reach the fusion parameters is to enlarge of the device dimensions. As a result, the thermoisolation necessary for reactor purposes can be achieved only in extremely large devices. Typical dimensions of an experimental tokamak-reactor under consideration are the following: major torus radius $R_0 = 8.14$ m, minor radius $a_{\mathrm{scp}} =$ 2.8 m [1].

A significantly improved confinement in a tokamak was gained with the High-confinement mode (H-mode) discovery [2]. The H-mode involves an essential reduction of turbulence at the plasma periphery. This suppresses transport on the periphery and, consequently, essentially increases τ_E. Since its discovery the H-mode attracted attention, being, firstly, a promising scenario of a reactor-tokamak, and, secondly, an interesting and complicated phenomenon which needs gross experimental and theoretical efforts to be well understood.

The H-mode was discovered in experiments on high-power auxiliary heating with neutral-atom beams in the ASDEX tokamak with a complicated magnetic configuration including a divertor [2]. Further, a similar improvement was revealed under RF heating by means of ion cyclotron [3] and electron cyclotron waves [4]. Our experiments [5] and also the experiments on the DIII-D tokamak [6] with ohmic heating without any auxiliary heat sources have shown that the possibility of the obtaining H-mode is not related to any specific features of the employed heating schemes.

In the cases mentioned above, excluding ours, improved confinement has been observed in tokamaks with a magnetic divertor which influences essentially the properties of the plasma periphery. The fact that the H-mode was obtained in the TUMAN-3 tokamak with no divertor evidences that magnetic configuration is not a crucial factor

Table 1. Main parameters of the TUMAN-3 tokamak.

	Parameter ranges	Typical parameters in ohmic heating regime
Major radius, R_0, m	0.44–0.64	0.53
Minor radius, a_l, m	0.11–0.24	0.22
Aspect ratio, $A = R_0/a_l$	2.3–5.8	2.4
Longitudinal magnetic field, B_t, T	0.4–1.2 (2.0†)	0.68
Plasma current, I_p, kA	40–180 (250†)	150
Discharge duration, τ_p, 10^{-3} s	40–80 (200†)	80
Average density, $\bar{n}$, 10^{19} m^{-3}	0.4–6.2	1.0–3.5
Central electron temperature, $T_e(0)$, eV	150–800 (1000‡)	500
Central ion temperature, $T_i(0)$, eV	80–200 (400‡)	180
Power of ion cyclotron heating generator, P_{RF}, MW	< 1.0	—
Parameters of injected pellets of H and LiD:		
velocity, V_{pel}, m s^{-1}	70–800	—
sore, d_{pel}, μm	100–1200	—

† Post modernization parameters of the tokamak (TUMAN-3M),
‡ Parameters obtained with auxiliary ion cyclotron heating.

Table 2. Main diagnostic tools in TUMAN-3.

Diagnostic tool	Parameters to be measured
Magnetic probes, loops and Rogovskii's coils	Loop voltage, U_p Plasma current, I_p Plasma loop position, ΔR Stored energy, W Magnetic field fluctuations, $\partial B/\partial t$
Microwave interferometer, $\lambda = 2$ mm, 10 channels	Radial distribution of electron density, $n_e(r)$
Thomson scattering of laser radiation	Radial distribution of electron temperature, $T_e(r)$
Neutral particle analyzer	Radial distribution of ion temperature, $T_i(r)$
Detectors of soft X-ray radiation	Temporal evolution of X-ray radiation and electron temperature, $I_{SXR}(t)$, $T_{e0}(t)$
Detectors of hard X-ray radiation	High energy electron spectrum, $f_e(v)$
Microwave reflectometer	Intensity of density fluctuations at the periphery, $\tilde{n}_b$
Microwave radiometer	Electron temperature, energy and density of runaway electrons, $T_e(t)$, $E_{tail}(t)$, $n_{tail}(t)$
Bolometer	Power of radiation losses, P_{RAD}
Spectroscopy of visual and ultraviolet radiation	Densities of working gas, impurities, effective charge of plasma, n_0, n_i, Z_{eff}
Multielectrode electrostatic probes	Plasma density, fluctuations and convection flux at the periphery, n_b, $\tilde{n}_b$, Γ_b
Heavy Ion Beam Probe diagnostics	Radial distribution of plasma potential, Φ_p

governing the transition, which in its turn, gives an additional evidence, in favor of the universality of the mechanism of confinement improvement in the H-mode. Experiments described here have been carried out on the TUMAN-3 tokamak working in the Ioffe Physico-Technical Institute for years [7]. The main parameters of the tokamak are shown in Table 1. An important feature of the device geometry which helps in H-mode investigations is the small aspect ratio $A = R_0/a_l$. It allows one to obtain relatively high plasma current I_p at low toroidal field B_t and, thus, to increase ohmic power input. Another feature of TUMAN-3 is a versatile power supply which allows operation scenarios with rapid changes in I_p and B_t and enhances the possibilities for influencing on the plasma periphery. The device is equipped with a proper set of diagnostic tools ensuring reliable measurement of a number of plasma parameters. The main diagnostic tools employed in the TUMAN-3 tokamak are listed in Table 2.

An experimental investigation of the ohmic H-mode in the TUMAN-3 tokamak revealed some specific features of the H-mode and elucidated the physical mechanisms underlying the confinement improvement effect. In particular, the key role of the radial electric field in the Low-High (L-H) transition, prerequisites to a spontaneous transition, location of reduced transport region, and parametric dependences of the energy confinement time have been established and various methods to trigger the H-mode have been proposed and realized.

2 Ohmic H-mode signatures

After the first experiment on H-mode triggering it was understood that the transition has evident features of bifurcation. Actually, under nearly the same conditions both low and high confinement regimes may be realized. In Fig. 1 typical temporal dependences of the main plasma parameters in a discharge with transition into the ohmic H-mode are

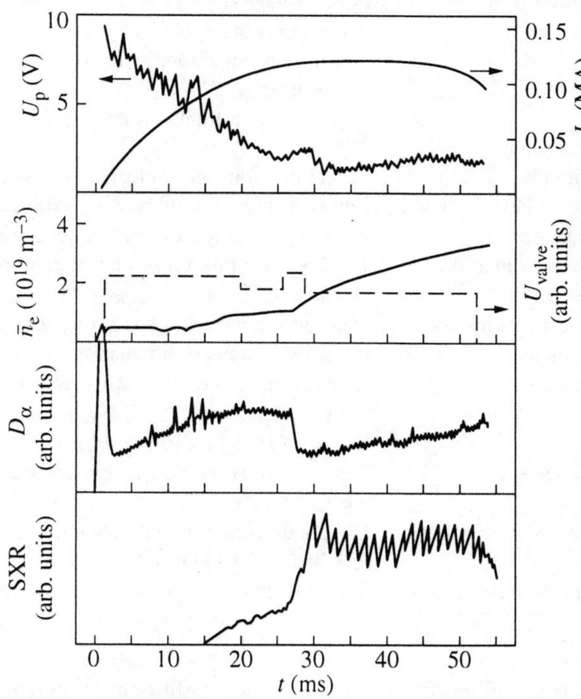

Fig 1. Behavior of the plasma current I_p, loop voltage U_p, average density of electrons $\bar{n}_e$, intensity of spectral line D_α, and soft X-ray emission SXR in a discharge with a transition to ohmic H-mode after 27 ms.

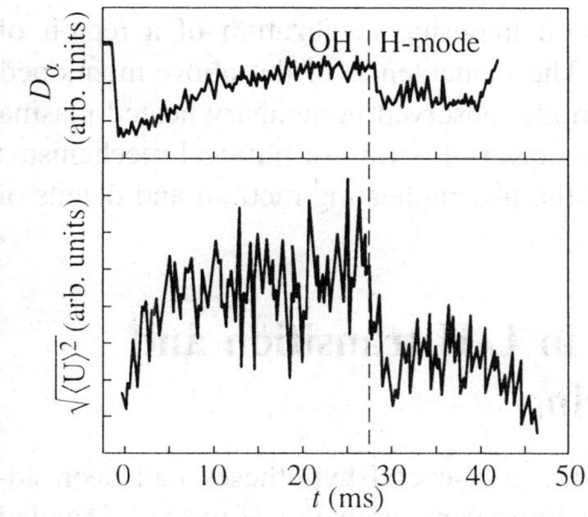

Fig. 2. Averaged amplitude of density fluctuations measured by reflectometer with probing frequency of 26 GHz in the $f > 200$ kHz part of the spectrum measured during transition into ohmic H-mode. In upper window—D_α radiation.

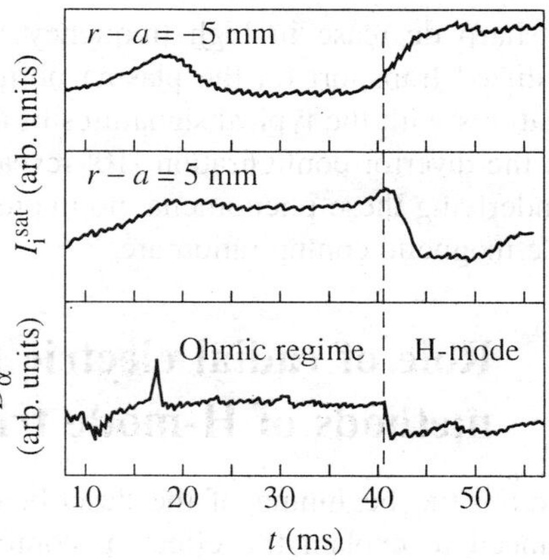

Fig. 3. Evolution of the ion saturation current onto electrostatic probe placed inside (upper window) and outside (middle window) the last closed magnetic surface ($I_i^{sat} \propto n_e$).

shown. After 27 ms of discharge a rapid growth of plasma density begins. Simultaneosly D_α radiation which is proportional to the source of charged particles drops drastically. These features indicate a rapid and large increase in the particle confinement time. This phenomenon, revealed in the first experiment in TUMAN-3, was interpreted as a transition into the ohmic H-mode [5]. However, even at very close initial parameters this transition may not arise. In this case, the discharge remains in the conventional (low) confinement mode. A transition having occurred the discharge, as a rule, proceeds in the H-mode until the density rises to limiting values.

The transition improves the confinement of not only particles but energy as well. As a result of the improvement of energy confinement the electron temperature increases after the transition, which is illustrated by an increase in the SXR radiation, shown in Fig. 1. Measurements of fluctuation intensities by reflectometry have shown that the transition is characterized by a sharp suppression of electrostatic turbulence on the plasma periphery. It should be noted that only high-frequency fluctuations with $f_{fluct} > 200$ kHz are suppressed, as no amplitude changes appear in the low frequency part of the spectrum [8].

One more signature of the discovered mode is that the confinement improved on the plasma periphery. This was established when measuring the plasma parameters with electrostatic probes. It was shown that a region of significant density gradient appears near the last closed flux surface (LCFS). The plasma density proportional to the ion saturation current, decreases outside and increases inside the LCFS [9]. The results of these measurements are presented in Fig. 3. By processing the above data, a density decay length was found to be 1.2 cm before the transition and to decrease to 0.3 cm in the ohmic H-mode. Measurements with electrostatic probes in this region confirmed the occurrence of the radial fluctuation driven flux decrease [9].

Thus, the first experiments with ohmic H-mode revealed the following features of this regime: bifurcation character, growth of the energy and particle confinement times, a sharp decrease in high frequency fluctuation intensity, localization of a region of reduced transport on the plasma periphery. The coincidence of the above-mentioned features with the typical signatures of the H-mode, observed in auxiliary heated plasma in the divertor configuration [10] revealed the universal nature of physical mechanisms underlying these phenomena, no matter what the plasma heating method and details of the magnetic configuration are.

3 Role of radial electric field in L-H transition and methods of H-mode triggering

Before the beginning of the described experiments, several hypotheses had been advanced to explain the effect of confinement improvement in the H-mode. Detailed analysis of the theoretical models is beyond the scope of the present paper. A critical review of various theoretical approaches can be found in [11]. Here, it is appropriate to describe in brief only the model which was, in our opinion, the most plausible and was confirmed in our further experiments. According to this model, at different rotation velocities of radially adjacent plasma layers their characteristic turbulence must be suppressed by decorrelation [12]. Such sheared rotation can arise in a plasma as result of a $\mathbf{E} \times \mathbf{B}$ drift. For a sufficient nonuniformity of the electric field in radial direction, the rotation appears to be essentially sheared and fluctuation spectra in the adjacent layers appear to be shifted, which results in decorrelation. As a reason for the appearance of sheared electric field, various physical mechanisms have been proposed, including orbit ion losses and nonambipolar radial drift in the periphery region [11].

To verify the above hypothesis of L-H transition, experiments on artificial generation of a radial electric field have been carried out on TUMAN-3. For this purpose an electrode with isolated head was inserted into a plasma to a distance of $\sim$ 4 cm inside the LCFS, so that radial field could be generated by applying a bias to this electrode [13]. The experiments have shown that the transition to the H-mode occurs when a sufficient bias of any polarity is applied to this electrode. The behavior of different plasma parameters with positive and negative electrical field excited by this electrode is shown in Fig. 4. One can see that for the negative (directed into plasma bulk) field the derivatives of density and soft X-ray signal are larger, which indicates a more pronounced improvement of energy and particle confinement in this case. Calculations of the particle confinement times confirmed the substantial increase in τ_p after artificially driven transition and the more pronounced effect of negative radial field on the transport. In the initial regime $\tau_p = 2.5$ ms, in the H^--mode (negative bias) $\tau_p = 7.9$ ms, and in the H^+-mode (positive bias) $\tau_p = 4.3$ ms.

A detailed comparison has shown that the main plasma parameters in the H^--mode and in ohmic H-mode are the same. This point became the basis to assume inward direction of radial field in the spontaneously arising ohmic H-mode. Further verification of this assumption was the termination of the H-mode by a positive bias, also realized in our experiments, see Fig. 5. In this case, the transition to enhanced confinement regime occurs spontaneously after 20 ms, then, after a positive biasing at 25 ms the density growth cuts off, D_α light emission increases, and soft X-ray intensity starts decrease: the ohmic H-mode is switched off.

Under rather different conditions, similar results have been obtained in the CCT [14]

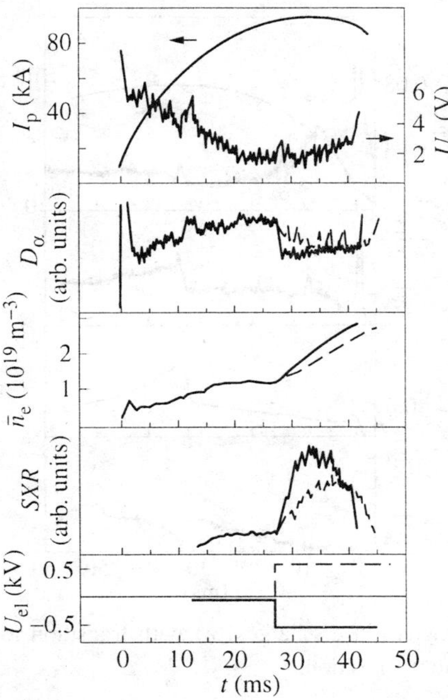

Fig. 4. Evolution of some plasma parameters during transition to the H-mode triggered by the bias with positive and negative electrode polarity.

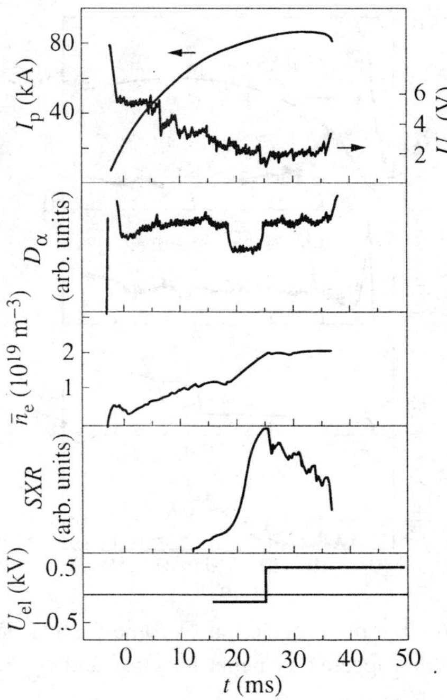

Fig. 5. Evolution of some plasma parameters in a shot with spontaneous ohmic H-mode terminated by applying a bias with positive electrode polarity.

and TEXTOR [15] tokamaks. Thus, the key role of the radial electric field was proven experimentally and the plasma confinement characteristics were shown to be well controllable by biasing an electrode inserted into plasma. Being efficient in small devices, the above method of control appears to be unsuitable for the bigger machines with hotter plasma, particularly, in a reactor, because any electrode will be destroyed by the enormous energy flux inherent in the fusion plasma. Hence, a noncontact way to generate essentially nonuniform radial fields on the periphery is needed. The search led us to develop a method to trigger the H-mode by means of pellet evaporation on the plasma periphery [16]. Theoretically, a significant density gradient is accompanied by electric field [17]:

$$E_r = \frac{T_i}{e}\left[\frac{\partial \ln n}{\partial r} + (1-k)\frac{\partial \ln T_i}{\partial r}\right].$$

The density gradient needed can be produced by pellet evaporation. In our case, the injection was accomplished with a pneumatic accelerator. Pellets were made of LiD. To ensure evaporation on the periphery, the minimum possible pellet velocity and size were chosen: ~ 150 m s^{-1} and 100 μm, respectively. With such a choice the injection triggers a transition to the H-mode, see Fig. 6(a) [16]. There, after a jump caused by pellet evaporation, the density continues to increase, while the D_α intensity drops. The adjacent Fig. 6(b) illustrates the evolution of some plasma parameters for a discharge with injection of a pellet of greater size and higher velocity. In this case, immediately after the jump, the density starts to decrease while the D_α intensity reflecting the plasma

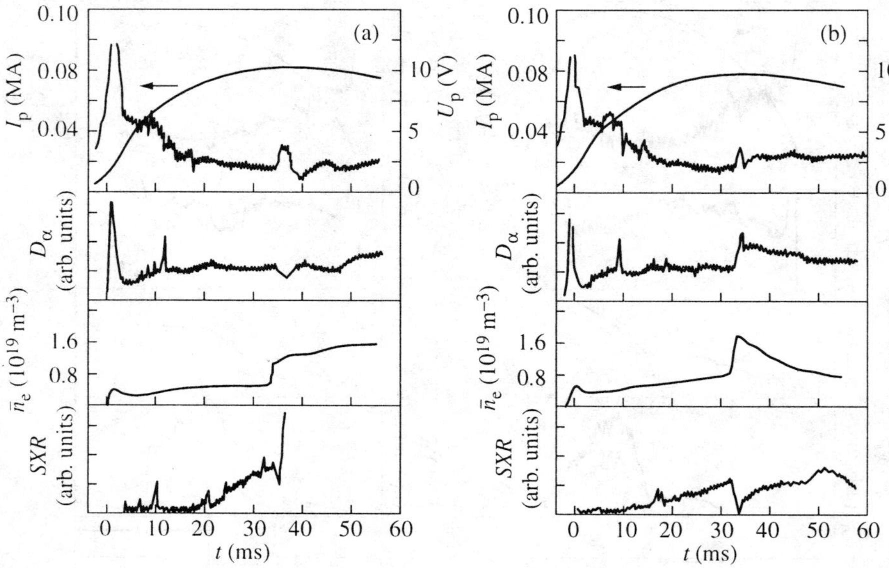

Fig 6. Temporal evolution of some plasma parameters in discharges (a) with transition to the H-mode triggered by pellet injection and (b) without transition.

flow to the wall increases significantly—there is no transition to the H-mode.

The method of triggering the H-mode by edge pellet ablation can be used in large devices with long discharge duration, because it requires no electrode insertion into plasma. Note that the H-mode in the T-10 tokamak could be triggered [18] by the technique proposed, and now experiments on the DIII-D are in due course.

Other methods to initiate the H-mode have also been demonstrated in our experiments, including pulsed gas puffing [16], minor radius magnetic compression [19], simultaneous current ramp-up, and magnetic compression [20].

4 Peculiarities of energy confinement in ohmic H-mode

A knowledge of parametric dependences of plasma confinement is of great importance for designing the next generation of fusion devices. The most important thing for a reactor is to make a prognosis of its energy confinement time. To predict the τ_E for future devices, empirical scalings obtained by statistical analysis of experimental data from the modern tokamaks are used. In view of the aforesaid, experimental studies of parametric dependences of τ_E are of great importance.

In our experiments the τ_E dependences on averaged density, plasma current, and input power were investigated in the ohmic H-mode [21]. In contrast to the scalings applicable to the ordinary ohmic regime, the energy confinement time in the ohmic H-mode appears to be practically independent of density and strongly increases with the plasma current. The maximum τ_E obtained in TUMAN-3 at high currents is as long as 30 ms, which essentially longer than the values predicted for the ordinary ohmic heating. Owing to the vessel wall boronization [22], the input power could be changed significantly in the ohmic heating regime without any changes in current. In this case, reducing the power significantly improved the confinement. The investigation performed

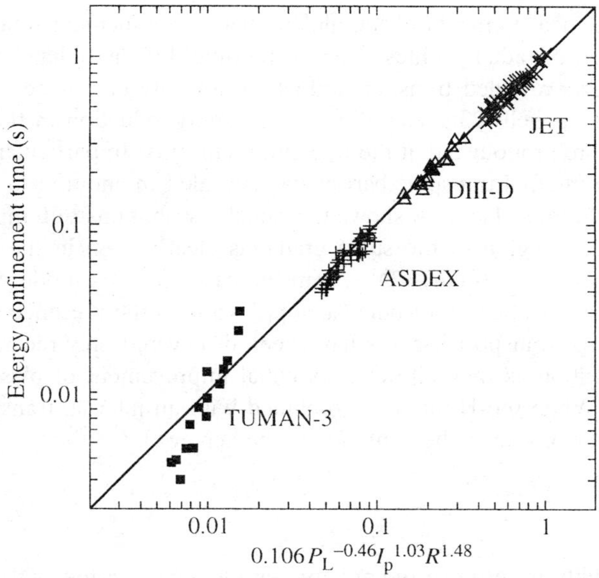

Fig 7. Comparison of experimentally measured energy confinement times with the prediction of the JET/DIII-D H-mode scaling.

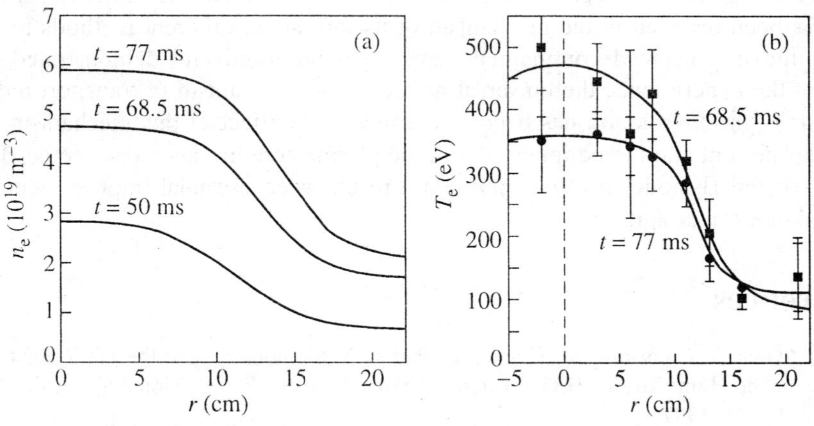

Fig 8. Radial distributions of the electron density and temperature in a discharge with internal transport barrier in the region 10 cm $< r <$ 16 cm.

confirmed the essential difference between the ohmic H-mode and the ordinary ohmic heating regime and the similarity of the former to the H-mode observed under auxiliary heating. This conclusion can be illustrated by a comparison of the energy confinement times measured in the ohmic H-mode with a scaling describing the confinement in the H-mode on large tokamaks ASDEX, DIII-D, and JET, presented in Fig. 7. The same figure shows the experimental points obtained on these large tokamaks and used to compile the scaling. Despite the absence of auxiliary heating and comparatively simple magnetic configuration, the energy confinement times in the H-mode in TUMAN-3 are in good agreement with the scaling.

In addition to the measurements of the global energy confinement times, measurements of the evolution of radial profiles effective thermal diffusivity have been carried out. The measurements revealed transport reduction not only on the periphery but in more interior regions as well [23]. The effect of transport reduction in the central regions of plasma is more pronounced at the maximum currents. In particular, in a recent experimental run, an internal transport barrier was revealed in addition to the external one inherent in the H-mode. Figure 8 shows the radial distributions of temperature and electron density, with a region of the steep gradients clearly seen in the mid part of cross section ($10 \text{ cm} < r < 16 \text{ cm}$) [24]. As in the case of the H-mode, the effect of transport reduction is explained by turbulence suppression in the presence of a sheared $\mathbf{E} \times \mathbf{B}$ rotation. Internal transport barriers have been discovered only recently [25], but attracted a great attention as they allow an essential improvement of plasma confinement in a tokamak. When the H-mode is combined with an internal transport barrier, an enhancement factor of 4 over the L-mode can be achieved.

Conclusion

Studies of the possibility to improve the energy confinement in a tokamak, carried out on the TUMAN-3 device have led to a discovery of the ohmic H-mode—a promising scenario of plasma heating. In the course of these experiments the properties of the discovered regime have been investigated in detail, the crucial role of the radial electric field has been revealed in the mechanism of its formation, different methods to initiate the regime of enhanced confinement have been proposed and demonstrated. As a result of the experiments, the universal nature of the mechanism of transport reduction by sheared plasma rotation has been established. The effect of this mechanism on the transport not only on the periphery but in the plasma core has been discovered. Further studies of the H-mode in a tokamak could result in an essential improvement of the tokamak-reactor design.

References

[1] R. Aymar, V. Chuyanov, M. Huguet, R. Parker, Y. Shimomura, and the ITER Joint Central Team and Home Teams, *Fusion Energy 1996*, XVI Conf. Proc., (Montreal), IAEA, Vienna, vol. 1, p. 3, 1997.

[2] F. Wagner, G. Becker, K. Behringer, et al., *Phys. Rev. Lett.* **49** 1408 (1982).

[3] K. Steinmetz, J.-M. Noterdaeme, F. Wagner, et al., *Phys. Rev. Lett.* **58** 124 (1987).

[4] J. M. Lohr, B. W. Stallard, R. Prater, et al., *Phys. Rev. Lett.* **60** 2630 (1988).

[5] A. N. Arbuzov, L. G. Askinazi, V. I. Afanas'ev, et al., *17th Eur. Conf. on Controlled Fusion and Plasma Heating* (Amsterdam, 1990), EPS, Geneva, vol. 14B, part I, p. 299, 1990.

[6] T. H. Osborne, N. H. Brooks, K. H. Burrell, T. Carlstrom, R. J. Groebner, D. Hill, W. Howl, A. Kellman, L. Lao, N. Ohyabu, M. Perry, and T. Taylor, *Nuclear Fusion* **30** 2023 (1990).

[7] G. M. Vorob'ev, V. E. Golant, S. V. Gornostaev, A. V. Grigor'ev, M. P. Gryaznevich, T. P. Evtushenko, V. A. Ipatov, A. I. Kislyakov, S. V. Lebedev, B. M. Lipin, R. N. Litunovsky, O. A. Minyaev, V. V. Rozhdestvensky, N. V. Sakharov, P. P. Teplov, A. A. Fedorov, K. G. Shakhovets, and S. G. Shchemelinin, *Sov. J. Plasma Phys.* **9** 65 (1983).

[8] L. G. Askinazi, V. V. Bulanin, D. O. Korneyev, and S. V. Lebedev, in *Magnetic Turbulence and Transport*, ed. by P. Hennequin and M. A. Dubois, Editions de Physicque, Orsay, p. 189, 1993.

[9] L. G. Askinazi, V. E. Golant, E. R. Its, D. O. Korneyev, S. V. Krikunov, B. M. Lipin, S. V. Lebedev, K. A. Podushnikova, V. V. Rozhdestvensky, N. V. Sakharov, S. P. Yaroshevich, L. Kryska, and P. Dvoracek, *18th Eur. Conf. on Controlled Fusion and Plasma Physics* (Berlin, 1991), EPS, Geneva, vol. 15C, part I, p. 401, 1991.

[10] ASDEX Team, *Nuclear Fusion* **29** 1959 (1989).

[11] R. J. Groebner, *Phys. Fluids B* **5** 2343 (1993).

[12] H. Biglari, P. H. Diamond, and P. W. Terry, *Phys. Fluids B* **2** 1 (1990).

[13] L. G. Askinazi, V. E. Golant, S. V. Lebedev, V. A. Rozhanskij, and M. Tendler, *Nuclear Fusion* **32** 271 (1992).

[14] R. J. Taylor, M. L. Brown, B. D. Fried, H. Grote, J. R. Liberati, G. J. Morales, P. Pribyl, D. Darrow, and M. Ono, *Phys. Rev. Lett.* **63** 2365 (1989).

[15] R. R. Weynants, R. J. Taylor, P. E. Vandenplas, F. Durodie, B. Schweer, D. Bora, T. Delvigne, J. Devos, K. H. Dippel, M. Gaugneaux, B. Geisen, Y. T. Lie, A. M. Messiaen, R. Moyer, U. Samm, R. P. Schorn, C. Stickelmann, R. Van Nieuwenhove, G. Van Oost, G. H. Wolf, and J. W. Yang, *17th Eur. Conf. on Controlled Fusion and Plasma Heating* (Amsterdam, 1990), EPS, Geneva, vol. 14B, part I, p. 287, 1990.

[16] L. G. Askinazi, V. E. Golant, S. V. Lebedev, L. S. Levin, V. A. Rozhanskij, and M. Tendler, *Phys. Fluids B* **5** 2420 (1993).

[17] V. A. Rozhanskij, and M. Tendler, *Phys. Fluids B* **4** 1877 (1992).

[18] V. G. Kapralov, V. A. Rozhansky, and K. V. Khlopenkov, *22nd EPS Conf. on Controlled Fusion and Plasma Physics* (Bournemouth, 1995), EPS, Geneva, vol. 19C, part I, p. 117, 1995.

[19] T. Yu. Akatova, L. G. Askinazi, V. I. Afanas'ev, V. E. Golant, V. K. Gusev, E. R. Its, S. V. Krikunov, S. V. Lebedev, B. M. Lipin, K. A. Podushnikova, G. T. Razdobarin, V. V. Rozhdestvensky, N. V. Sakharov, A. S. Tukachinsky, A. A. Fedorov, F. V.Chernyshev, S. P. Yaroshevich, L. Kryska, and P. Dvoracek, *Plasma Physics and Controlled Nuclear Fusion Research*, 13th Conf. Proc. (Washington), IAEA, Vienna, vol. 1, p. 509, 1991.

[20] L. G. Askinazi, T. Yu. Akatova, V. E. Golant, M. P. Gryaznevich, V. K. Gusev, E. R. Its, S. V. Krikunov, S. V. Lebedev, B. M. Lipin, K. A. Podushnikova, G. T. Razdobarin, V. V. Rozhdestvensky, N. V. Sakharov, A. S. Tukachinsky, A. A. Fedorov, F. V. Chernyshev, K. G. Shakhovets, and S. P. Yaroshevich, *17th Eur. Conf. on Controlled Fusion and Plasma Heating* (Amsterdam, 1990), EPS, Geneva, vol. 14B, part I, p. 411, 1990.

[21] M. V. Andreiko, L. G. Askinazi, V. E. Golant, V. A. Kornev, S. V. Lebedev, L. S. Levin, G. T. Razdobarin, V. V. Rozhdestvensky, A. S. Tukachinsky, and S. P. Yaroshevich, *Plasma Phys. Control. Fusion* **36** A165 (1994).

[22] V. M. Sharapov, S. V. Mirnov, S. A. Grashin, I. A. Kovan, A. V. Krasilnikov, V. A. Krupin, S. V. Lebedev, L. S. Levin, A. N. Romannikov, and A. P. Zakharov, *J. Nucl. Mat.* **220-222** 730–735 (1995).

[23] M. V. Andrejko, L. G. Askinazi, V. E. Golant, S. V. Lebedev, and A. S. Tukachinsky, in *Proc. Int. Workshop on Transport Studies in Fusion Plasmas*, Varenna, p. 57, 1993.

[24] S. V. Lebedev, M. V. Andreiko, L. G. Askinazi, V. E. Golant, V. A. Kornev, S. V. Krikunov, L. S. Levin, V. V. Rozhdestvensky, A. S. Tukachinsky, and S. P. Yaroshevich, accept. for publ. in *Plasma Phys. Control. Fusion* **40** (1998).

[25] F. M. Levinton, M. C. Zarnstorff, S. H. Batha, M. Bell, R. E. Bell, R. V. Budny, C. Bush, Z. Chang, E. Fredrickson, A. Janos, J. Manickam, A. Ramsey, S. A. Sabbagh, G. L. Schmidt, E. J. Synakowski, and G. Taylor, *Phys. Rev. Lett.* **75** 4417 (1995).

Two topics of astrophysics-and-physics overlap: Fundamental constants and Neutron stars

D. A. Varshalovich, A. Y. Potekhin, A. V. Ivanchik, and D. G. Yakovlev

Introduction

Many subjects have been studied at the Department of Theoretical Astrophysics of the Ioffe Institute in the past decades. For this paper, we have selected two of them: fundamental constants and neutron stars. Each of them reveals a deep connection between the fundamental theoretical physics and the astrophysics.

Contemporary endeavors to build the Grand Unification Theory (GUT) have led to the prediction that the fundamental physical constants should have changed during the history of the Universe (some 10 billion years). To check this exciting prediction, one should "measure" the constants in the early cosmological epoches. This can be done only by astrophysical methods. The first section of the present review, written by D.A.V., A.Y.P., and A.V.I., describes the work in this direction.

The second section, written by D.G.Y., is also devoted to exciting circle of problems related to very remarkable astronomical objects—the neutron stars. These stars are characterized by extreme conditions: superhigh densities, magnetic fields, gravitational and centrifugal accelerations not available in terrestrial laboratory. The studies of neutron stars present a challenge to the contemporary physics.

1 Measuring the fundamental constants in the early Universe

Every physical theory uses some parameters determined either from experiment or from a more fundamental theory. The most fundamental parameters are called fundamental physical constants (e.g., the electron charge e, the Planck constant $\hbar$, the speed of light c, the masses of elementary particles, etc.). However a numerical value of any dimensional parameter depends on arbitrary choice of physical units. Only dimensionless combinations are truly fundamental, such as the fine-structure constant $\alpha = e^2/(\hbar c)$, the proton-to-electron mass ratio $\mu = m_p/m_e$, etc. The current theories cannot derive their numerical values. Hopefully these values will be provided by the future theory, which has not been created yet. Attempts to construct such a theory have led to a number of models, claiming that the fundamental constants are, in fact, not true constants but slowly varying functions of time.[1] Clearly, a discovery of these variations would be a

[1] The prediction of the theory that the fundamental constants depend on energy transfer has been confirmed in experiment. In this paper, we consider only the space-time variability of their low-energy limits.

great step in our understanding of Nature. Even a reliable upper bound on a possible variation rate of a fundamental constant presents a valuable tool for selecting viable theoretical models.

Historically, a hypothesis that the fundamental constants may depend on the *cosmological time* t (that is the age of the Universe) was first discussed by Milne [1] and Dirac [2]. The latter author proposed his famous "large-number hypothesis" and suggested that the gravitational constant was directly proportional to t. Teller [3] showed, however, that such variation was inconsistent with geological and paleontological findings. Later the variability of fundamental constants was analyzed, using different arguments, by such distinguished physicists as Gamow [4], Dyson [5], and others. The interest in the problem has been revived due to recent major achievements in the GUT and superstring models (e.g., [6]).

Presently, the fundamental constants are being measured with a relative error of $\sim 10^{-8}$. These measurements obviously rule out considerable variations of the constants on a short time scale, but do not exclude their changes over the entire lifetime of the Universe, $\sim 1.5\times10^{10}$ years. Moreover, one cannot rule out the possibility that the constants differ in widely separated regions of the Universe; this could be disproved only by experiments and observations.

It is extremely difficult to increase the accuracy of the present-day experiments and trace possible variation of a fundamental constant during the entire history of the Universe. Fortunately, Nature has provided us with a tool for direct measuring the physical constants in the early epoches. This tool is based on observations of quasars, the most powerful sources of radiation. Many quasars belong to most distant objects we can observe. Light from the distant quasars travels to us about 10^{10} years. This means that the quasar spectra registered now were formed $\sim 10^{10}$ years ago. Because of the expansion of the Universe, distant objects recede from us at a great speed. As a result, the wavelengths of the lines observed in spectra of these objects ($\lambda_{\rm obs}$) increase compared to their laboratory values ($\lambda_{\rm lab}$) in proportion $\lambda_{\rm obs} = \lambda_{\rm lab}(1+z)$, where the *cosmological redshift* z can be used to determine the age of the Universe at the line-formation epoch. In some cases, the redshift is as high as $z \sim 3$–5, so that the intrinsically far-ultraviolet lines are registered in the visible range. The examples are demonstrated in Fig. 1. Analyzing these spectra we, like a time-machine, may study the epoch when the Universe was several times younger than now.

Here we review briefly the studies of the space-time variability of the fine-structure constant α and the proton-to-electron mass ratio μ. The studies are based on the quasar spectra analyses carried out at the Department of Theoretical Astrophysics of the Ioffe Institute. This activity has been reviewed in detail in [7, 8].

1.1 Fine-structure constant

The name of the fine-structure constant α was introduced by Sommerfeld in 1916 because the constant defined the fine splitting of atomic energy levels. Now we know that α is the fundamentally parameter in quantum electrodynamics.

The strongest bound on the possible time-variation rate of α was derived in 1976 by Shlyakhter [9] and recently, from a more detailed analysis, by Damour and Dyson [10], who obtained $|\dot{\alpha}/\alpha| < 0.7\times10^{-16}$ yr^{-1}, The analysis was based on measurements of isotope ratios in the Oklo site in Gabon, where a unique natural uranium nuclear fission reactor had operated 1.8 billion years ago. The isotope ratios of samarium produced

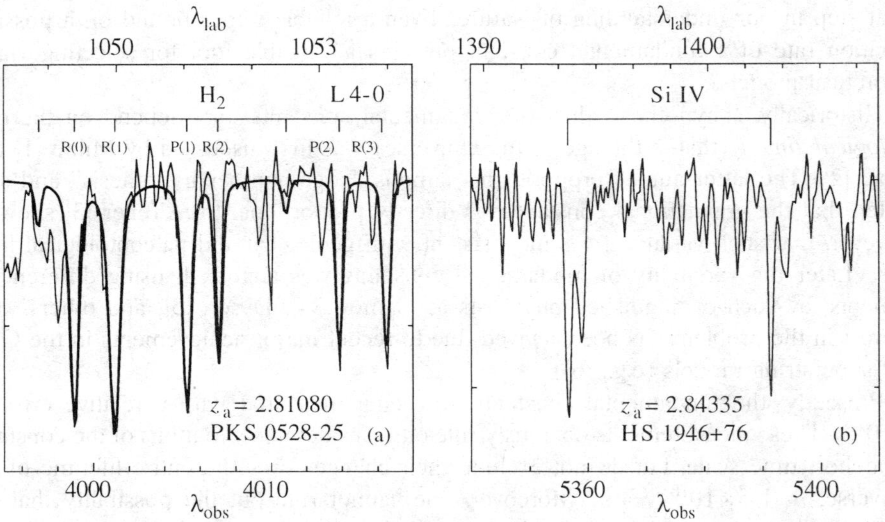

Fig 1. Portions of quasar spectra which show the absorption lines of H_2 and Si IV with large redshifts, $z = (\lambda_{obs} - \lambda_{lab})/\lambda_{lab} \approx 2.8$. The lower horizontal axis gives the wavelengths in the observer's frame (λ_{obs}) and the upper axis gives the wavelengths in the quasar's frame λ_{lab} (in Å). (a) The spectrum (thin line) of the quasar PKS 0528–250, obtained with the 4-meter CTIO telescope (Chile), containing H_2 lines which belong to the L 1–4 branch of the spectrum; thick line plots the spectral fit. (b) The spectrum of the quasar HS 1946+76, obtained with the 6-meter SAO telescope (Russia), containing Si IV doublet lines which correspond to the ${}^2S_{1/2} \rightarrow {}^2P_{3/2}$ and ${}^2S_{1/2} \rightarrow {}^2P_{1/2}$ transitions.

in this reactor by the neutron capture reaction ${}^{149}\mathrm{Sm} + \mathrm{n} \rightarrow {}^{150}\ \mathrm{Sm} + \gamma$ would be completely different, if the energy of the nuclear resonance responsible for this capture were shifted at least by 0.1 eV.

Another strong bound, $|\dot{\alpha}/\alpha| < 5\times10^{-15}$ yr^{-1}, was obtained by Dyson [5] from an isotopic analysis of natural radioactive decay products in meteorites.

A weak point of these tests is their dependence on the model of the phenomenon, fairly complex, involving many physical effects. In addition, the tests are associated with the Solar system. They cannot be extended to distant space regions and to the early Universe: the law of possible space-time variation of α is unknown *a priory*. It is the extragalactic astronomy that allows us to study these remote regions of spacetime.

To find out whether α changed over the cosmological time, we have studied the fine splitting of the doublet lines of Si IV, C IV, Mg II, and other ions, observed in the spectra of distant quasars. According to quantum electrodynamics, the relative splitting of these lines $\delta\lambda/\lambda$ is proportional to α^2 (neglecting very small corrections). Consequently, if α changed with time, then $\delta\lambda/\lambda$ would depend on the cosmological redshift z. This method of measuring α in distant regions of the Universe had been first suggested by Savedoff [11] and was used later by other authors. For instance, Wolfe *et al.* [12] derived an estimate $|\dot{\alpha}/\alpha| < 4\times10^{-12}$ yr^{-1} from an observation of the Mg II absorption doublet at $z = 0.524$.

Many high-quality quasar spectra measured in the last two decades have enabled us to significantly increase the accuracy of determination of $\delta\lambda/\lambda$ at large z. The first estimates [13] indicated that α might be changing at the fractional rate $\sim 3\times10^{-12}$ yr^{-1}

at the 3σ confidence level. This exciting result prompted us to undertake a more thorough study.

At the first stage of the work, it was reanalyzed all the published data on the fine-splitting doublet lines observed in quasar spectra. We considered a total of about 1500 pairs of doublet lines with redshifts between 0.2 and 3.7 and compiled a catalog of the wavelengths [14]. An analysis revealed no statistically significant deviation of the ratio $\delta\lambda/\lambda$ from the present value. The wealth of observational material collected in our catalog allowed us to study whether α is different in causally disconnected regions of the Universe. Many regions of formation of the spectral lines, observed at large redshifts in different directions in the sky, had been causally disconnected at the epoches of line formation. Thus, no information could have been exchanged between these regions of the Universe and, in principle, the fundamental constants could be different there. However, a weak dependence of the fine doublet splitting on the direction in the sky found in [7] lies within the 3σ relative error $|\Delta\alpha/\alpha| < 3\times10^{-4}$.

Using these results, we optimized the strategy of studying the time-dependence of α. At the second stage of the project [15], we realized a special program of spectroscopic observations of several quasars with the 6-m telescope of the Special Astrophysical Observatory (SAO) of the Russian Academy of Sciences. Our aim was to achieve the highest possible accuracy of the fine-splitting measurements at large redshifts. An example of the spectra observed is shown in Fig. 1. As a result, we obtained a new constraint on the possible deviation of the fine-structure constant at $z = 2.8$–3.1 from its present ($z = 0$) value: $|\Delta\alpha/\alpha| < 1.6\times10^{-4}$. The corresponding upper limit of the α variation rate averaged over $\sim 10^{10}$ yr is $|\dot{\alpha}/\alpha| < 2\times10^{-14}$ yr^{-1}. This constraint is much more stringent than those obtained from previous astronomical observations.

According to our analysis, some theoretical models are inconsistent with observations [14, 7, 15]. For example, a power law $\alpha \propto t^n$ could be reconciled with observations if only $n < 10^{-4}$. This inequality eliminates a broad class of models. In particular, it is in conflict with the Gamow's hypothesis [4] that $n = 1$ and with the theoretical laws with $n = 1, -1/4$, and $-4/3$ published by various authors in the 1980s. Moreover, the Teller–Dyson's hypothesis on the logarithmic dependence of α on t [3, 5] has also been shown to be inconsistent with observations.

1.2 Proton-to-electron mass ratio

The dimensionless constant $\mu = m_p/m_e$ approximately equals the ratio of the constants of strong interaction $g^2/(\hbar c) \sim 14$ and electromagnetic interaction $\alpha \approx 1/137.036$, where g is the effective coupling constant calculated from the amplitude of nucleon-π-meson scattering.

In order to check the cosmological variability of μ we have used high-redshift absorption lines of molecular hydrogen H_2 in the spectrum of the quasar PKS 0528–250. This is the first (and, in a sense, unique) high-redshift system of H_2 absorption lines discovered in 1985 [16][2]. It was the discovery of a new class of the extragalactic objects—discrete clouds of molecular hydrogen with large cosmological redshifts. A study of these objects yields information of paramount importance on the physical conditions $\sim 10^{10}$ years ago, particularly, on variability of the fundamental constants.

[2]For this work, Levshakov and Varshalovich were awarded with the Belopolsky Prize of the Russian Academy of Sciences in 1990.

It is important that the electronic, vibrational, and rotational energies of a molecule depend on its reduced mass in different ways. Comparing ratios of wavelengths λ_i of various H_2 electron-vibration-rotational lines in a quasar spectrum at some redshift z and in laboratory (at $z = 0$), we can trace variation of μ. The method had been used previously by Foltz *et al.* [17], whose analysis was corrected later in our papers [18, 7, 19]. In the latter papers, we calculated the sensitivity coefficients K_i of the wavelengths λ_i with respect to possible variation of μ and applied a linear regression analysis to the measured redshifts of individual lines z_i as function of K_i. If the proton mass in the epoch of line formation were different from the present value, the measured z_i and K_i values would correlate. However, no such correlation was found within the statistical errors, and we obtained the upper limit on the deviation of μ at $z = 2.811$ (i.e., $\sim 10-12$ billion years ago) from the present value: $|\Delta\mu/\mu| < 0.002$.

We have made even more refined analysis using the modern high-resolution spectrum of the same object obtained by Lanzetta *et al.*. It has enabled us to identify several tens of the H_2 lines; a portion of the spectrum is shown in Fig. 1. From this spectrum we have obtained [20] the restriction: $\Delta\mu/\mu = (-1.15 \pm 0.76)\times 10^{-4}$ at $z = 2.8108$. It corresponds approximately to $|\dot{\mu}/\mu| < 2\times 10^{-14}$ yr^{-1}.

1.3 Conclusions

We have demonstrated how the extragalactic astronomy provides an insight into what the Universe had looked like at early cosmological epoches and in distant regions of space. We have given only two examples, concerning the bounds on possible variation rates of the fine-structure constant α and the proton-to-electron mass ratio μ. The extragalactic astronomy can also give the bounds on other fundamental constants or their combinations. For example, we have shown [7, 8, 19] that not only α and μ but also the ratio of the atomic mass unit to m_e did not change in $\sim 10^{10}$ years by more than a few parts in 10^4. This means that the coupling constants of electromagnetic and strong interactions did not change to a high accuracy. Moreover, they were the same in causally disconnected regions of the Universe.

The derived constraints give effective criteria for selecting theoretical models of elementary interactions which predict variations of the physical constants over the cosmological time scale. In particular, we were the first who showed that the Teller–Dyson's hypothesis on the logarithmic time-dependence of the fine-structure constant was incompatible with observational data.

The work was supported by the Russian Foundation for Basic Research (grants 93-02-02958 and 96-02-16849), the International Science Foundation (grant NUO-300), and by the Russian State Programme "Fundamental Metrology". The results were awarded with the Scientific Council Prize of the Ioffe Institute in 1993 and the Main Prize of the International Academic Publishing Company "Nauka" in 1996.

2 Neutron stars

"Neutron stars are superstars," David Pines once said. And right he was. These tiny stars (of radius $\sim$ 10 km and mass $\sim 1.4 M_{\odot}$) are the champions among the stars. They contain superdense matter, show superfast rotation, have superstrong magnetic fields, operate as superpowerful accelerators of superenergetic particles. They entertain astronomers behaving as radio pulsars, X-ray pulsars, X-ray bursters, soft gamma repeaters, and the sources of quasiperiodic X-ray oscillations. Their birth in supernova explosions produces strong neutrino outbursts (detected once from the supernova 1987A). Hopefully, they are the best emitters of gravitational waves.

Here is a history of the neutron-star physics in the Department of Theoretical Astrophysics of the Ioffe Institute from its beginning till February 1998. The first article [21] was written by Yu. N. Gnedin, A. Z. Dolginov, and A. I. Tsygan in 1969, two years after the discovery of pulsars, to model radiation formation in the X-ray source Sco-X-1. The history involves about 250 articles in the refereed journals and conference proceedings (about 50 written before 1980, about 70 from 1980 till 1989, and about 130 later). Their subject is mainly theory of all corners of neutron stars and nearby space related to observations. The most enthusiastic patriarches were Yu. N. Gnedin (left the department in 1984) and A. I. Tsygan. The second generation started working later: N. A. Silant'ev—since 1965 (left in 1985), G. G. Pavlov, A. D. Kaminker, Yu. A. Shibanov—since 1967, D. G. Yakovlev—since 1971, I. G. Mitrofanov—since 1972 (left in 1981), V. A. Urpin—since 1973 (on leave since 1996). A. G. Muslimov was employed in 1980, converted into the neutron star physics in 1984 and left in 1992. The next generation came in the middle of the 1980s: V. G. Bezchastnov in 1986, D. A. Shalybkov in 1989, A. Y. Potekhin and K. P. Levenfish in 1992, V. D. Pal'shin and D. Y. Konenkov in 1995, and D. A. Baiko in 1996. We organized four neutron-star workshops in Leningrad (St. Petersburg) in 1988, 1992, 1995, and 1997. You can find our neutron-star information in Internet (http://www.ioffe.rssi.ru/nsgroup.html). Our activity is summarized in four sections.

2.1 Radiation from neutron stars

The neutron stars are surrounded by atmospheres, where their radiation is formed. You cannot interpret the radiation without atmospheric models. You cannot borrow the models from other stars which are not superstars. New general theory of radiative transfer in anisotropic media was developed mainly by the efforts of Gnedin, Pavlov, Silant'ev, and Dolginov (Department's head till 1986) in the 1970s. Basic formulation of the radiative transfer problem in a strongly magnetized plasma [22] has become the foundation of the new field – the radiative transfer in magnetized compact objects. The results were summarized in a monograph [23] and its authors received a prestigious Belopolsky Prize in Russia in 1987.

The electron cyclotron lines in the spectra of magnetized compact objects have three fathers. Gnedin and Sunyaev [24] predicted the lines theoretically. The third father, J. Trümper, and his colleagues discovered the lines in 1976 in the spectrum of Her X-1. Since then observations of the electron cyclotron lines have allowed astronomers to measure the magnetic fields of X-ray pulsars.

The radiative transfer theory in a hot, fully ionized, and strongly magnetized atmosphere of a neutron star was completed [25] in 1983. More than that, high magnetic fields were shown [26] to affect the radiative transfer by polarizing the electron-positron vacuum. Strong anisotropy of the radiation and the vacuum polarization effects explain [25] the main features of the pulse shapes and their variations with photon energy in some X-ray pulsars.

It was a tricky job to construct the atmosphere models [27] of X-ray bursters: the atmosphere expands enormously during a burst with nearly Eddington luminosity. The job led to useful constraints on the parameters (masses, radii, distances, chemical compositions of the atmospheres) of some X-ray bursters [28].

One cannot build the atmosphere models without knowing the physics of atmospheric plasma. This plasma can be partially ionized and non-ideal; atoms and ions can be distorted by huge magnetic fields. The first realistic study [29] of ionization in the magnetized hydrogen atmospheres was done in 1974. A self-consistent model equation of state of non-magnetized hydrogen plasma constructed recently [30] takes account of the effects of pressure ionization and plasma non-ideality. The structure and radiative transitions of the hydrogen atoms moving in a strong magnetic field was a hard problem to solve (see, e.g., [31]): one should take into account strong electric fields induced in an atom-comoving reference frame. Similar but more complicated problem for the He^+ ion is being under study now [32].

Recent achievements are the realistic models [33] for the hydrogen, helium, and iron atmospheres of the isolated neutron stars with low magnetic fields $B < 10^9$–10^{10} G. The radiation spectra from the H and/or He atmospheres are harder than the blackbody spectrum with the same effective temperature. Thus, the traditional blackbody interpretation of thermal radiation from the neutron stars may overestimate the effective temperature by a factor of 1.5–3 and lead to wrong conclusions on neutron star structure and evolution. The spectra from the iron atmospheres are closer to the blackbody but contain many spectral lines and photoionization jumps.

Many efforts were made to compose the first models [34, 35] of hot (the effective temperature $> 10^6$ K) hydrogen atmospheres of isolated neutron stars with strong magnetic fields $B \sim 10^{11}-10^{13}$ G. The fields make the spectra softer than in the nonmagnetic stars but harder than the blackbody. The ion cyclotron absorption line can be generated in optics or UV. The radiation is highly polarized, peaked sharply along the magnetic field and has a fan-like beaming across the field. It is no problem now to calculate the radiative flux from a rotating neutron star with the nonuniform surface temperature and magnetic field distributions including the gravitational lensing near the star. All the atmosphere models are available in Internet (http://stella.ioffe.rssi.ru/Stars) and widely used by the astrophysical community.

Needless to say, the models were applied for interpretation of the spectra and light curves of the soft X-ray emission from several isolated neutron stars. The interpretation of observations of the Geminga pulsar [36] shows that the Geminga's surface can be very nonuniform. The fits of the X-ray spectrum of the Vela pulsar (detected with the ROSAT orbital observatory) with the hydrogen atmosphere models [37] give the distance to Vela consistent with other independent observations, contrary to the blackbody fits. The same is true for a few other neutron stars. The new generation of sensitive detectors and advanced atmosphere models will open a new era for the studies of mysterious surfaces and interiors of neutron stars.

2.2 Neutron star magnetospheres

The magnetosphere lies above the atmosphere. It is a place of violent processes: energetic particles are accelerated, radio and gamma-ray emission is formed, etc. In 1974 Tsygan proposed a model of generation of cosmic gamma-bursts [38] in the magnetospheres of vibrating neutron stars. The vibration energy was taken from the elastic stress-energy ($\sim 10^{47}$ erg) of the solid neutron-star core (such cores were popular in the 1970s). Detailed theory of the electric fields around vibrating stars was developed later [39]. Now the vibrational model is being discussed in the literature as the model of the soft gamma repeaters.

X-ray emission from a hot spot on the surface of an X-ray pulsar is a good accelerator of electrons in the pulsar's magnetosphere [40]. Interaction of electrons with radiation near the magnetized neutron-star surface is affected by the electron cyclotron resonance. The resonance can strongly reduce the critical luminosity of the neutron stars [41] and modify the magnetospheric particle acceleration [42].

One needs the quantum relativistic theory to describe the cyclotron radiation (with possible maser effects) in a strongly magnetized plasma around the neutrons stars [43]. The plasma emits also the electron–positron annihilation radiation (see [44] and references therein) that depends on distribution functions of electrons and positrons.

Who could have imagined that the electric fields near a rotating neutron star are controlled by the general relativistic effect of drag of reference frame? But the effect is crucial [45] for the existence of radiopulsars with the spin periods about 1 s and the magnetic fields $\sim 10^{12}$ G. The electric fields are sufficient to accelerate the electrons, emitted from hot polar caps of radio pulsars, up to Lorentz factors $\sim 10^6$–10^7. The electrons generate gamma-quanta and the electron–positron pairs. The spectrum and beaming of such gamma-rays, the inverse current of positrons, the temperature of the hot polar caps and the efficiency of the electron–positron avalanche were modelled numerically [46].

The emission of particles and gamma-quanta by the radiopulsars, and (possibly) radioemission can be governed by an external magnetic field [47]. For a pulsar with the spin period of 1 s and the polar magnetic field $\sim 10^{12}$ G, the effect becomes important if the external field is about 0.5–3 G. For instance, such a field can be provided in a binary with a degenerate dwarf possessing an intrinsic field $\sim 10^8$ G.

A neutron star which enters an X-ray binary and has a non-dipolar magnetic field, emits asymmetric X-ray radiation and produces an accelerated particles [48]. After averaging over the neutron star spin, the force component along the spin axis survives. It may exceed gravitational attraction of the binary to the Galaxy. The effect is mostly pronounced in the low-mass X-ray binaries during the intense accretion onto the neutron stars. These systems may form the Galactic halo and escape partly into intergalactic space.

2.3 Internal structure and evolution

The internal structure is still a mystery. What is the composition of superdense neutron star cores? Nucleons and hyperons? Or maybe kaon condensates or strange quark matter? What is the equation of state: soft or stiff?

We started with the correct basic formalism for the electron thermal and electric conduction in degenerate surface layers of neutron stars without [49] and with [50] strong magnetic fields. These results were updated recently [51, 52, 53]. Thermodynamic properties of the surface layers and the plasma screening of nuclear reactions in these layers were also studied [54, 55].

Transport properties in the neutron star cores have been analyzed too, particularly, the electron and muon thermal conductivities [56]. It was a surprise to find out (see, e.g., [57, 58]) that the electrical resistance in the nonsuperfluid cores across strong magnetic fields can be greatly enhanced, accelerating ohmic decay of the internal magnetic fields.

The quantized Feynman–Onsager vortices of neutrons in the superfluid and magnetized cores of the pulsars interact with magnetic proton fluxoids. This is a fantastic picture. While the pulsar spins down, the vortices are expelled from the core into the crust dragging the fluxoids. This triggers the glitches, and possibly reduces the pulsar magnetic moment [59].

Now a few words about the neutrino generation. Neutrinos can easily escape even from the superstars and are important cooling agents. Various neutrino energy loss mechanisms in the neutron star crusts and cores without and with strong magnetic fields were examined (see, e.g., [55]). One of them is the neutrino generation due to the electron bremsstrahlung by atomic nuclei in the neutron star crusts [60].

It was a tedious job to study the suppression of various neutrino reactions and heat capacity by neutron and/or proton superfluid in the neutron star cores (e.g., [61] and references therein). The suppression by the strong superfluidity was shown to be several orders of magnitude weaker than thought before. The neutrino-pair emission due to the Cooper pairing of nucleons in the neutron-star cores and crusts has been analyzed [62].

Cooling of the neutron stars depends on their internal structure and is a good tool to study this structure. Our recent cooling calculations for the stars with superfluid cores [61, 62] include properly the simultaneous effects of the neutron and proton superfluids. As a result, we have obtained preliminary constraints on the superfluid critical temperatures of neutrons and protons in the cores of PSR 0656+14, Geminga and several other pulsars by comparing the theory with the observational data on X-ray thermal radiation of these stars.

2.4 Neutron star magnetic fields

The magnetic fields, especially the superstrong ones, are always of interest. What is their origin? How do they evolve?

We have already described the studies of fields in the neutron star cores. But the fields can be confined in the crust, being independent of the core fields. The crustal fields can be generated by thermomagnetic effects in young neutron stars [63]. Urpin and coauthors [64, 65, 66, 67] elaborated the theory of the ohmic decay of the crustal fields in the cooling isolated neutron stars assuming the fields are supported by the crustal electric currents. The electric conductivity in a young star increases with its cooling and slows the decay. In a colder star (age $t > 10^5$ yr), the conductivity becomes temperature independent (determined by the electron–impurity scattering) and the field decay becomes especially slow. The decay depends on the cooling scenario (standard or rapid cooling) and on the equation of state in the stellar core (the stiffer the equation of state, the larger the crust and slower ohmic dissipation). The Joule heat can be sufficient to control the temperature of an old star ($t > 10^7$ yr). The field decay affects spin-down of the pulsars and, hence, their spin-down age. Observations of pulsar magnetic fields are more consistent with the standard cooling of neutron stars composed of matter with stiff equation of state. The theory can explain larger magnetic fields of the young pulsars in supernova remnants.

Another problem is the ohmic decay of the magnetic fields in the crusts of accreting neutron stars (see [66] and references therein). The accreted matter, which sinks within the crust, reheats the star and carries the field inside. The evolution of the magnetic field and spin period is associated with the accretion regime. The theory [66] considers the magneto-rotational evolution of neutron stars in the binaries with low-mass and massive counterparts. In particular, it gives a natural explanation of the existence of the millisecond pulsars.

The author of the second section of this article (D.G.Y) is grateful to Yu. N. Gnedin, I. G. Mitrofanov, G. G. Pavlov, A. Y. Potekhin, Yu. A. Shibanov, and A. I. Tsygan for useful comments.

References

[1] E. Milne, *Proc. R. Soc.* **A 158** 324 (1937).
[2] P. A. M. Dirac, *Nature* **139** 323 (1937).
[3] E. Teller, *Phys. Rev.* **73** 801 (1948).
[4] G. Gamow, *Phys. Rev. Lett.* **19** 759 (1967).
[5] F. J. Dyson, *Aspects of Quantum Theory*, ed. A. Salam and E. P. Wigner, Cambridge Univ., Cambridge, p. 213, 1972
[6] T. Damour and A. M. Polyakov, *Nucl. Phys.* **B 423** 532 (1994).
[7] D. A. Varshalovich and A. Y. Potekhin, *Space Sci. Rev.* **74** 259 (1995).
[8] D. Varshalovich, A. Potekhin, and A. Ivanchik, *J. Journals* **1** 7 (1997).
[9] A. I. Shlyakhter, *Nature* **25** 340 (1976).
[10] T. Damour and F. J. Dyson, *Nucl. Phys.* **B 480** 37 (1996).
[11] M. P. Savedoff, *Nature* **264** 340 (1956).
[12] A. M. Wolfe, R. L. Brown, and M. S. Roberts, *Phys. Rev. Lett* **37** 179 (1976).
[13] S. A. Levshakov, *Vistas Astron.* **37** 535 (1993).
[14] A. Y. Potekhin and D. A. Varshalovich, *Astron. Astrophys. Suppl. Ser.* **104** 89 (1994).
[15] D. A. Varshalovich, V. E. Panchuk, and A. V. Ivanchik, *Pis'ma Astron. Zh. (Engl. transl.: Astron. Lett.)* **22** 8 (1996).
[16] S. A. Levshakov and D. A Varshalovich, *Monthly Not. R. Astron. Soc.* **212** 517 (1985).
[17] C. B. Foltz, F. H. Chaffee, and J. H. Black, *Astrophys. J.* **324** 267 (1988).
[18] D. A. Varshalovich and S. A. Levshakov, *JETP Lett.* **58** 237 (1993).
[19] D. A. Varshalovich and A. Y. Potekhin, *Pis'ma Astron. Zh. (Engl. transl.: Astron. Lett.)* **22** 3 (1996).
[20] A. Y. Potekhin, A. V. Ivanchik, D. A. Varshalovich, K. M. Lanzetta, J. A. Baldwin, G. M. Williger, and R. F. Carswell, *Astrophys. J.* in press (1998).
[21] Yu. N. Gnedin, A. Z. Dolginov, and A. I. Tsygan, *Pis'ma Zh. Exper. Teor. Fiz.* **10** 441 (1969).
[22] Yu. N. Gnedin and G. G. Pavlov, *Sov. Phys.–JETP* **38** 903 (1973).
[23] A. Z. Dolginov, Yu. N. Gnedin, and N. A. Silant'ev, *Propagation and Polarization of Radiation in Cosmic Media* (Nauka, 1979; Gordon and Breach, 1995).
[24] Yu. N. Gnedin and R. A. Sunyaev, *Astron. Astrophys.* **36** 379 (1974).
[25] A. D. Kaminker, G. G. Pavlov, and Yu. A. Shibanov, *Astrophys. Space Sci.* **91** 167 (1983).
[26] G. G. Pavlov and Yu. N. Gnedin, *Sov. Sci. Rev.*, ed. R. A. Sunyaev, Harwood Acad. Publ., vol. 3, p. 197, 1984.
[27] V. E. Zavlin and Yu. A. Shibanov, *Sov. Astron.* **135** 499 (1991).
[28] A. D. Kaminker, G. G. Pavlov, Yu. A. Shibanov *et al.*, *Astron. Astrophys.* **220** 117 (1989).
[29] Yu. N. Gnedin, G. G. Pavlov, and A. I. Tsygan, *Zh. Exper. Teor. Fiz.* **66** 421 (1974).
[30] A. Y. Potekhin, *Phys. Plasmas* **3** 4156 (1996).
[31] A. Y. Potekhin and G. G. Pavlov, *Astrophys. J.* **483** 414 (1997).
[32] V. G. Bezchastnov, G. G. Pavlov, and J. Ventura, *Phys. Rev. A* **58** 180 (1998).
[33] V. E. Zavlin, G. G. Pavlov, and Yu. A. Shibanov, *Astron. Astrophys.* **315** 141 (1996).
[34] G. G. Pavlov, Yu. A. Shibanov, V. E. Zavlin, and R. J. Meyer, *The Lives of the Neutron Stars*, ed. A. Alpar *et al.*, NATO Kluwer, Dordrecht, ASI Ser. C, vol. 450, p. 71, 1995
[35] Yu. A. Shibanov and V. E. Zavlin, *Astron. Lett.* **21** 38 (1995).
[36] D. Page, Yu. A. Shibanov, and V. E. Zavlin, *Astrophys. J.* **301** 247 (1995).
[37] D. Page, Yu. A. Shibanov, and V. E. Zavlin, *Röntgenstrahlung from the Universe*, ed. H. U. Zimmermann *et al.*, MPE Report, vol. 263, p. 173, 1996.

[38] A. I. Tsygan, *Astron. Astrophys.* **44** 21 (1975).
[39] A. G. Muslimov and A. I. Tsygan, *Astrophys. Space Sci.* **120** 27 (1986).
[40] A. I. Tsygan, *Astrophys. Space Sci.* **77** 187 (1981).
[41] I. G. Mitrofanov and G. G. Pavlov, *Mon. Not. R. Astron. Soc.* **200** 1033 (1982).
[42] I. G. Mitrofanov and A. I. Tsygan, *Astrophys. Space Sci.* **84** 35 (1982).
[43] V. G. Bezchastnov and G. G. Pavlov, *Astrophys. Space Sci.* **178** 1 (1991).
[44] A. D. Kaminker, P. G. Mamradze, and G. G. Pavlov, *Astrophys. Space Sci.* **192** 175 (1992).
[45] A. G. Muslimov and A. I. Tsygan, *Mon. Not. R. Astron. Soc.* **255** 61 (1992).
[46] V. D. Pal'shin and A. I. Tsygan, *Astron. Rep.* **40** 385 (1996).
[47] A. I. Tsygan, *Mon. Not. R. Astron. Soc.* **292** 317 (1997).
[48] V. D. Pal'shin and A. I. Tsygan, *Astron. Lett.* **24** 131 (1998).
[49] D. G. Yakovlev and V. A. Urpin, *Sov. Astron.* **24** 303 (1980).
[50] D. G. Yakovlev, *Astrophys. Space Sci.* **98** 37 (1984).
[51] D. A. Baiko and D. G. Yakovlev, *Astron. Lett.* **22** 708 (1996).
[52] A. Y. Potekhin and D. G. Yakovlev, *Astron. Astrophys.* **314** 341; **327** 442 (1996).
[53] A. Y. Potekhin, G. Chabrier and D. G. Yakovlev, *Astron. Astrophys.* **323** 415 (1997).
[54] D. G. Yakovlev and D. A. Shalybkov, *Sov. Sci. Rev.*, ed. R. A. Sunyaev, Harwood Acad. Publ., vol. 7, p. 311, 1989.
[55] D. G. Yakovlev and A. D. Kaminker, *Equation of State in Astrophysics*, ed. G. Chabrier and E. Schatzman, Cambridge Univ. Press, Cambridge, p. 214, 1994.
[56] O. Y. Gnedin and D. G. Yakovlev, *Nucl. Phys.* **A 582** 697 (1995).
[57] D. G. Yakovlev, *Strongly Coupled Plasma Physics*, ed. H. M. Van Horn and S. Ichimaru, Univ. Rochester Press, Rochester, NY, p. 157, 1993.
[58] V. A. Urpin and D. A. Shalybkov, *Astron. Astrophys.* **294** 117 (1995).
[59] G. Srinivasan, D. Bhattacharya, A. G. Muslimov, and A. I. Tsygan, *Current Sci.* **59** 31 (1990).
[60] P. Haensel, A. D. Kaminker, and D. G. Yakovlev, *Astron. Astrophys.* **314** 328 (1996).
[61] K. P. Levenfish and D. G. Yakovlev, *Astron. Lett.* **22** 56 (1996).
[62] D. G. Yakovlev, A. D. Kaminker, and K. P. Levenfish, *Neutron Stars and Pulsars*, ed. N. Shibazaki, N. Kawai, S. Shibata, and T. Kifune, Universal Acad. Press, Tokyo, p. 195, 1998.
[63] V. A. Urpin, S. A. Levshakov, and D. G. Yakovlev, *Mon. Not. R. Astron. Soc.* **219** 703 (1986).
[64] V. A. Urpin and A. G. Muslimov, *Mon. Not. R. Astron. Soc.* **256** 261 (1992).
[65] V. A. Urpin and K. A. Van Riper, *Astrophys. J.* **411** L87 (1993).
[66] V. A. Urpin and D. Y. Konenkov, *Neutron Stars and Pulsars*, ed. N. Shibazaki, N. Kawai, S. Shibata, and T. Kifune, Universal Acad. Press, Tokyo, p. 171, 1998.
[67] J. Miralles, V. A. Urpin, and D. Y. Konenkov, *Astrophys. J.* in press (1998).

Cosmic gamma-ray bursts

E. P. Mazets

1 Introduction

High-intensity pulsed hard-photon fluxes with photon energy in the range from ~ 1 KeV to tens of MeV propagate in the interstellar space of our Galaxy. When such a very thin wave of photon density runs through the Solar system and encounters a gamma-ray detector aboard a spacecraft the observer sees what is known as a cosmic gamma-ray burst (GRB). With their discovery and first observations in the early 70s [1], the GRBs became subject of intense study to find out from where they come, where and how they are produced.

Research on GRBs involves recording the time history of events, measuring energy spectra and their variations, mapping the sources and inspecting them at other wavelengths during active and quiescent periods, evaluating statistics.

Historically, several important periods in studying the GRBs can be pointed out after their discovery by the Vela spacecrafts in 1967–73. Thus, the launches of the Venera 11, 12 missions with high sensitivity gamma-burst *KONUS* [2] and *SIGNE* detectors [3] allowed moving from a few events recorded per year to an event detected and studied every 2–3 days. The results obtained by Venera 11, 12, 13, 14 missions in 1978–83 gave rise to the present observational database of GRB parameters [2, 4]. A number of types of bursts with different time history and duration distribution were studied. Short bursts were indicated as a separate class. Sources of soft repeating bursts were observed. The energy spectra of bursts were investigated. A rapid spectral variability was found. Absorption and emission features were discovered in the spectra of some bursts. Localization of 160 sources revealed their isotropic distribution on the celestial sphere. On plotting the frequency of burst appearance against their intensity, deviations from the $-3/2$-power law valid for a uniform 3D distribution of burst sources were detected in the region of weak bursts.

Investigation of the GRBs entered a new phase with the BATSE experiment aboard the Compton Gamma-Ray Observatory (CGRO) launched in a near-earth orbit in 1991 [5]. The BATSE[1] high-sensitivity detectors measure, on the average, one burst per day. New data confirmed and developed the results obtained with the Veneras. Location of the sources of many hundred bursts confirmed their isotropic angular distribution with a high statistical accuracy.

Starting in November 1994, precise investigations of the bursts have proceeded aboard the *Wind* space station launched in an orbit very suitable for their observations [6]. More simple detectors of gamma-bursts placed on a number of other space-

[1] Burst and Transient Source Experiment.

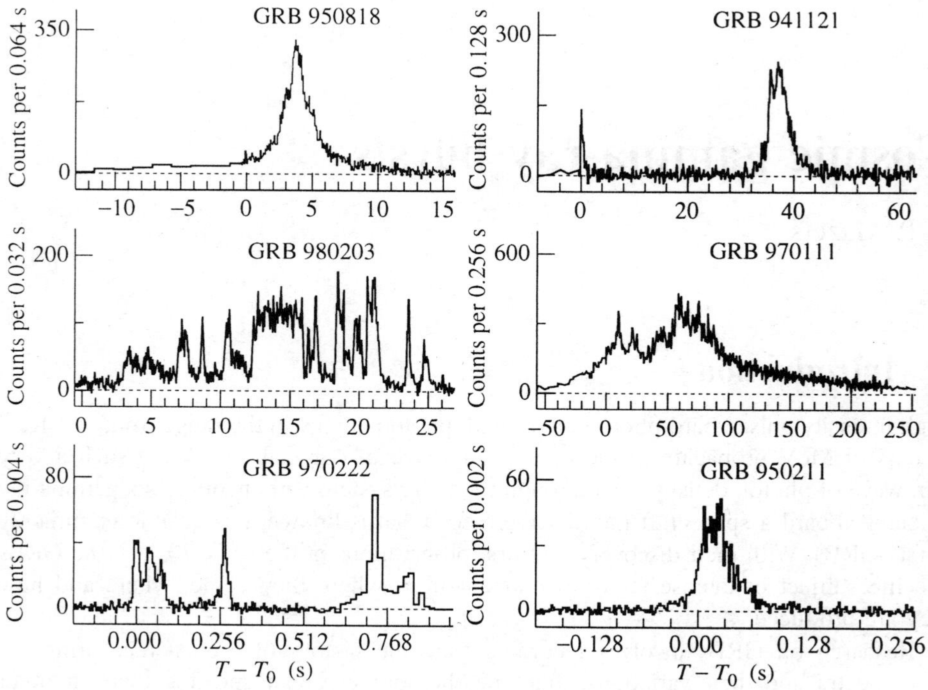

Fig 1. Several characteristic types of brightness curves of cosmic GRBs.

crafts, also efficiently contribute to the overall database. The most valuable results were obtained with interplanetary stations, such as PVO, Ulysses, Mars. These space stations together with the near-earth satellites gave rise to an interplanetary triangulation network whose long bases enable a high-accuracy location of gamma-bursts by a relative time-lag in their arrival at different spacecrafts [7].

Finally, with the launch of the BeppoSAX satellite in the mid-1996, the long-standing researchers' dream of precise and, principally, fast location of bursts, which would allow optical detection of a burst source with large ground-based telescopes within a short time after the event, started to come into being [8].

Thus, during the last years much has been learned, the number of publications on bursts exceeds two thousands [9], but non-controversial generally accepted explanation of the bursts' origin is still lacking. Tens of different hypotheses and models of the phenomenon are in competition, which evidences that the epithets mysterious and enigmatic, as applied to gamma bursts, are not figure of speech.

2 Time history of GRBs

The duration of the bursts varies over a wide range from tens of milliseconds to several minutes. Time histories of the bursts are of remarkable variability. A number of attempts have been undertaken to rank the events according to their morphology. As an example, some brightness curves measured in the *KONUS* experiment aboard the *Wind* station are shown in Fig. 1 for the energy range 50–200 KeV.

First of all, mention should be made of the existence of two classes of events. Short bursts (GRB970222, GRB950211, see Fig. 1) strongly differ from longer events in their rise time ~ 10–100 ms in addition to their small duration < 0.5–1 s. At least two groups of events can be isolated among the ordinary longer ones. Simple single pulsed bursts (GRB950818) last for 5–15 s. In some cases, their duration increases severalfold and, then, they begin to look as long structureless events (GRB970111). Bursts with a complex multipulse time profile are the most numerous. The number of individual peaks observed in a burst profile may vary significantly and be occasionally as great as several tens (GRB980203). Frequently these peaks form quasi-periodic series. However, there is no strictly regular periodicity in the burst profiles with one exception (the event of March 5, 1979). As a rule, the rise and fall times of separate peaks in complex bursts are shorter than those of individual pulsed events. In many cases, precursors have been observed in the profiles separated from the main phase of a burst by several tens of seconds (GRB941121). Obviously, observations of weak precursors are limited by instrument sensitivity.

Unquestionably, the time structure of a burst reflects the most inherent properties of the emission process in its source. An explanation of the temporal behavior of bursts seems to be the main purpose of any source model. Unfortunately, up to now most of the burst models under development deal with source energetics and energy spectra, but pay only minor attention to the time structure.

3 Energy spectra

The energy spectra constitute a more uniform pattern compared with the variety of the burst time structures. Smooth continuous spectra measured with practically achievable precision can be equally well represented, e.g., by the spectral distributions of thermal bremsstrahlung or synchrotron thermal radiation in the optically thin case. In practice, a combination of exponential and power distributions has been often used for data analysis. The actual mechanism of radiation is unclear as yet. With thermal origin of radiation supposed, the estimates of temperature in different burst sources fall within 10^9–10^{10} K. Representative examples of the observed spectra are presented in Fig. 2.

The problem of the existence of spectral features in burst spectra remains under discussion. For the first time features of two types were noticed in the *KONUS* experiment aboard Venera 11, 12 missions [11]. Absorption features were observed over the

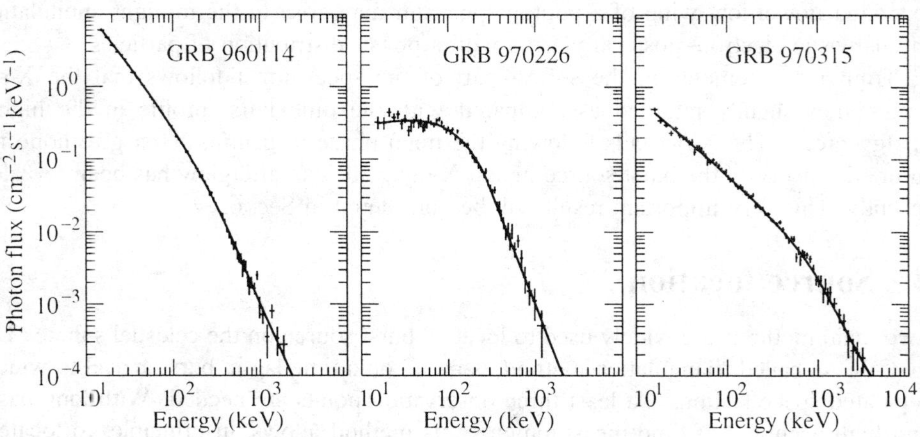

Fig 2. Examples of burst energy spectra.

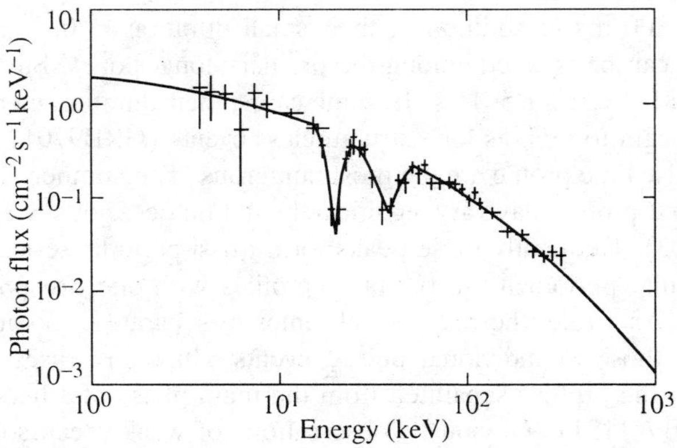

Fig 3. Absorption lines in the GRB880205 spectrum [12].

energy range 30–100 KeV. They were assumed to be consistent with electron cyclotron absorption at a magnetic field strength B of $(2-8) \times 10^{12}$ G. Results obtained on the Ginga spacecraft are a prominent example of absorption features [12] (Fig. 3).

Emission features lie in the range 350–450 KeV and higher. It is highly probable that these features are consistent with the gravitational shift of electron–positron pair annihilation radiation. The occurrence of these two types of features may indicate burst generation in strongly magnetized neutron stars.

No spectral features of bursts could be detected in observations with CGRO [13]. Meanwhile, new evidence of their existence has been found. Some of the results obtained with the *Wind* station are presented in Fig. 4.

Energy spectra exhibit strong and rapid variability [4, 15]. Continuous spectra measured over different phases of the bursts, as a rule, differ drastically. The hardness of spectra may change as fast as the intensity of radiation does. GRB970226 profiles measured in different energy ranges G_1 and G_2 and the related variations of spectral hardness characterized by a ratio of intensities N_2/N_1 are shown in Fig. 5. It is worth noting that the most rapidly changing the spectral features. Observations aboard the SMM satellite have shown for the first time that the hard radiation of burst may extend to energies of tens and hundreds of MeV [16]. It is possible that such hard tails in the spectra are related to annihilation peculiarities, being their continuation [17], since the extended power-low wing of a photon spectrum may arise in the in-flight annihilation in relativistic electron–positron plasma with a power distribution of particles.

From measurements in the softest part of the spectrum it follows that the X-ray emission is slightly retarded as compared with the time burst profile in the higher energy range. The X-ray tails following the main phase of gamma burst give hope for future discovery of the burst source in soft X-rays. Such an afterglow has been revealed recently. This very important result will be considered in Sec. 6.

4 Source location

Two main methods are widely used to locate a burst source on the celestial sphere. The first, the so-called triangulation method, is based on a time-lag in burst arrival to widely separated space stations. At least three observation points are needed. With long bases and high accuracy of time measurements this method allows, in principle, to locate a

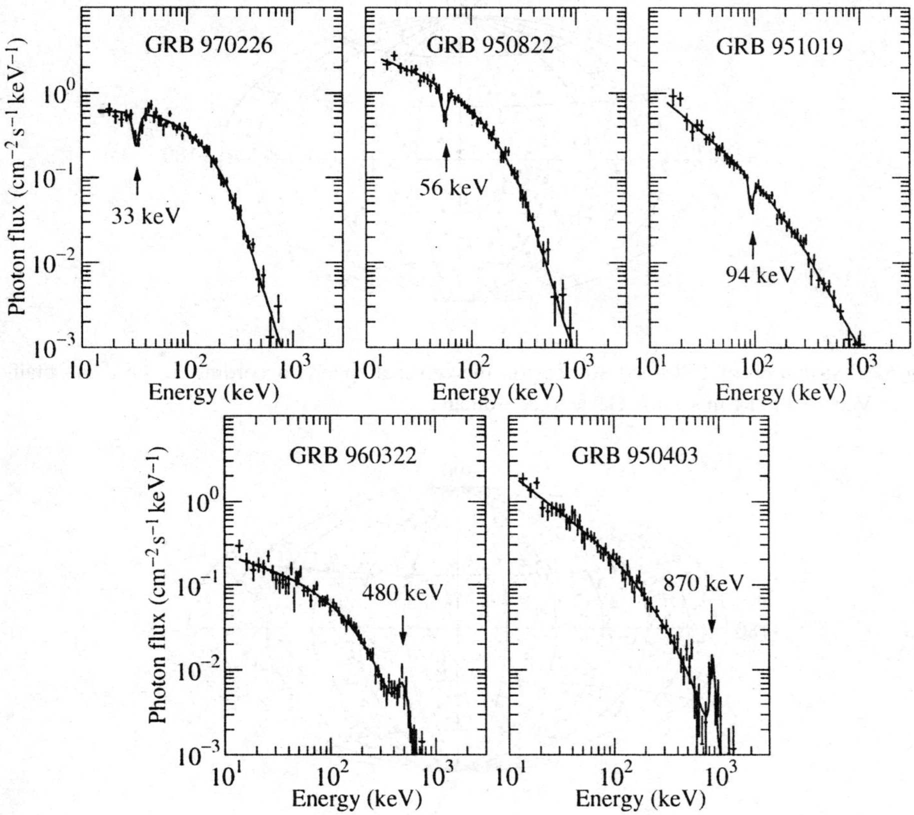

Fig 4. Examples of peculiarities in burst spectra (*KONUS-Wind*).

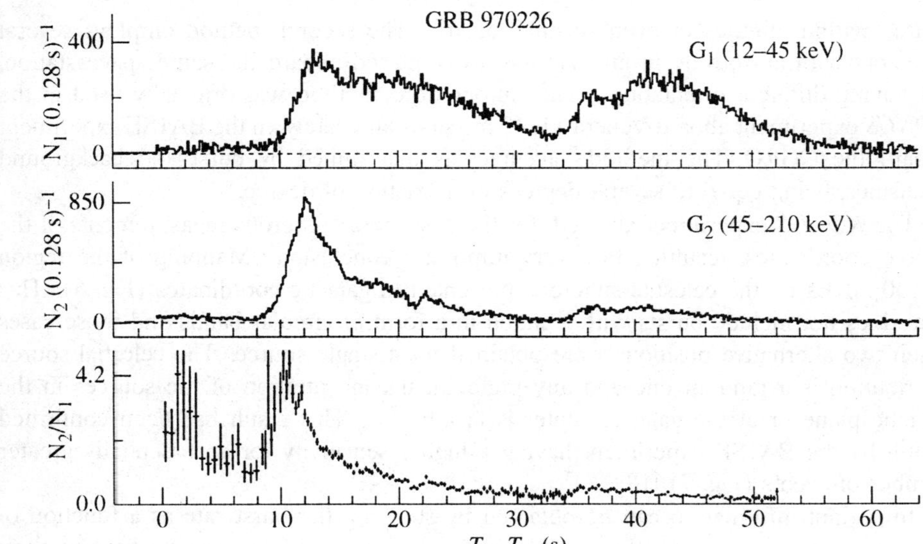

Fig 5. Distinctions between brightness curves taken in different energy ranges are determined by spectral variability of radiation.

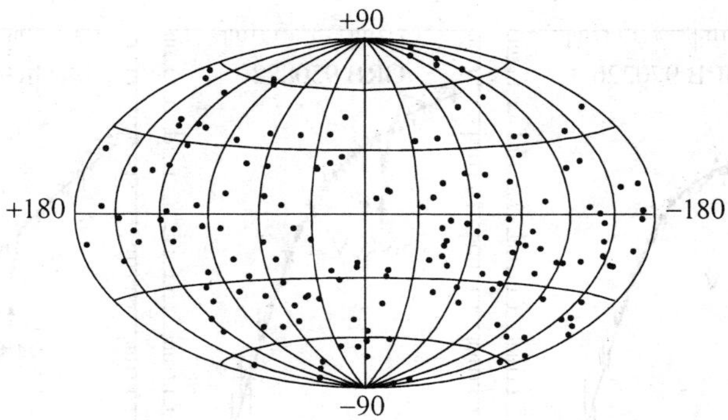

Fig 6. Distribution of 168 burst sources on the celestial sphere, according to the data obtained in the Venera 11–14 missions. Galactic coordinates.

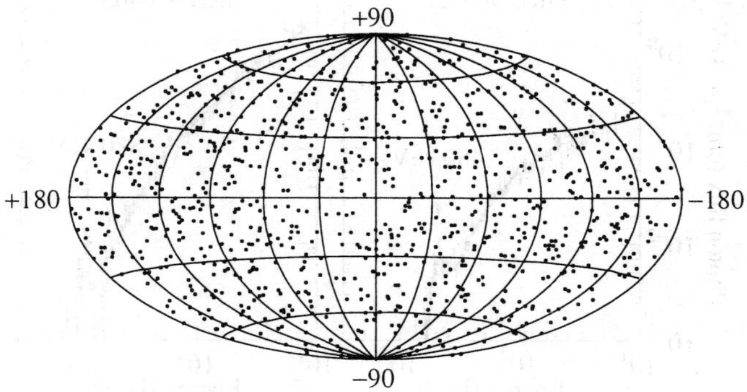

Fig 7. Distribution of 1005 sources on the celestial sphere. The BATSE observation [18].

source within minutes or even seconds of arc. The second method employs several detectors with anisotropy in angular sensitivity placed aboard the same space station, but having different orientations. This autonomous method was originally used in the *KONUS* experiment aboard Venera 11, 12 missions and, lately, in the BATSE experiment aboard the CGRO. The method's accuracy is determined by burst and background intensities, being equal to several degrees or a fraction of degree.

The *KONUS* experiment allowed, for the first time, numerous measurements of the source coordinates, resulting in a very important conclusion. Mapping of the region of 160 GRBs on the celestial sphere is presented in galactic coordinates (Fig. 6). This map does not include bursts with location in a form of circular bands and those cases when two alternative positions were obtained for a single source. The celestial source distribution is a random one and any preferential concentration of the sources in the galactic plane or at the galactic center is lacking [4]. This result has been confirmed in full by the BATSE experiment having a higher sensitivity for an essentially greater number of events (Fig. 7) [18].

Important information can be obtained by studying the burst rate as a function of burst intensity *I*. Naturally, the smaller bursts appear, on the average, to have as their origin more distant sources. As a measure of intensity is taken the total energy fluency *S*, erg cm^{-2}, or a peak power of a burst *P*, erg cm^{-2} s, or, finally, the maximal counting

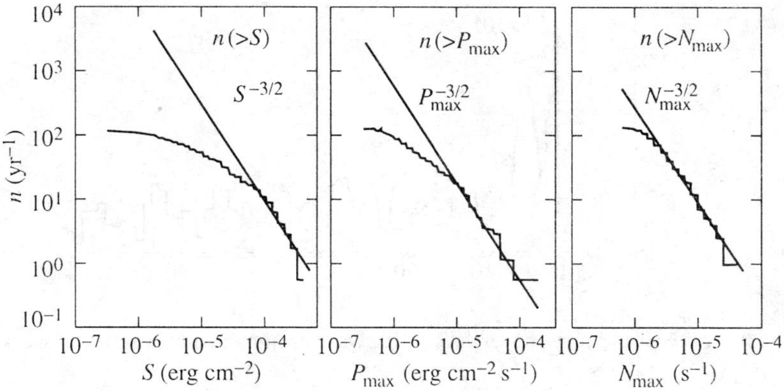

Fig 8. Frequency of burst events n against the total energy flux S, luminosity in maximum P_{max}, and maximum counting rate N_{max}.

rate in a burst peak N_{max}, s^{-1}. The cumulative GRB intensity distribution $n(>I)$ is commonly considered in the form $\lg n - \lg I$. For a uniform spatial distribution of bursts such a function must obey the $-3/2$-power law. The observed distributions reveal very strong deviations from this law in the region of weak bursts (Fig. 8). Part of weak bursts is lost because of the observational selection effects. The smallest selection distortions in a distribution $n(>I)$ take place when N_{max} is chosen as intensity measure. However, the observed distortions in this distribution are too strong, and the deficiency of weak events is real. What this means is that the spatial source density falls off at long distances. Such a shape of $n(>I)$ may be consistent with the isotropic source distribution over the sky in the following three cases: (i) the sources are in the neighborhood of the Sun at distances less than the interstellar ones, (ii) the sources are in an extended galactic halo with dimensions essentially greater than the galaxy disk diameter, and (iii) the sources are at cosmologically long distances. Lately the cosmological models have received the strongest observational support.

5 Repeating soft short bursts

Among a number of GRB observations a small peculiar class of events stands out whose sources have been named the soft gamma-repeaters (SGRs). Their main distinctive feature is their recurrence. Generation of a burst by a source may repeat many times. This excludes a large class of those models related to disruption or drastical rearrangement of a source. Up to now four such sources have been discovered and located. The most peculiar was the burst of March 5, 1979 [19]. It started with a very intensive initial burst with a hard radiation component and an annihilation line. Then, soft pulsed radiation was observed, typical of an the X-ray pulsar (Fig. 9). The most probable explanation of the periodic structure of this burst is the neutron star rotation with a period of 8 ± 0.05 c.

A prominent peculiarity of the GRBS0526-66 was a series of short repeating soft bursts observed in the *KONUS* experiment. A total of 16 repeating bursts were recorded during 687 days of observations in 1979 (Venera 11, 12) and 1981–83 (Venera 13, 14), i.e., on the average, one burst in 48 days. Soon a repetition of bursts from the B1900+14 (also in the *KONUS* experiment) and also in SGR1806-20 was observed. In the follow-

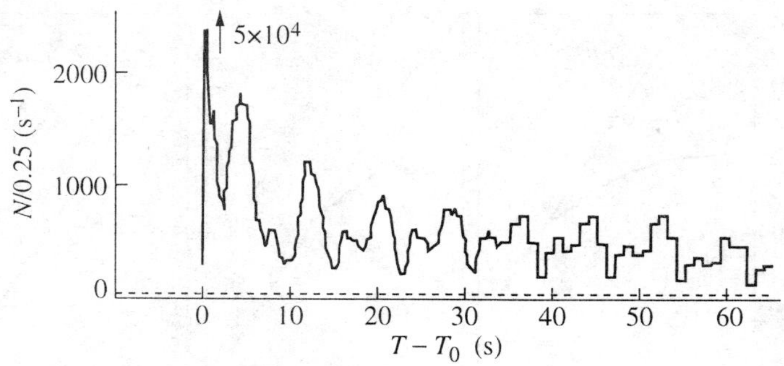

Fig 9. The most powerful burst of March 5, 1979, was accompanied by a series of weak repeated bursts in 1979–82.

ing years, periods of new activity of B1900+14 were observed with the BATSE, and new repeating bursts in SGR1806-20 were detected with the BATSE, *KONUS-Wind*, and Ulysses instruments. On June 29, 1997, these instruments discovered and located SGR1806-14, the fourth source of repeating bursts.

The energy spectra of all the repeating bursts are very soft. Their approximation by

$$\frac{dN}{dE} \sim E^{-1} \exp\left(-\frac{E}{kT}\right)$$

estimates kT at 20–30 KeV for all the four sources. Evidently, the soft repeating bursts are emitted owing to nonstationary sporadic accretion onto a magnetized neutron star. In this connection it is interesting to note that the regions of location of three SGRs are in close proximity with supernovae remnants [20].

6 Attempts of identification

Reliable identification of the GRB sources with one of the known types of astrophysical objects observed and studied in other spectral regions would be a cardinal result. The characteristic scale of distances to sources would be determined and, hence, their natural energetics.

The functioning of the interplanetary triangulation network of satellites in the 80s allowed determining the positions of several burst sources to within error limits of the order of one arcminute. It was revealed that these areas do not contain any bright objects in radio-, optical, or X-ray ranges. Thorough search with highest sensitivity revealed there only a few objects of weak brightness. It should be noted that the inspection of those areas was performed long after a GRB. The afterglow of sources, if any, might fall below the threshold of detectability.

The interest in observations at other wavelengths increased significantly after remnants of optical bursts in certain regions of their location were revealed on archival astronomical plates. This might mean that those bursts are recurrent, the same sources were active in the past, and X-ray radiation was accompanied by optical photon emission. Pitifully, all the efforts to reveal any GRB sources in active or quiescent state at other wavelengths failed to produce unambiguous conclusions and until very recently [21] this situation remained uncertain. It changed drastically after a launch of the Italo-Dutch BeppoSAX satellite in mid-1996. To record the GRB, an all-sky monitor of bursts was

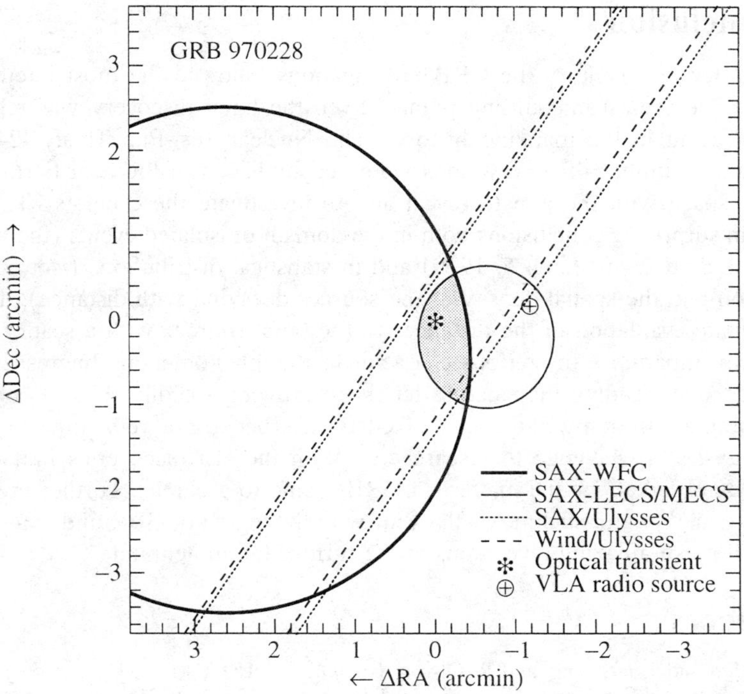

Fig 10. Location of GRB970228—an interception of the larger circle with triangulation band. Position of decaying X-ray source is the smaller circle. Optical transient is an asterisk [8].

mounted aboard this satellite. If a burst falls within the field of vision of a wide-angle camera it can be located with an accuracy of several minutes of arc. Then, the satellite can be reoriented with the aim to direct its narrow-angle X-ray telescope to the burst region. Such a sequence of observations took place for a GRB of February 28, 1997 [8]. Large ground-based telescopes were activated as well as the space telescope "Hubble." As a result, a weak decayed X-ray and optical source was discovered within a small region of burst location (Fig. 10). This is in all probability the same object and, hence, for the first time an afterglow of a GRB source was discovered in the X-ray region as well as in the visible. It is notable that this decayed optical source coincides in position with a distant galactic (probable value of $Z = 0.2$–2). A new success was the observation of the GRB970508. X-ray and optical decayed afterglow were discovered a redshift $Z \sim 0.8$ [22, 23].

These estimates of Z correspond to a distance of $\sim 10^{28}$ cm and a released energy of $\sim 10^{51} - 10^{52}$ ergs. The relevant cosmological models assume that the burst origin could be a merger of two neutron stars or of a neutron star and a black hole, and the pattern of gamma-radiation of a burst and X-ray and optical afterglow should be determined by ultrarelativistic expansion of the resulting fireball. Thus, observations at other wavelengths gave, as expected, urgently wanted results very important for elucidating the origin of the GRBs. These observations surely require further development and continuation. Efforts of burst hunters will be aimed in this direction for the nearest future.

7 Conclusion

For a quarter of a century the GRB investigations followed the most interesting path with dramatic events and turning points. Even the burst discovery was related to an important event in the mankind history—The Nuclear Test-Ban Treaty [24]. During these years a number of observations were accomplished in the near-Earth and interplanetary space with the aim to reveal and to investigate these bursts. The attempts resulted in surprising conclusions both in the domain of isolated events (the most bright example is a burst of March 5, 1979) and in statistical distributions (isotropic angular distribution and the spatial density of the sources decaying with distance). The lack of any adequate evaluation of the distances to the burst sources with a scatter from 10^{14} to 10^{28} cm imparted a mysterious character to this phenomenon. Interest in studying them and understanding their nature has been growing steadily. First a concrete task of the gamma-astronomy, the origin of GRBs has become a very important problem of astrophysics, a challenge to researchers. With the last discoveries initiated by the BeppoSAX observations the interest in GRBs came to a climax. Further investigations will show whether an anticlimax will occur with the region of GBR understood or these bursts will make amazing revelations at the turn of the millennium.

References

[1] R. Klebesadel, I. Strong, and R. Olson, *Astrophys. J.* **182** L85 (1973).
[2] E. Mazets and S. Golenetskii, *Astrophys. Space Sci.* **75** 47 (1981).
[3] C. Barat, G. Chambon, K. Hyrley, *et al.*, *Astrophys. Space Sci.* **75** 83 (1981).
[4] E. Mazets and S. V. Golenetskii, *Sov. Sci. Rev. E. Astrophys Space Phys.* **6** 281 (1988).
[5] C. Meegan, G. Fishman, R. Wilson, *et al.*, *Gamma Ray Bursts, AIP Conf. Proc.* vol. 307, N.Y., p. 3, 1994.
[6] R. Aptekar, D. Frederiks, S. Golenetskii, *et al.*, *High Velosity Neutron Stars and Gamma-Ray Bursts, AIP Conf. Proc.* vol. 336, N.Y., p. 158, 1996.
[7] K. Hurley and T. Cline, *Gamma-Ray Bursts, AIP Conf. Proc.*, vol. 307, N.Y., p. 653, 1994.
[8] J. van Paradijs, P. Groot, T. Galama, *et al.*, *Nature* **387** 783 (1997).
[9] K. Hurley, *Gamma Ray Bursts, AIP, Conf. Proc.* vol. 384, N.Y., p. 985, 1996.
[10] C. Kouveliotou, *Ap. J. Suppl. Ser.* **92**, 637 (1994).
[11] E. Mazets, S. Golenetskii, R. Aptekar, *et al.*, *Nature* **290** 378 (1981).
[12] T. Murakami, M. Fujii, K. Hayashida, *et al.*, *Nature* **335** 234 (1988).
[13] D. Palmer, B. Teegarden, B. Schaefer, *et al.*, *Gamma-Ray Bursts, AIP Conf. Proc.* vol. 307, N.Y., p. 247, 1994.
[14] S. Golenetskii, R. Aptekar, D. Frederiks, *et al.*, *4th Gamma Ray Bursts Symposium, AIP Conf. Proc.*, vol. 428, N.Y., p. 284, 1998.
[15] L. Ford, D. Band, and J. Matteson, *Gamma-Ray Bursts, AIP Conf. Proc.* vol. 307, N.Y., p. 298, 1995.
[16] P. Nolan, G. Share, and S. Matz, *High Energy Transients in Astrophysics, AIP Conf.* vol. 115, N.Y., p. 399, 1984.
[17] S. Golenetskii, E. Mazets, R. Aptekar, *et al.*, *Astrophis and Space Sci.* **124** 243 (1986).
[18] M. Briggs, W. Paciesas, and G. Pendelton, *Ap. J.* **45** 40 (1996).
[19] E. Mazets, S. Golenetskii, V. Iluinskii, *et al.*, *Nature* **282** 587 (1979).
[20] R. Rothschild, *High Velosity Neutron Stars and Gamma-Ray Bursts, AIP Conf. Proc.* vol. 366, N.Y., p. 51, 1996.
[21] B. Schaefer, *Gamma-Ray Bursts, AIP Conf. Proc.* vol. 307, p. 382, 1994.
[22] S. Djorgovski, M. Metzger, S. Kulkarni, *et al.*, *Nature* **387** 876 (1997).
[23] M. Metzger, S. Djorgovski, S. Kulkani, *et al.*, *Nature* **387** 878 (1997).
[24] J. Bonnell and R. Klebesadel, *Gamma-Ray Bursts, AIP Conf. Proc.* vol. 384, N.Y., p. 977, 1996.

Development of new methods of mass spectrometry. Basic research and discoveries

B. A. Mamyrin

1 Introduction

Nowadays mass spectrometry is one of the most universal methods for determining the composition of substances. This is due to the fact that the substances being analyzed may be in any aggregative state: solid, fluid, gas, vapor, or plasma, with atoms, molecules, clusters, ions of any natural or artificially created substances analyzed. The concentration of components in mixtures of substances may vary over extremely broad limits, from nearly 100 to 10^{-10}% (without preliminary concentration of minor impurities).

Therefore, it is difficult to name an area of science or technology where mass spectrometry would not be applied: physics, chemistry, biology, geochemistry, space research, medicine, criminology, ecology; technology of a variety of productions, including metal manufacture, fuel and power industry, production of pure and highly pure substances, food-processing industry, etc.

Regarding the whole set analytical parameters of mass-spectrometric analyzers, dynamic mass spectrometers, in which ions are separated under the action of variable electrical fields, have indisputable advantage over static devices.

The obtained record-breaking values of basic analytical parameters of dynamic mass spectrometers are listed in the table; they exceed parameters of static devices $10^3 - 10^6$ times. It is mass analyzers of this kind that were being developed at the Laboratory of Mass Spectrometry of the Ioffe Institute. The high analytical parameters of our instruments allowed us to perform a number of important fundamental and applied investigations and to obtain some unexpected results, which gave rise to three fundamental discoveries (Secs. 4–6).

2 Development of resonance mass spectrometers with magnetic focussing of the time of flight of ions

One of the first dynamic mass spectrometers developed in the laboratory was the magnetic resonance mass spectrometer (MRMS) with focussing of the time of flight of ions packets along a circular orbit in a magnetic field. A detailed description of how the instrument operates is given in [1, 2].

Table 1. Record-breaking values of analytical parameters of dynamic mass spectrometers (M, ion mass; P, gas pressure in ion source).

Range of masses of singly-charged ions (amu)	$1–10^7$
Resolution ($M/\Delta M$)	up to 10^7
Absolute sensitivity $P_{\min}$ (Torr) or minimum number of atoms in probe	10^{-16} tens of atoms
Relative sensitivity ($P_{\min}/P_{\max}$)	10^{-12}
Minimum time of registering the whole mass spectrum (s) (One laser pulse ~ 2 ns)	10^{-6}
Error of ion mass measurement (amu)	up to 10^{-9}

Principal analytical features of an MRMS having dimensions, weight, and power consumption acceptable for a laboratory device are the following:

– resolution as high as several hundred thousand [2],

– absolute sensitivity of up to several thousand atoms in a sample,

– double sequential separation of ions and high dispersion of hundreds of millimeters per atomic mass unit produce an extremely good shape of mass spectrum peaks, at which isotope ratios as low as 10^{-11} can be measured [3].

Several modifications of the device with pulsed and sinusoidal feed of the modulator have been developed in the laboratory. One type is produced commercially (MI-9303) [3] (Fig. 1).

In Secs. 3–7 the principal results obtained with MRMS are presented.

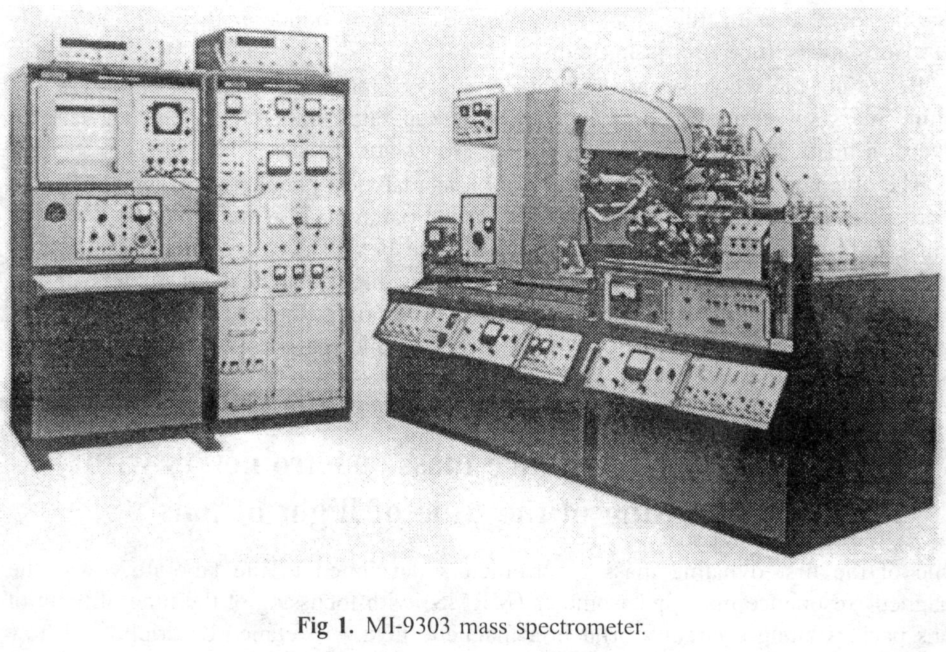

Fig 1. MI-9303 mass spectrometer.

3 Study of helium isotopes in nature

$^3He/^4He$ isotope ratios lie in an extremely wide range from 1 to 10^{-10}. Such isotope ratios could not be measured with static mass spectrometers, and, prior to the creation of our device, data on the isotopy of helium had been very scarce. Our investigations have shown that helium from the Earth's atmosphere has an isotope ratio $^3He/^4He \sim 1.4 \times 10^{-6}$, whereas helium from deep wells shows an isotope ratio $^3He/^4He$ which is the smaller, the deeper is the sampling point located. On the average, for the deep-seated helium of the Earth's crust $^3He/^4He \approx 3 \times 10^{-8}$. This brought up the "paradox of atmospheric helium"—whence comes the excess ^3He? We started to investigate helium from the deepest "wells"—volcanoes (their active zones lie at a depth of several hundred kilometers!). To our great surprise, it was found that they contain helium with an isotope ratio $^3He/^4He$ 1000 times larger than that in gases from the terrestrial crust [5]. Calculations have shown that the "output" of volcanoes and egress of ^3He into space just ensure the $^3He/^4He$ ratio measured by us in the atmospheric helium $\sim 1.4 \times 10^{-6}$. Based on numerous researches performed in cooperation with geochemists, we obtained for the first time the general pattern of He isotope distribution throughout the Earth and near space (Fig. 2).

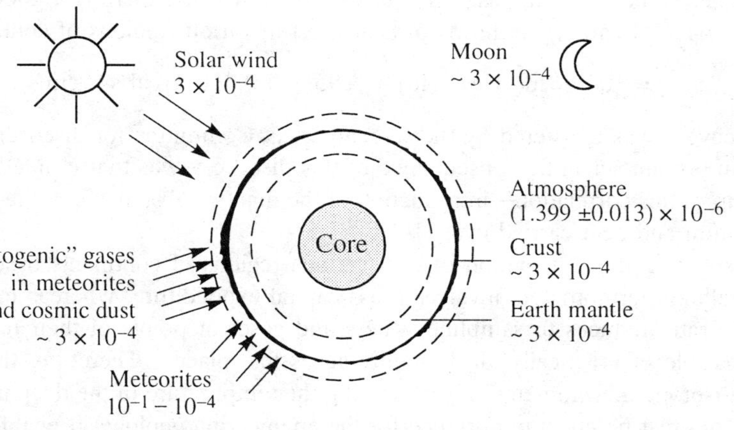

Fig 2. $^3He/^4He$ isotope ratios in geospheres and near space.

4 The importance of discovery of helium with anomalously high ^3He content in the Earth's mantle

The discovery in the Earth's mantle of helium with an anomalously high $^3He/^4He$ ratio of $\sim 3 \times 10^{-5}$ is explained by the capture during Earth formation of the so-called "protogenic" helium contained in many meteorites, cosmic dust, lunar soil and solar wind (Fig. 2). In all objects of near space $^3He/^4He \approx 3 \times 10^{-4}$. The protogenic helium, mixed with radiogenic helium, is responsible for the ratio $^3He/^4He \approx 10^{-5}$ in the Earth's mantle, with this value observed with a surprising persistence all over the globe—in volcanic gases of all continents, all islands and all oceans.

This discovery turned upside down all geochemical concepts about the degassing of the Earth's deep interior: it had been believed that inert gases had abandoned the deep interior in the stage of the "hot" phase of Earth formation, being preserved partly in rock and mainly in the atmosphere. Since ^{3}He, the most mobile of all inert gases, was preserved in the mantle, then, naturally, all other inert gases [6] were preserved too. This was an important argument in favor of the hypothesis of "cold" accretion of the Earth during its formation.

Investigations have shown that deep breaks of the terrestrial crust (invisible, being concealed by sedimentary rock) in which the mantle approaches the surface of the Earth can be readily traced by high values of the ^{3}He/^{4}He ratio. This is of great practical importance for the prospecting geology, since it is in the zones of deep breaks that many types of mineral resources are concentrated.

The discovery of helium with anomalously high isotope ratio in the Earth's mantle was registered by the Russian Committee for Discoveries and Inventions [7].

5 Discovery of the correlation between the ^{3}He/^{4}He ratio and the intensity of heat fluxes in the upper stratum of the Earth's crust under the continents

Studies of the distribution of ^{3}He/^{4}He in the surface stratum of the continental crust and the heat flux (q) in the same regions have shown that there is a good correlation between ^{3}He/^{4}He and q (in terms of heat flux) in various regions of continents [8, 9]

$$q = [0.166\ln(^3\mathrm{He}/^4\mathrm{He}) + 3.95 \pm 0.03]10^{-6}\ \mathrm{cal\ s^{-1}\ cm^{-2}}.$$

The discovery was registered by the Czechoslovak Committee for discoveries [10].

An important scientific consequence of this discovery was that it aided in revealing the origin of this correlation—introduction of the mantle substance into the Earth's crust with helium and heat carried over [10].

Investigating the distribution of q in various regions of continents by drilling wells and installing thermometers involves gross capital expenditure, whereas measuring the ^{3}He/^{4}He ratio reduces to sampling waters and gases at points of their natural egress, being feasible in practically all difficultly accessible places. Therefore, the developed "helium-isotope thermometer" for measuring the temperature of the deep interior of the Earth is of great practical importance for the engineering geology. It enables predicting temperature in designing deep mines and wells, which is of paramount importance, since the possible temperature at the same depth in various regions of continents may vary by more than a factor of 5, for example, from 30 to 150 °C!

As a result of the opened up possibility of studying the isotopy of natural helium, a new branch of geochemistry has originated—the geochemistry and space chemistry of helium, which has changed much the understanding of processes occurring on the Earth and in near space. Some important data, essentially changing the concepts of some natural processes, have been obtained. It has been found that the amount of space substance falling onto the Earth (as cosmic dust and micrometeorites) is $\sim 2 \times 10^4$ ton yr^{-1} [11], rather than 2×10^8 ton yr^{-1} (!), as it was believed in [12]; the "paradox" of growth of oceanic concretions has been solved, with the average time of their growth found to be $\sim 10^4$ yr [13], instead of 5×10^7 yr (!) [14]; a novel model for calculating the rate of accumulation of Earth's oceanic deposits has been substantiated. Helium-isotope searching criteria for use in prospecting of uranium-thorium and lithium-beryllium deposits have been substantiated; it has been shown that the ^{3}He/^{4}He ratio in the Earth's atmosphere is very stable up to heights of $\sim$ 10 km, being equal to $(1.399 \pm 0.013) \times 10^{-6}$ [15], this value is used as reference all over the world in studying the isotopy of helium.

The results of our investigations of the isotopy of helium (in cooperation with geochemists) were summed up in two monographs [16, 17].

6 Discovery of the dislocation-dynamic diffusion

A study of the strength properties of LiF crystals at extremely low temperatures in the laboratory of physics of profiled crystals at the Ioffe Institute has revealed specific features indicating the influence of the density of the medium (^{4}He in liquid or gaseous state). According to the formulas of classical thermal diffusion, atoms of helium could not penetrate into the samples under study at liquid helium temperature. However, MRMS measurements have shown a huge amount of helium (10^{10}–10^{13} atoms per 1 cm^2 of crystal surface) is evolved upon heating from crystals deformed in liquid ^{3}He. Control tests have shown that helium atoms are captured only by strained crystals and the amount of captured atoms is proportional to the number of dislocations formed. This was the reason for naming the discovered phenomenon the dislocation-dynamic diffusion (DDD) [18]. In the case of the dislocation-dynamic diffusion:

(1) the number of atoms of the medium penetrating into the crystal lattice of a sample is independent of temperature over an extremely wide range from 1.6 to 300 K,

(2) the captured helium is firmly retained in the crystal lattice under normal conditions (for many months) and is evolved in a narrow interval of temperatures corresponding to fast annihilation of dislocations on gradually elevating the temperature to $\sim 300\,°$C,

(3) the DDD processes are common to all tested monocrystals and metals (LiF, NaCl, KBr, CsI, Cu, Rb, Sn, Al, In, Cd, Zn, Ti, F) in various media (He, Ne, N_2, H_2O, etc.) [19],

(4) control tests have shown that DDD is not related to microcrack formation.

The DDD phenomenon is not only of scientific importance as a new mechanism of penetration of atoms of the medium into the crystal lattice of solids, but also of practical value, since in some cases it essentially modifies the strength properties of materials. This discovery was registered [20].

7 Development of a special MRMS for precision measurement of fundamental physical constants

The present-day table of exact fundamental physical constants (FPCs) includes more than 130 constants (Planck, Faraday, and Avogadro constants, optical and physicochemical constants, charges, masses and magnetic moments of electron, proton, deuteron, muon, etc.). However most of them are calculated using fundamental relations of physics from values of several (of order 10) quantities, which are some combinations of FPCs, measured experimentally with high accuracy. By the moment of beginning of our studies, one of the constants preventing the accuracy of the whole system of FPCs from being improved was the ratio of the magnetic moment of proton to nuclear magneton μ_p/μ_N. It can be readily shown that $\mu_p/\mu_N = \omega_N/\omega_c$, where ω_N is the frequency of nuclear magnetic resonance of proton and ω_c is the frequency of cyclotron resonance of proton, measured in the same magnetic field (the absolute value of the magnetic field is of no importance). Nowadays ω_N can be measured with very high accuracy by NMR spectroscopy; therefore, measuring ω_c and reducing the magnetic field strength to the same value for ω_c and ω_N present major difficulties. We succeeded in overcoming these experimental and calculation difficulties using an installation developed by us on the basis of a special MRMS. The value obtained by us, $\mu_p/\mu_N = 2.7927729 \pm 0.0000012$ [21], is until now the world's most accurate directly measured value of this quantity. This has resulted in an essential adjustment of such important quantities as Planck and Faraday constants, proton mass and charge of electron, magnetic moments of nuclei and other constants, in expressions for which these constants appear as factors.

The importance of our measurement of μ_p/μ_N can be judged from the following situation related to the book of American scientists, B. N. Taylor, W. H. Parker, and D. N. Langenberg "The Fundamental Constants and Quantum Electrodynamics" [22]. In the book, a thorough analysis was given of all experimental data on FPCs (measured before our basic work), and experimental values of the fine structure constant α and other quantities in the expressions for which it appears were compared with theoretical values of α calculated by methods of the quantum electrodynamics. The basic conclusion of the authors of the book is the following: "We believe there is some cause for concern in the present unsatisfactory state of agreement between quantum electrodynamic theory and experiment" (see p. 299 in [22]) and "In any case it appears that all is not well with the quantum electrodynamics" (see p. 306 in [22]).

The authors of the subsequent adjustment of FPCs rejected some μ_p/μ_N values measured in the US, UK, and FRG and also the Faraday number determined in NBS, and accepted our μ_p/μ_N value, determined much more accurately, so that the quantum electrodynamics was completely "rehabilitated." As a result, such important quantities as Planck and Faraday constants, charge of electron, magnetic moments of particles, etc. were changed by more than $20 \times 10^{-4}\%$, which is almost 3 times larger than the errors reported previously.

At present, an installation for further measuring of the FPCs and atomic masses is being assembled in our laboratory on the basis of a unique special-purpose MRMS with a resolution exceeding $\geq 10^6$.

8 Determination of the half-life of tritium and the influence of the electronic environment of the nucleus on this parameter

The accurate measurement of the half-life of tritium $(T_{1/2})$ is of paramount scientific importance, since the characteristics of this process: E_0, upper limit to the energy of β-electrons, their average energy, and $T_{1/2}$ are important quantitative characteristics for calculating some weak interaction constants, solving conclusively the problem of existence of the rest-mass of neutrino, evaluating quantitatively the influence of the electronic environment of the nucleus on the β-particle disintegration of tritium and estimating the lifetime of free neutron [23].

The concept of the helium-isotope method for determining $T_{1/2}$ [24], proposed and developed in our laboratory, consists in that the isotope ratios $^3He/^4He$ are measured after accumulation of ^{3}He in different amounts of the same mixture of tritium with ^{4}He during two different time intervals. As shown in [24], this obviates the necessity for measuring the initial absolute amounts of T and ^{4}He in the mixture of gases being used, accounting for leakages of ^{3}He and T and measuring temperatures and pressures, inherent in all the previously known methods for $T_{1/2}$ measurement.

The data on $T_{1/2}$ available in the literature disagree by more than 1.5%. In some cases measuring $T_{1/2}$ required up to 10 years and more for a single experiment. Our method furnished a value $T_{1/2} = 12.279 \pm 0.033$ years [25]. There is good reason to believe that the error can be reduced to 0.1% and lower.

Theoretical calculations show that the half-life of tritium nuclei must be influenced by modification of the electronic environment of the nucleus [26, 27].

To check this phenomenon experimentally, we developed in our laboratory a differential helium-isotope method for measuring the difference between $T_{1/2}$ for different electronic environments of the tritium nucleus [28]. This method allows obtaining in a relatively short time much more accurate values of the $T_{1/2}$ difference, than does the direct comparison of independent $T_{1/2}$ values for different electronic environments of the nucleus.

The measuring installation created in our laboratory in accordance with [28] furnished a $T_{1/2}$ ratio for T atoms and T_2 molecules:

$$\frac{(T_{1/2})_{\text{mol}}}{(T_{1/2})_{\text{at}}} = 1.0026 \pm 0.0004$$

$$\Delta T_{1/2} = 11.7 \pm 1.8 \text{ days.}$$

Theoretical estimates of the difference of half-lives of molecular and atomic tritium, obtained using various calculation techniques, lie in the interval 9.0–22.4 days [26, 27].

The proposed technique for measuring $\Delta T_{1/2}$ can furnish, on being further developed with the use of various chemical compounds, important data for judging about the structure of the atomic phase space accessible to a β-electron, and may become, with a representative body of data on $\Delta T_{1/2}$ for various tritium compounds accumulated, a basis for a new method for identification of electron states of hydrogen-containing substances.

It is interesting to note that Misra and Sudarshan [29] analyzed the so-called "quantum Zeno effect." The data obtained in our experiment are the first experimental corroborative evidence for the occurrence of this effect.

9 Development of nonmagnetic pulsed time-of-flight mass spectrometers. Mass reflectrons

The world's first pulsed nonmagnetic time-of-flight mass spectrometer with analytical performance acceptable to laboratory practice was developed in our laboratory and was the only dynamic mass spectrometer exhibited in the "Atomium" pavilion where scientific achievements were demonstrated at the World exhibition in Bruxelles.

Improvements of ion sources and detectors, made in our laboratory and other laboratories all over the world, allowed achieving a resolution of several hundred for devices of this kind. Increasing the resolution further was hindered by a strong initial energy

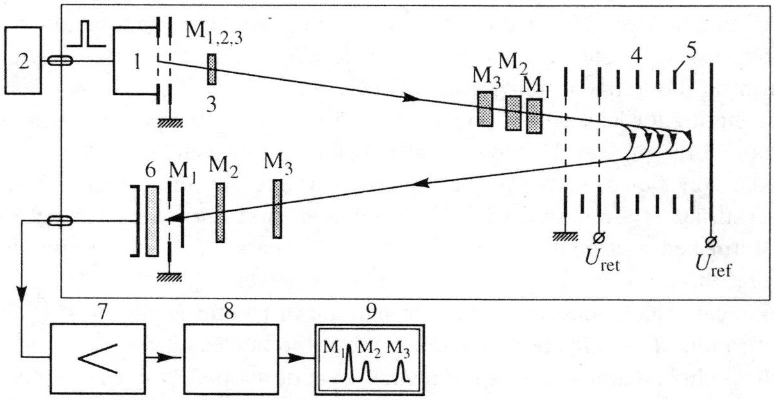

Fig 3. Schematic of mass reflectron. 1, pulsed ion source; 2, power supply of the source (pulsed generator or laser); 3, packet of ions emerging from the source; 4, two-sectional ion reflector (U_{ret}, retarding potential; U_{ref}, maximum potential of the reflector); 5, electrodes creating a uniform electric field; 6, secondary-electron multiplier; 7, amplifier; 8, computer; 9, display.

spread of ions, inevitably originating when ions are formed in a source in any ionization modes.

A dramatic improvement of the resolution of time-of-flight devices could be achieved by using the new scheme of a time-of-flight analyzer "mass reflectron," developed by us. This name was given to the new device in view of the fact that reflection of ions from a two-stage planar electrostatic mirror (see inventor's certificates and patents [30–32]) underlies the operation of the device.

In the mass reflectron scheme (Fig. 3), the second order of the time of flight of ion packets from the source to detector is focussed on the basis of energy and vertical and horizontal angles of ion take-off from the source, which allows obtaining, at acceptable dimensions of modern mass reflectrons, a resolution as high as 40 000.

Mass reflectrons with a laser source of ions have acquired particular importance (see review [33]). Laser evaporation and ionization of a substance being analyzed in the course of several nanoseconds furnishes a means of obtaining ion pulses of extremely short duration, evaporating and ionizing in undestroyed state heavy clusters and thermally labile organic molecules, peptides, and proteins, since, during a laser pulse of 10^{-10}–10^{-12} s, the probability of excitation of electron bonds in large molecules to critical state, resulting in their disintegration, is small. This circumstance caused rapid development of investigations with the use of laser mass reflectrons in biophysics, biochemistry, biology, medicine and gene engineering. Laser mass reflectrons allowed analysis of heavy clusters and thermally labile organic molecules (including polypeptides and proteins) with a molecular weight $\geq$ 500 000 amu [33]. Prior to the creation of laser mass reflectrons this had been practically impossible.

The number of works employing mass reflectrons, published in various journals and conference proceedings, has reached several thousand [34]. Manufacturing companies all over the world produce laser mass reflectrons (Bruker, Leybold–Heraeus, Finnigan Mat, etc.), in many laboratories mass reflectrons were fabricated for special physical and chemical investigations [34, 35].

In our laboratory studies of fullerenes and metallofullerenes, including those in coop-

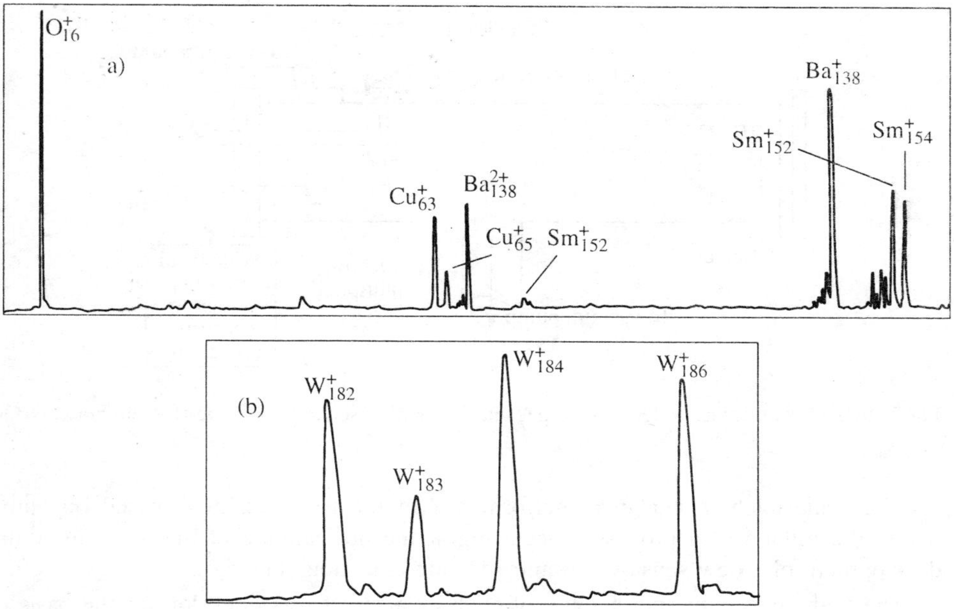

Fig 4. Mass spectra taken with the help of a laser mass reflectron in a single laser pulse. (a) HTSC ceramics $SmBa_2Cu_3O_6$: ion energy range 120–180 eV, $P_{las} = 1-10^{10}$ W cm^{-2}, and spot diameter 12 μm. (b) Tungsten isotopes: ion energy range 320–400 eV and $R_{0.5} = 1000$.

eration with many laboratories in St. Petersburg and Moscow, are being performed using mass reflectrons with thermal and laser evaporation of samples; laser mass reflectrons with high power density focussed on a sample $\geq 10^9$ W cm^{-2} allow analyzing the elemental and isotopic composition of practically any substances (metals, semiconductors, dielectrics) (Fig. 4) [33].

10 Use of mass reflectrons for monitoring technological processes at research institutes and in the industry

In many experiments in physics and chemistry and in monitoring various technological processes it is necessary to analyze complex mixtures of gases.

The automated mass reflectron FTIAN-3 system, developed in the laboratory [35] was produced commercially by the industry and is now used in many laboratories at research institutes and metallurgical integrated iron- and-steel works of Russia, the countries of the former USSR, Cuba, Bulgaria, and China for monitoring fast processes in manufacture of steel and non-ferrous metals.

In recent years, we proposed, developed, and investigated new principles for construction of gas-analyzing systems, which increased by five orders of magnitude the sensitivity and drastically improved the stability of gas analysis in the presence of chemically active gases (O_2, H_2, SO_2, CO, etc.) in a mixture (see patent [36]). This enabled us to create an automated system FTIAN-5, tested under industrial conditions. Documentation for commercial manufacture of this system has been prepared.

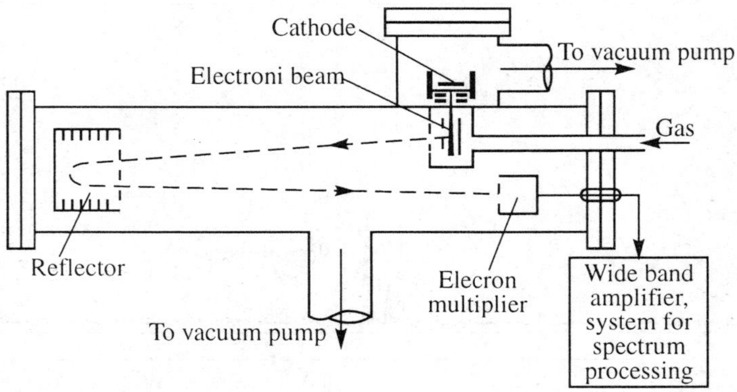

Fig 5. Schematic of mass reflectron for gas analysis with a separate chamber for the hot cathode.

The achieved high analytical performance of the new system is based on eliminating the influence of processes occurring at the hot cathode of the ion source and development of a new sensitive circuitry for ion detection (Fig. 5).

The high sensitivity and stability of analysis allowed us to develop on the basis of FTIAN-5 a fundamentally new method for determining with high accuracy the temperature of steel without interrupting the fast converter process of steel melting (see patent [37]) and methods for determining microimpurities in pure gases used in the industry and in scientific laboratories (O_2, N_2, Ar).

Four FTIAN-5 systems have operated successfully for two years at the Severstal' integrated iron-and-steel works (Cherepovets), monitoring the operation of high-power steel-melting aggregates of the converter shop and of the installation for vacuum refining of special-type steels.

11 Conclusion

We briefly reviewed main directions of research performed in the laboratory of mass spectrometry. All of them are being developed now, being associated with new fundamental or practically important results. To these belong, in the first place, research aimed at adjusting the fundamental physical constants, studies concerned with the isotopy of inert gases in the nature and investigations which produced for the first time experimental proof that modification of the electronic environment of tritium nuclei affects their decay period.

Naturally, a great number of particular results having independent importance (methods of vacuum obtaining, special electronic devices, methods of accurate measurement of the uniformity of magnetic field, obtaining tritium (hydrogen) atoms in a closed volume for a long time, etc.) were obtained during creation of large experimental installations and development of techniques for precision measurements. Since all the fundamental results were based on achieving high analytical parameters of the created dynamic mass spectrometers, this has resulted in their application (especially in the case of mass reflectrons) in many physical, chemical, and biological researches all over the world [34].

References

[1] B. A. Mamyrin and B. N. Shustrov, *USSR Inventor's Certificate* No. 167673, *BI*, **2**, p. 8, 1965.

[2] B. A. Mamyrin, S. A. Alekseenko, and N. N. Aruev, *Sov. Phys. JETP* **80** 2125 (1981).

[3] *Mass Spectrometer MI 9303* Prospekt SKB AP AN SSSR 1982.

[4] B. A. Mamyrin, L. V. Khabarin, and V. S. Yudenich, *Prib. Tekh. Exper.* **2** 173 (1985).

[5] G. S. Anufriev, B. A. Mamyrin, I. N. Tolstikhin, and I. L. Kamenskii, *Dok. Akad. Nauk SSSR* **184** 1197 (1969).

[6] G. S. Anufriev, *Dok. Akad. Nauk* **250** 672 (1980).

[7] B. A. Mamyrin, G. S. Anufriev, and L. V. Khabarin, *Fundamental Aspects of Helium Isotope Distribution over the Earth* Diplom na otkrytie, No. 253, 1982.

[8] B. A. Mamyrin, B. G. Polyak, V. I. Kononov, I. N. Tolstikhin, and L. V. Khabarin, *Proc. XVI General Assambly of the Int. Union Geophysics* (Grenoble) p. 17, 1975.

[9] B. A. Mamyrin, B. G. Polyak, V. I. Kononov, and L. V. Khabarin, *Dok. Akad. Nauk* **247** 1220 (1979).

[10] B. A. Mamyrin, B. G. Polyak, V. I. Kononov, V. N. Yakutseni, E. M. Prasolov, and V. R. Chermak, *Relation between Heat Flux and Helium Isotope Ratio* Diplom na otkrytie No. 68, 1992.

[11] G. S. Anufriev, A. Ya. Krylov, V. P. Pavlov, and T. I. Mazina, *Dok. Akad. Nauk* **237** 284 (1977).

[12] E. V. Sabotovich, G. N. Bondarenko, and T. I. Koromyslichenko, *Cosmic Substance in Oceanic Deposits and Glacial Rock*, Kiev, Naukova Dumka, p. 194, (in Russian) 1978.

[13] G. S. Anufriev and B. S. Boltenkov, *Lithology and Mineral Resources* **5** 451 (1997).

[14] S. Krishnaswami and J. K. Cohran, *Earth Planet Sci. Lett.* **4** 45 (1978).

[15] B. A. Mamyrin, G. S. Anufriev, I. L. Kamenskii, and I. N. Tolstikhin, *Dok. Akad. Nauk* **195** 188 (1970).

[16] B. A. Mamyrin and I. N. Tolstikhin, *Isotopy of Helium*, Moscow, Atomizdat, p. 222, (in Russian) 1981.

[17] B. A. Mamyrin and I. N. Tolstikhin, *Helium Isotopes in Nature*, Oxford–Amsterdam–New York–Tokyo, Elsevier, p. 273, 1984.

[18] O. V. Klyavin, B. A. Mamyrin, L. V. Khabarin, Yu. M. Chernov, and V. S. Yudenich, *Fiz. Tverd. Tela* **18** 1281 (1976).

[19] O. V. Klyavin, B. A. Mamyrin, L. V. Khabarin, Yu. M. Chernov, and V. S. Yudenich, *Fiz. Tverd. Tela* **24** 2001 (1976).

[20] O. V. Klyavin, B. A. Mamyrin, L. V. Khabarin, Yu. M. Chernov, V. S. Yudenich, and G. I. Agafonov, *The Phenomenon of Dislocation-Dynamic Diffusion* Diplom na otkrytie, No. A-066, 1997.

[21] B. A. Mamyrin, N. N. Aruev, and S. A. Alekseenko, *Sov. Phys. JETF* **36** 1 (1973); *ibid.* **57** 1152 (1983).

[22] B. N. Teilor, V. M. Parker, and D. N. Langenberg, *The Fundamental Constants and Quantum Electrodynamics* New York–London, Academic Press, p. 353, 1969.

[23] J. Byrne, *Nature* **310** 212 (1984).

[24] Yu. A. Akulov, B. A. Mamyrin, L. V. Khabarin, and V. S. Yudenich, *Zh. Tekh. Fiz.* **57** 291 (1987);
Yu. A. Akulov, B. A. Mamyrin, L. V. Khabarin, and V. S. Yudenich, *USSR Inventor's Certificate* No. 1295948, *BI*, **4**, p. 265, 1988.

[25] Yu. A. Akulov, B. A. Mamyrin, L. V. Khabarin , and V. S. Yudenich, *Pis'ma Zh. Tekh. Fiz.* **14** 940 (1988).

[26] E. G. Drukarev and M. I. Strikman, *Phys. Lett.* B **186** 1 (1987).

[27] B. Budick, *Phys. Rev. Lett.* **51** 1034 (1983).

[28] Yu. A. Akulov, B. A. Mamyrin, and P. M. Shikhaliev, *Pis'ma Zh. Tekh. Fiz.* **19** 72 (1993).

[29] B. Misra and E. C. G. Sudarshan, *J. Math. Phys.* **18** 756 (1977).

[30] B. A. Mamyrin, *USSR Inventor's Certificate* No. 198034, *BI*, **13** 148 (1967).

[31] B. A. Mamyrin, V. I. Karataev, D. V. Shmikk , and V. A. Zagulin, *Sov. Phys. JETP* **37** 45 (1973).

[32] B. A. Mamyrin, V. I. Karataev, and D. V. Shmikk, *US Patent* No. 4072862, 1978; *UK Patent* No. 1474149, 1977; *French Patent* No. 7530831, 1978; *FRG Patent* No. 2532552, 1980.

[33] B. A. Mamyrin, *Int. J. Mass Spectrom. Ion Process.* **131** 1 (1994).

[34] *Int. J. Mass Spectrom. Ion Process.* Special issue: Time-of-flight Mass Spectrometry and Its Application **131** (1994);
Proc. 43rd ASMS Conf. Mass Spectrom. Appl. Topics (Atlanta) 1995.

[35] *Prospectus of the FTIAN-3 System*, PO "Elektron", Sumy, 1988.

[36] M. A. Ivanov, B. A. Mamyrin, and I. L. Fedichkin, *Mass Spectrometer for Gas Analysis* RF Patent No. 169023, 1993.

[37] M. A. Ivanov, V. G. Smorodin, and V. V. Chifonenko, *A Method for Measuring the Temperature of Fluid Medium* RF Patent No. 2063005, 1996.

Part IV

Institute's Research Divisions

This part presents general information on Institute's structure, personnel, and officers and also briefly describes main research areas of Institute's scientific units. Nowadays, the Institute consists of five large divisions (see below) and has 1218 professional researchers, of which 862 hold scientific degrees of candidate and doctor of sciences[1]. The divisions are subdivided into 73 research units headed by prominent scientists in the relevant areas.

Since 1989, Academician Zhores I. Alferov has been heading the Institute.

Physics of Semiconductor Heterostructures

Director: *Zh. I. Alferov*, Academician.
Staff: 211 researchers, including 138 with scientific degrees.
Structure: 12 research departments.

- **Quantum-Size Heterostructures** *P. S. Kop'ev*
  MOCVD and MBE technology and physics of III–V and II–VI low-dimensional heterostructures.
  Optoelectronic devices based on low dimensional heterostructures.
- **Integrated Optics Based on Heterostructures** *E. L. Portnoi*
  Optical phenomena in semiconductor heterostructure waveguides.
  Terahertz frequency optoelectronics.
- **Photovoltaic Convertors** *V. M. Andreev*
  Heterostructure solar cells.
  Thermophotovoltaic converters.
- **Optoelectronic Phenomena in Heterostructures** *V. I. Korolkov*
  Electrical and photoelectric phenomena in heterostructures and high-speed electronic devices.
- **Physical and Functional Microelectronics** *R. P. Seisyan*
  Optical and magnetooptical spectroscopy of electron and exciton states in low-dimensional semiconductor heterosystems.
  High-resolution image formation by short-wave electromagnetic radiation.
- **Semiconductor Luminescence and Laser Diodes** *I. S. Tarasov*
  High power semiconductor lasers with low internal loss.
  Self-organized periodic nanoheterostructures based on multicomponent solid solutions.
- **Characterization of Electronic Materials** *S. G. Konnikov*
  Characterization methods based on interaction with particle beams.
  Complete in-chain characterization of new materials, structures, and devices of SSE.
  Materials science and characterization of electronic materials, especially III–V and II–VI semiconductor heterostructures.
- **Semiconductor Quantum Electronics** *S. A. Gurevich*
  Structural properties, conductivity, and charge-memory effects in composites with nanosize metallic clusters.
  Laser diodes and semiconductor optical amplifiers for high-speed optical communication systems.

[1] See relevant footnote in Introduction.

- **Electronic Semiconductors** *A. A. Rogachev*
  Exciton molecules captured by quantum dots in multivalley semiconductors.
  Self-consistent quantum wells for electrons and holes at heterointerface in III–V semiconductors.
- **Theory of Optical and Electrical Phenomena in Semiconductors** *V. I. Perel*
  Optical orientation and alignment in semiconductor nanostructures.
  Transport phenomena in semiconductor nanostructures.
- **Theory of Quantum Coherent Phenomena in Solids** *E. L. Ivchenko*
  Resonant optical phenomena in semiconductor nanostructures.
  Weak localization in semiconductor low-dimensional systems.
- **Infrared Optoelectronics** *Yu. P. Yakovlev*
  Narrow-gap III–V semiconductor heterostructures and tunable lasers.
  Light-emitting diodes and photodiodes for 2–6 μm spectral range.

Plasma Physics, Atomic Physics, and Astrophysics

Director: *V. E. Golant*, Academician.
Staff: 317 researchers, including 213 with scientific degrees.
Structure: 19 research departments.

- **Experimental Astrophysics** *E. P. Mazets*
  Gamma and X-ray astronomy.
- **Space Spectrometry** *V. A. Romanov*
  Antiproton, deutron, and positron fluxes in cosmic rays.
  Search for antihelium in primary cosmic radiation.
  Modeling of comet phenomena.
- **Physics of High-Temperature Plasma** *E. Z. Gusakov*
  Plasma processes in tokamaks in improved energy confinement regimes.
  Spherical tokamak GLOBUS-M.
  Diagnostics of high-temperature plasma.
  High-frequency plasma heating in tokamaks.
  Interaction of electromagnetic waves with plasma.
- **Physics of Low-Temperature Plasma** *G. A. Dyuzhev*
  Near-electrode processes in gas discharge.
  Sources of negative hydrogen ions.
  Fullerene formation in low-temperature plasma.
- **Physical Gas Dynamics** *S. V. Bobashev*
  External and internal flows of gas and plasma, magnetohydrodynamic interaction.
  Polarized laser spectroscopy of atoms and molecules.
  Plasma diagnostics in soft X-ray spectral range.
  Nonlinear structures in bounded plasma and high-efficiency thermionic converters.
- **Dense Plasma Dynamics** *E. M. Drobyshevskii*
  Electromagnetic hypervelocity acceleration of solid bodies.
  Origin of the solar system and its minor bodies.
  Search for objects of dark matter in the Universe.
- **Plasma/Gas Dynamics** *A. P. Bedin*
  Physics and gas dynamics of super- and hypersonic motion of bodies.
  Flows around bodies and properties of shock waves in low-temperature nonequilibrium plasma.

- **Nuclear Astrophysics** *G. E. Kocharov*
  Paleoastrophysics.
  Nuclear astrophysics.
  Solar physics.
- **MHD Phenomena** *K. B. Abramova*
  Gas discharge, plasma technology, dense plasma dynamics.
  MHD-phenomena (shock waves, light emission from metals).
  Solar physics.
- **Cyclotron** *V. O. Naidenov*
  Nuclear structure.
  In-beam nuclear spectroscopy.
  X-ray and gamma-ray diagnostics of fusion plasma.
  Application of cyclotron beams to studies of solids.
- **Cosmic Rays** *V. A. Dergachev*
  Heavy nuclei of cosmic rays in Earth's magnetosphere.
  Large-scale changes of solar activity, cosmic rays, geomagnetic fields, and climate.
  X-ray polarization of solar flares.
  Spectrometry of solar neutrons.
- **Physics of Atomic Collisions** *V. V. Afrosimov*
  Elementary processes at atomic and molecular particle interaction.
  Neutral-particle diagnostics of high-temperature plasma.
  Condensed matter structure diagnostics.
  Ion and electron optics.
- **Atomic Collisions in Solids** *Yu. S. Gordeev*
  Ion-atom and ion-molecule collisions.
  Ion beam–surface interaction.
  Photoelectron and Auger electron spectroscopy of solids.
- **Mass Spectrometry** *B. A. Mamyrin*
  Dynamic spectrometers.
  Precise measurements of electromagnetic physical constants.
  Beta decay of tritium.
  Isotopes of rare gases in nature.
- **Physics of Adsorption-Desorption Processes** *V. N. Ageev*
  Electron-stimulated, field-induced, and thermal desorption.
  Surface chemical compounds and 2D graphite films on metals.
  Photoinduced processes at solid surfaces and interfaces.
- **Physics of Elementary Structures of Surface** *V. N. Shrednik*
  Emission-related and surface effects in high electric field.
  Electron and ion spectroscopy and structural analysis of surface and interface.
  Nonequilibrium surface ionization.
- **Theoretical Astrophysics** *D. A. Varshalovich*
  Quasar spectroscopy and cosmology.
  Neutron stars.
  Interstellar and intergalaxy matter.
- **Computer Simulation** *Yu. P. Golovachov*
  Supersonic flows of homogeneous and reacting media.
  Computational fluid dynamics of heterogeneous media.
  Nonequilibrium nonstationary processes in semiconductor structures.
  Semiconductor growth processes.

- **Atomic Radiospectroscopy** *R. A. Zhitnikov*
  Spin-dependent interactions in gases and plasma.
  Excited states and impurity paramagnetic centers in cryocrystals of solid gases.

Physics of Dielectrics and Semiconductors

Director: *R. V. Parfen'ev.*
Staff: 193 researchers, including 129 with scientific degrees.
Structure: 13 research departments.

- **Physics of Ferroelectricity** *V. V. Lemanov*
  Ferroelectric phase transitions.
  Relaxors and incipient ferroelectrics, ferroelectric thin films.
- **Physics of Magnetism** *V. A. Bokov*
  Physics of magnetism, domain walls, spin waves.
- **Neutron Researches** *N. M. Okuneva*
  Neutron scattering in condensed matter.
  Critical dynamics and evolution of microstructure in disordered systems.
- **Acoustics of Solids** *A. B. Sherman*
  Solid state acoustics.
- **Wide-Bandgap Semiconductors Electronics** *Yu. A. Vodakov*
  Wide-bandgap semiconductors with high-bonding energy—SiC, GaN, AlN.
- **Diffusion and Creation of Defects in Semiconductors** *V. I. Sokolov*
  Physics of diffusion and defect structure evolution.
  Reliability of semiconductor devices.
- **Kinetic Phenomena in Solids at Low Temperatures** *R. V. Parfeniev*
  Quantum phenomena in solids at low temperatures.
  High-pressure physics of semiconductors.
- **Physics of Thermoelements** *M. V. Vedernikov*
  Physics of thermoelectrics.
  New methods of thermoelectric energy conversion.
  Transport properties of solids at high temperatures.
- **Physics of Rare-Earth Semiconductors** *I. A. Smirnov*
  Physics of rare-earth semiconductors and semiconductor compounds.
  Thermal properties of solids.
- **Physics of Anisotropic Materials** *V. N. Bogomolov*
  Ultradispersed and low-dimensional systems based on nanoclusters.
- **Optical Phenomena in Ferroelectric and Magnetic Crystals** *R. V. Pisarev*
  Optics of solid state, magnetooptics, and electrooptics.
- **Diffraction Methods for Investigations of Real Crystal-Structures** *L. M. Sorokin*
  Characterization of single crystals, heterostructures, and superlattices by high-resolution diffraction methods.
  Origin of structural defects and mechanisms of their generation in electronic materials.
- **Theory of Semiconductors and Dielectrics** *Yu. A. Firsov*
  Hopping conductivity, multiphonon processes in semiconductors.
  Critical properties of layered superconductors and weakly disordered systems.
  Critical states of Josephson and CDW structures, superconductivity in heavy fermion systems.

Solid State Electronics

Director: *R. A. Suris*, Corresponding Member of the Russian Academy of Sciences.
Staff: 248 researchers, including 175 with scientific degrees.
Structure: 14 research departments.

- **Power Semiconductor Devices** *I. V. Grekhov*
  Physical processes in semiconductors at ultrahigh current densities and high fields.
  Impurities and defects in semiconductors.
  New techniques for semiconductor device fabrication.
  New microelectronic devices based on multilayer thin film HTSC-ferroelectric structures.
- **Fast Semiconductor Devices** *A. F. Kardo-Sysoyev*
  Physical processes in fast power high-response semiconductor devices.
- **Physics of Semiconductor Devices** *V. E. Chelnokov*
  High temperature, microwave and short wave SiC-based optoelectronics.
  Amorphous Si microelectronics.
  Defect formation in Si.
- **Physical Phenomena in Epitaxial Semiconductor Structures** *Yu. V. Zhilyaev*
  High-temperature fast power III–V semiconductor electronics.
- **Physics of Cluster Structures** *A. Ya. Vul'*
  Formation of heavy fullerenes.
  Polymerization of fullerenes.
  Physical properties of fullerene-containing polymers.
- **Nanoelectronics** *I. L. Saveliev*
  Quantum galvanomagnetic effects in low-dimensional structures.
  Technology of quantum heterostructures, MBE of group III nitrides.
  Structures with quantum wells, collective effects and intersubband transitions in semiconductor heterostructures.
- **Amorphous Semiconductors** *V. G. Golubev*
  Thin-film structures from amorphous, nano- and microcrystalline Si and C.
  Amorphous-nanocrystalline Si doped with rare-earth metals.
  Composite III–V compounds with built-in system of nanoscale atomic clusters.
  Direct induction melting of refractory oxides in cold crucible.
  Levitation melting of refractory metals and alloys.
- **Photoelectric and Nonlinear Optical Phenomena in Semiconductors** *E. V. Beregulin*
  Fundamental properties of various GaN crystal modifications.
  Optical and luminescent properties of porous silicon.
  Interaction of FIR laser radiation with semiconductors and low-dimensional structures.
  Nonlinear picosecond spectroscopy of semiconductors.
- **Nonequilibrium Processes in Semiconductors** *A. G. Zabrodskii*
  Interaction between semiconductors and electromagnetic radiation of various nature and frequency.
  Quantum electronic transport, carrier interactions, Coulomb gap formation and metal-insulator phase transitions.
  Defect formation in semiconductors during growth and under influence of nuclear radiation.
  Theory of spontaneous ordering and self-organization in composition-modulated structures and nanostructures.

- **Optoelectonics** *Yu. N. Denisyuk*
  Holographic and optoelectronic methods of recording, processing, and visualization of information.
- **Applied Mathematics and Mathematical Physics** *E. A. Tropp*
  Numerical treatment of 3D convection in rectangular enclosure in rotating coordinate system.
  X-ray photoemission in nonhomogenous semiconductor structures.
  New methods for conversation of heat to work by thermal nonhomogenous and non-equilibrium systems.
- **Solid State Theory** *O. V. Konstantinov*
  Scanning near-field optical microscopy.
  Nanostructure self-organization in epitaxial growth.
  Many-electron correlations in atoms and clusters.
- **Theoretical Foundations of Microelectronics** *R. A. Suris*
  Electronic and optical processes in semiconductor quantum wells and quantum dots.
  Optoelectronic devices based on quantum wells and quantum dots.
  Epitaxial growth of nanostructures.
  Electronic structure, optical and electronic properties of carbon.
  Noise properties of HTSC structures.
- **Quantum Electronics** *M. P. Petrov*
  Holographic investigations in photorefractive materials.
  Nonlinear phenomena in optical fibers.
  High temperature superconductivity.

Solid State Physics

Director: *B. P. Zakharchenya*, Academician.
Staff: 249 researchers, including 192 with scientific degrees.
Structure: 15 research departments.

- **Optics of Semiconductors** *B. P. Zakharchenya*
  Optical properties and spectroscopy of semiconductors (photoluminescence, optical orientation of charge carriers).
  Magnetooptical effects.
- **Phase Transitions in Solids** *Yu. M. Baikov*
  Optical and electrical properties of transition-metal oxides.
  Protonic conductivity in perovskite materials.
- **Microwave Spectroscopy of Crystals** *P. G. Baranov*
  ODMR and EPR studies of dielectrics and semiconductors.
- **Spectroscopy of Solids** *A. A. Kaplyanskii*
  Electronic properties and lattice dynamics of semiconductors, insulators, HTSC materials and their heterostructures.
  Condensed-matter spectroscopy.
- **Surface Optics** *A. N. Titkov*
  Scanning tunneling microscopy and spectroscopy; atomic force microscopy; semiconductor nanostructures and surfaces.
- **Photoelectric Phenomena in Semiconductors** *V. I. Ivanov-Omskii*
  Amorphous and glassy semiconductors.
  Low-dimensional structures.
  IR-luminescence.

- **Physicochemical Properties of Semiconductors** *A. A. Vaipolin*
  Synthesis and growth of compound semiconductors.
  Anisotropy of electrical, photoelectric, and optical properties of diamond-like compounds.
  Dynamics of interatomic interactions in crystal structures.
  Formation of structures with random arrangement of atoms.
- **Physics of Electrical Phenomena in Disordered Systems** *A. N. Ionov*
  Conductivity of semiconductors and insulators; polymers and organic semiconductors and compounds.
- **Physical Kinetics and Electroacoustics** *V. L. Gurevich*
  Kinetic phenomena in condensed matter (electronic and nonelectronic transport, noise phenomena, mesoscopic systems).
- **Physics of Profiled Crystals** *S. P. Nikanorov*
  Growth, structure, micro- and macroplasticity of crystals.
- **Material Dynamics** *A. B. Sinani*
  Mechanics of fracture and deformation of solids under shock-wave loading.
- **Physics of Plastic Deformation** *B. I. Smirnov*
  Defects and physicomechanical properties of solids.
- **Physics of Strength of Composites** *A. M. Leksovskii*
  Dynamics of micro-cracks and fracture of composites.
  Thermal stability of thin polymer composite films.
- **Prognosis of Fracture** *V. S. Kuksenko*
  Fatigue, brittle fracture, and cracks.
  Physicomechanical model of micro- to macrofracture transition as basis for prediction of fracture.
- **Elementary Processes of Fracture** *A. I. Slutsker*
  Kinetics of fracture at microlevel.
  Evolution of defects; fluctuation dynamics.
  Strength enhancement and lifetime prolongation.

Part V

Personalia

Directors of the Institute and Institute's Divisions

Abram F. Ioffe (1880–1960)

Institute's Founder and Director (1918–1950)

Abram F. Ioffe was born in Romny (Poltava Gubernia, Russia) on October 29, 1880. He graduated from St Petersburg Institute for Technology in 1902 and postgraduated from University of Munich in 1905 (*Pd. D. cum laude* under W. C. Röntgen). He took his magister degree (St Petersburg University) in 1913 and doctor of physics (Petrograd University) in 1915. He was with his Physico-Technical Institute in 1918–1952, being also a founder and director of a number of institutions under auspices of the Physico-Technical Institute; including the Special Laboratory of Semiconductors (1952–1954). From 1955 onwards, he was director of the Institute of Semiconductors.

His outstanding achievements were in the domain of mechanical (Ioffe effect related to mechanical strength of solids) and electrical properties of dielectrics (discovery of conductivity laws in ionic crystals), semiconductors (theory of rectification at metal-semiconductor contact, thermoelectric generators and associated inventions, Ioffe–Regel' rule for the short-range order of composition), measurement of the electron charge, cathode- and X-rays, experimental verification of quantum theory of photoeffect.

From 1906, Abram F. Ioffe had been with the St Petersburg (Petrograd, Leningrad) Polytechnic Institute, as extraordinary professor since 1913 and professor ordinarius from 1915 to 1948, being a founder and the first Dean of the Physico-Mechanical Faculty in 1919–1947. He was also a lecturer at the St Petersburg Institute of Mining and other institutions in 1908–1914 and a visiting professor at MIT (USA) and University of California in 1927.

Abram F. Ioffe was elected a member of the Russian Academy of Sciences in 1920 (being a corresponding member since 1918). He was Vice-President of the USSR Academy of Sciences in 1926–1929 and in 1942–1945. For his research and organizing activities he was awarded the highest honors:

- Ivanov Award (Russian Academy of Sciences, 1915),
- Honored Science Worker of the USSR (1933),
- Stalin Prize (USSR, 1942),
- Hero of Socialist Labor (USSR, 1955),
- Lenin Prize (USSR, 1961).

Among his many distinctions were:

- President of the Association of Soviet physicists (1920–1930),
- Corresponding Member of the Gettingen Academy of Sciences (1924–1938 and since 1946),
- Doctor of Law *honoris causa* from the University of California (1927),
- Corresponding Member of Berlin Academy of Sciences (1928–1938 and since 1956),
- Honorary Member of Boston Academy of Arts and Sciences (1929),
- Honorary Doctor of Sorbonne (Paris, 1946),
- Honorary Doctor of Metallurgy of the Graz Polytechnic Institute (Austria, 1948),
- Honorary Doctor of Bucharest University (Romania, 1948),
- Honorary Member of the Chinese Physical Society (1949),
- Honorary Member of the All-Union Lenin Academy of Agricultural Sciences (Moscow, 1956),
- Honorary Member of the French Physical Society (1957),
- Honorary Member of the Indian Academy of Sciences (1958),
- Honorary Member of the Leopoldina Academy of Natural History (Germany, 1958),
- Honorary Member of *Academia dei Lincei* (Italy, 1959),
- Member of F. and I. Joliot-Curies' Society (France, 1960).

He raised a number of scientific editions and periodicals existing until the present and was president and executive officer of a number of conferences and congresses held in the USSR and abroad.

A. F. Ioffe is a prominent author of the classical monographs on semiconductors, manuals on general physics, and books on the history of physics. But the Ioffe's most distinguished service to the Russian and World's science is the School of Physics raised by him.

Anton P. Komar (1904–1985)
Institute's Director (1950–1957)

Anton P. Komar was born in Berezna (Kiev Gubernia, Russia) on January 30, 1904. In 1930, he graduated from the Kiev Polytechnic Institute in metal processing. In 1930–1936, he was a member of the Institute's staff and then was transferred to the Ural Physico-Technical Institute in Sverdlovsk (now, Ekaterinburg). Since 1948 till 1950 he had been with the Lebedev Physical Institute, Moscow. Joining the Institute again in 1950 as director, he held office till 1957. In 1957–1972, he headed a department of the Institute's branch for nuclear research in Gatchina (Konstantinov Institute of Nuclear Physics since 1972) and rested with it as a scientific adviser.

Anton P. Komar was a prominent researcher and inventor in nuclear physics and particle accelerators. His activities in these fields included creation of the first Soviet betatron (1946), a large synchrotron with electron beam of 100 MeV (1953), and studies on photonuclear reactions. Under his directorship a new nuclear center of the Institute was founded.

He was inventor of the first Soviet field-ion microscope (the second in the world) with a single-atom resolution. In 1961, A. P. Komar solved the problem of image uniformity for a number of organic molecules having proposed a waveguide origin of their images and advancing a new nontrivial confirmation of wave nature of electron.

In 1937–1947, he held a chair at the Ural University, Sverdlovsk and, in 1951–1969, a chair at the Leningrad Polytechnic Institute.

As an eminent man of physics A. P. Komar was appointed a member of the Ukrainian Academy of Sciences in 1948.

Among his distinctions was that of the Stalin Prize in 1951.

Boris P. Konstantinov (1910–1969)

Institute's Director (1957–1967)

Boris P. Konstantinov was born in St Petersburg on July 6, 1910. In 1926–1929, he studied at the Leningrad Polytechnic Institute (courses unfinished). Since 1927 he had been a staff member of the Physico-Technical Institute where he held successively the following positions: laboratory assistant, engineer, engineer-researcher, senior researcher, head of laboratory, director (1957–1967), and head of department (from 1967). In 1932–1940, he also worked at the Electrophysical Institute (Leningrad). He earned scientific degrees: a candidate of sciences in physics and mathematics in 1942 and a doctor of sciences in physics and mathematics in 1943, both from the Physico-Technical Institute. He took a professorship at the Leningrad Machine-Tool Institute in 1945 and at the Leningrad Polytechnic Institute in 1947. He held chairs of experimental nuclear physics (since 1947) and isotopes (since 1951) and, in 1964–1966, was the Dean of Physico-Mechanical Faculty at the Leningrad Polytechnic Institute.

Boris P. Konstantinov is an author of distinguished works on acoustics (reverberation, sound reflection, musical acoustics, nonlinear acoustics). He headed special studies of light isotopes' separation (method of electrophoresis), resulting in the most advanced technology of thermonuclear fuel processing and in establishment of a fuel industry (methods of fuel refining and analysis, chemical reactors). His pioneering works in corpuscular and optical plasma diagnostics gave impetus to a new field of research at the Institute. He brought up experimental astrophysics, search for antimatter, initiated work on holographic data and information processing, and was one of the beginners of gamma-astronomy.

For his distinguished scientific services, he was elected a corresponding member of the USSR Academy of Sciences in 1953 and a full member in 1960. In 1967, he was elected Vice-President of the USSR Academy of Sciences.

In recognition of his labors, he was awarded the highest state honors:

- Stalin Prize (USSR, 1953),
- Hero of Socialist Labor (USSR, 1954),
- Lenin Prize (USSR, 1958).

He was Editor-in-Chief of a Russian journal, *Zhurnal Tekhnicheskoi Fiziki* (1959–1969) and a member of editorial boards of many scientific journals.

Vladimir M. Tuchkevich (1904–1997)

Institute's Director (1967–1987)

Vladimir M. Tuchkevich was born in Yanoutsy (Ukraine) on December 29, 1904. In 1928 he graduated from the Kiev University. From 1935 onward, he had been a member of the scientific staff of the Ioffe Institute where he was successively: senior scientist, head of laboratory (since 1947), director (1967–1987), honorary adviser and head of laboratory (since 1987 onward). Earlier, he worked at the Ukrainian Institute of Metrology and All-Ukrainian Institute of Roentgenology (1928–1935) and Radium Institute in Leningrad in 1935. He took the degree of doctor of sciences in physics and mathematics in 1955 from the Ioffe Institute.

V. M. Tuchkevich's first works were in the area of X-ray physics. During the war years he bestirred himself to organization and development of ship degaussing. Then, there was arduous work on isotope separation. Most of Vladimir M. Tuchkevich's scientific contributions were in the field of semiconductor physics and technology: high-power semiconductor devices, first planar transistor in the USSR, thyristors, single crystals of pure Ge, Ge rectifiers with record parameters, etc.

Since 1936, Vladimir M. Tuchkevich had delivered lectures at the Leningrad Polytechnic Institute (now St Petersburg Technical University) and, combined with his work as director, guidance of higher educational bodies. Many of his pupils and co-workers at the Institute became prominent physicists.

Vladimir M. Tuchkevich was elected a full member of the USSR Academy of Sciences in 1970, being admitted a corresponding member in 1968.

Among his many distinctions and awards were:

- Stalin Prize (USSR, 1942),
- Lenin Prize (USSR, 1966).
- Gold Medal "In recognition of services to science and humanity" (Czechoslovakian Academy of Sciences, 1977)
- Hero of Socialist Labor (USSR, 1984)

V. M. Tuchkevich was Editor-in-Chief of a Russian journal, *Zhurnal Tekhnicheskoi Fiziki.*

Zhores I. Alferov

Institute's Director (1987–present)
Director of the Division of Physics of Semiconductor Heterostructures (1989–present)

Zhores I. Alferov was born in Vitebsk, Belorussia, USSR, on March 15, 1930. In 1952, he graduated from the Department of Electronics of V. I. Ulyanov (Lenin) Electrotechnical Institute in Leningrad. Since 1953 he has been a staff member of the Physico-Technical Institute where he held consecutively the following positions: junior researcher (1953–1964), senior researcher (1964–1967), head of the laboratory (1967–1987), director (1987–present). He earned scientific degrees: a candidate of sciences in technology in 1961 and a doctor of sciences in physics and mathematics in 1970, both from the Ioffe Institute.

Since 1962 he has been working in the area of III–V semiconductor heterostructures. His outstanding contributions to physics and technology of III–V semiconductor heterostructures, especially investigations of injection properties, development of lasers, solar cells, LED's, and epitaxy processes have led to the creation of modern heterostructure physics and electronics.

In 1973 Zh. I. Alferov took over the chair of optoelectronics at the St Petersburg State Electrotechnical University (former V. I. Ulyanov (Lenin) Electrotechnical Institute) and in 1988 he was appointed to Dean of the Faculty of Physics and Technology at the St Petersburg Technical University.

He was elected a corresponding member of the USSR Academy of Sciences in 1972 and Academy's full member in 1979. From 1989 onward, he has been Vice-President of the USSR (Russian) Academy of Sciences and President of its St Petersburg Scientific Center.

For his research Professor Zh. I. Alferov was awarded a number of national and international prizes:

- Ballantyne Medal of the Franklin Institute (USA, 1971),
- Lenin Prize (USSR, 1972),
- Hewlett-Paccard Europhysics Prize (1978),
- State Prize (USSR, 1984),
- GaAs Symposium Award and H. Welker Medal (1987),
- Karpinskii Prize (FRG, 1989),

- Ioffe Prize (Russian Academy of Sciences, 1996).

and honorary memberships:

- Life Fellow of the Franklin Institute (USA, 1971),
- Foreign Member of the German Academy of Sciences (1987),
- Honorary Professor of the Havana University (Cuba, 1987),
- Foreign Member of the Polish Academy of Sciences (1988),
- Foreign Associate of the National Academy of Engineering (USA, 1990),
- Foreign Associate of the National Academy of Sciences (USA, 1990),
- Honorary Member of the Metrological Academy (St Petersburg, 1994),
- Foreign Member of the Belorussian Academy of Sciences (1995),
- Foreign Member of the Korean Academy of Science and Technology (1995),
- Honorary Member of the Pakistan Society for Semiconductor Science and Technology (1996),
- Honorary Academician of the International Academy of Refrigeration (St Petersburg, 1997),
- Academician of the International Academy of Ecology, Man and Nature Protection Sciences (St Petersburg, 1998),
- Honorary Doctor of the Humanity University (St Petersburg, 1998).

He is Editor-in-Chief of a Russian journal, *Pis'ma v Zhurnal Tekhnicheskoi Fiziki* (English-language version—*Technical Physics Letters*) and a member of the Editorial Board of a Russian journal *Nauka i Zhizn'* (Science and Life).

Zh. I. Alferov is author of 3 books, 400 articles, and 50 inventions on semiconductor technology.

Victor E. Golant

Director of the Division of Plasma Physics, Atomic Physics, and Astrophysics (1989–present)

Victor E. Golant was born in St Petersburg on January 14, 1928. He graduated from the Leningrad Polytechnic Institute in 1950. In 1950–1961, he was an engineer at the *Svetlana* factory in Leningrad (till 1956) and then a lecturer at the Leningrad Polytechnic Institute. From 1961 onward, he is with the Ioffe Institute, holding successively the following position: senior researcher, head of laboratory (since 1965), and director of division (since 1989). He received his degrees: candidate of sciences in physics and mathematics from the Leningrad Polytechnic Institute in 1955 and doctor of sciences in physics and mathematics from the Ioffe Institute in 1964.

The main contributions made by V. E. Golant were in fields of plasma physics and thermonuclear fusion, including wave–plasma interaction, high-frequency plasma heating, microwave plasma diagnostics, plasma transport and improved confinement regimes in tokamaks.

In 1967, V. E. Golant became a professor of the Leningrad Polytechnic Institute (now, St Petersburg Technical University) where he was later appointed to the chair of plasma physics at the Faculty of Physics and Technology.

In recognition of his merits, Professor V. E. Golant was elected a corresponding member (1984) and then a full member (1990) of the USSR Academy of Sciences. Since 1993, he has also been a foreign member of the American Physical Society.

V. E. Golant was awarded:

- State Prize (USSR, 1983),
- Artsimovich Prize (Russian Academy of Sciences, 1998).

He is Editor-in-Chief of a Russian journal, *Zhurnal Tekhnicheskoi Fiziki* (English-language version—*Technical Physics*), a member of the Editorial Board of *Nuclear Fusion*, and a member of the International Advisory Panel of *Plasma Physics and Controlled Fusion.*

He is an author of about 200 articles and 3 monographs. Among them should be mentioned the monograph “Fundamentals of Plasma Physics” (written with co-authors) translated into English, Japanese, Italian, and Chinese.

Robert V. Parfeniev

Director of the Dielectrics and Semiconductors Division (1993–present)

Robert V. Parfeniev was born in Arkhangelsk, USSR, on May 29, 1937. In 1961, he graduated with honor from the Radioelectronics Faculty of the Leningrad Polytechnic Institute and entered a post-graduate studentship of the Institute of Semiconductors, USSR Academy of Sciences, in Leningrad. Since termination of his post-graduate term in 1964, he has been a member of the scientific staff of this Institute and then of the Ioffe Institute where he was, successively, junior scientist, senior scientist, head of sector, head of laboratory (1985–present), deputy director (1989–1993) and then director of division (1993–present). He received his scientific degrees: candidate of sciences (in physics and mathematics) from the Institute of Semiconductors in 1966 and doctor of sciences (in physics and mathematics) from the Ioffe Institute in 1980. In 1987, he was given a professorship.

The main field of his reseach interest is low-temperatures physics of semiconductors, including quantum transport effects in magnetic field, and superconductivity. In his laboratory there was created a multi-purpose experimental base for investigation of stationary and pulsed high magnetic fields at low and ultralow temperatures, using high pressure of up to 40 kbars in autonomous pressure chambers.

Since 1984, Professor R. Parfeniev has been a lecturer at the St Petersburg Technical University.

In 1966 he was awarded Ioffe Prize of the Academy of Sciences for young scientists for experimental demonstration of the magnetophonon resonance effect predicted theoretically by V. Gurevich and Yu. Firsov. The title "Outstanding Inventor of the USSR" has been conferred on R. V. Parfeniev for his contribution to physics and technology.

He is involved in a work of a number of bodies of the Russian Academy of Sciences and international institutions such as:

- Scientific Council on Low-Temperature Physics of the Russian Academy of Sciences,
- Committee on Advanced Materials' Processing of the International Academy of Astronautics,
- "Cosmic Materials' Processing" section of the Council on Cosmos of the Russian Academy of Sciences.

Professor Robert Parfeniev is author of about 200 publication and 5 inventions.

Robert A. Suris

Director of the Solid State Electronics Division (1993–present)

Robert A. Suris was born in Moscow on December 31, 1936. In 1960, he graduated with honor from the Institute of Steel and Alloys (Moscow) in theory of metallurgy. From 1960 till 1988 he was a staff member of research centers for electronic industry in Moscow, where he held consecutively the following positions: engineer, researcher, senior researcher, head of laboratory, head of division. In 1988, he joined the Ioffe Institute to become head of laboratory and then director of division. He received his candidate of sciences (in physics and mathematics) degree from the Department of Physics of the Moscow State University in 1964 and a doctor of sciences (in physics and mathematics) degree from the Institute of Radio Engineering and Electronics (Moscow) of the USSR Academy of Sciences in 1974.

The main contributions to physics made by Robert A. Suris were in the domain of semiconductors, opto- and microelectronics, including invention of DFB laser and quantum cascade laser, and developememt of theory of sequential electron resonance tunneling and light amplification in semiconductor superlattices, MOS- and heterostructures, discovery of the trap recharge waves in semiconductors, etc.

R. Suris was an associate and then a full professor at the Quantum Electronics Department of the Moscow Physico-Technical Institute in 1970–1988 and in 1988 he became a full professor and then was appointed to the chair of solid state physics at the Faculty of Physics and Technology of the St Petersburg Technical University.

In 1995–1998, Professor R. Suris served as a member of the Editorial Board of *Semiconductor Science and Technology* and currently is a member of the Editorial Boards of Russian journals: *Zhurnal Tekhnicheskoi Fiziki* (English-language version—Technical Physics), *Pis'ma v Zhurnal Tekhnicheskoi Fiziki* (Technical Physics Letters), and *Microelectronics.*

In recognition of his services to the scientific development, Professor R. Suris was elected a corresponding member of the Russian Academy of Sciences in 1997.

He was recipient of Russian presidential award for outstanding scientists (1995) and Rank Prize for Optoelectronics (1998).

He held offices as a chair or a member of scientific councils of a number of institutions and international committees.

Professor R. Suris is author of about 200 publications and 20 inventions.

Boris P. Zakharchenya

Director of the Solid State Physics Division (1989–present)

Boris P. Zakharchenya was born in Orsha, Belorussia, USSR, on May 1, 1928. In 1952, he graduated from the Leningrad State University. Since 1952, he has been a member of the Physico-Technical Institute taking the following positions: post-graduate (1952–1955), junior researcher (1955–1960), senior researcher (1960–1969), head of laboratory (1969–present), director of division (1989–present). He earned scientific degrees: a candidate of sciences in physics and mathematics in 1955 and a doctor of sciences in physics and mathematics in 1966, both from the Physico-Technical Institute. Since 1972 he has been a professor of physics.

Among the major achievements of B. P. Zakharchenya in physics are the discoveries of:

- two exciton series in the CuO crystal, which proved the existence of the spin-orbital splitting in semiconductors (1952),
- Zeeman and Stark splitting of exciton lines in semiconductors (1954),
- a diamagnetic shift of exciton lines in magnetic field; this proved for the first time the existence of gigantic orbits in the Vannier–Mott exciton (1955)
- the fundamental magneto-optical effect in semiconductors (spectral oscillations of the absorption coefficients), due to optical transitions between Landau subbands; this was the first experimental confirmation of the existence of the so-called magneto-excitons (quasi-one-dimensional excitons), (1956),
- the effect of magnetic field inversion in the exciton spectrum; this was the first experimental demonstration of the motion of this quasiparticle (1960).

In 1970, B. Zakharchenya with co-workers gave birth to a new trend in the physics of semiconductors—studies of the optical orientation of electrons and nuclei. Together with V. Fleisher and R. Dzhioev he discovered the effect of optical cooling (to microkelvins) of the nuclear spin system. In studies of optical orientation and alignment of excitons in quantum-sized structures, a number of new effects were found.

B. Zakharchenya together with D. Mirlin observed alignment of angular momenta of hot electrons by linearly polarized light in bulk semiconductors in 1976 and in quantum-sized structures in 1989. Based on this phenomenon, a method was created for measurement of ultrafast processes of electron relaxation in semiconductors under steady-state excitation by light.

Professor B. P. Zakharchenya was elected a corresponding member of the USSR Academy of Sciences in 1976 and a full member of the Russian Academy of Sciences in 1992. He is a foreign member of the American Physical Society.

Other honors awarded to Professor B. P. Zakharchenya include:

- Lenin Prize (USSR, 1966),
- State Prize (USSR, 1976),
- Lebedev Gold Medal (Russian Academy of Sciences, 1996),
- Aronov Prize (Foundation for Intellectual Collaboration, Russia, 1998).

Professor B. P. Zakharchenya is Editor-in-Chief of a Russian journal *Fizika Tverdogo Tela* (English-language version—*Physics of the Solid State*). He is author and co-author of more than 200 scientific publications.

Winners of Scientific Awards

Enumerated below are the members of the Institute's staff, recipients of the most prestigious international (Nobel) and national awards for scientific works accomplished at the Institute. The national awards included are state awards and nominal prizes and medals established by the USSR (Russian) Academy of Sciences. Under each heading the entries are arranged in chronological order. In accordance with the purpose of this edition, its compilers take it reasonable to make no mention non-Institute's members who are participants of the honored works, and give abridged versions of the nomination formulas.

Nobel Prize in Chemistry

The Nobel Prize in Chemistry is one of the prizes awarded from a fund established under the will of Alfred Bernhard Nobel who specified that the awards should be annually given "to those who… shall have conferred the greatest benefit on mankind." According to Nobel's will the awarder of the Prize in Chemistry is the Royal Swedish Academy of Sciences. Distribution of this prize was started on December 10, 1901, the fifth anniversary of the death of the founder.

N. N. Semenov (1956)
Work on the kinetics of chemical reactions[1].

National Awards

Lenin Prize

Lenin Prize had been adjudged for outstanding achievements in science and technology, literature, and arts. It was first established by the Soviet Government in 1925 and was awarded till 1935. It was reestablished in 1957, but its awarding was again terminated after dissolution of the Soviet Union. The Prize was awarded by decision of the State Committee on Lenin and State Prizes, in accordance with its charter, and given only once during a recipient's lifetime.

A. A. Chernyshev (1930)
High-voltage and high-frequency technology.

B. A. Gaev, B. P. Konstantinov (1958)
Development of industrial methods to produce light isotopes.

Yu. A. Dunaev (1961)
Creation of high-temperature materials to protect landing spacecrafts.

A. F. Ioffe (1961)
Theoretical and experimental investigations of semiconductors and development of a theory of thermoelectric cells.

D. N. Nasledov, A. A. Rogachev, S. M. Ryvkin, B. V. Tsarenkov (1964)
Fundamental investigations which gave rise to semiconductor quantum generators.

E. F. Gross, A. A. Kaplyanskii, B. P. Zakharchenya (1966)
Theoretical and experimental investigations of excitons in crystals.

[1] These studies were carried out by N. N. Semenov at the Physico-Technical Institute mostly in the 1920s.

V. E. Chelnokov, I. V. Grekhov, V. B. Shuman, V. M. Tuchkevich (1966)
Development of germanium and silicon diodes and transistors and high-power semiconductor devices.

Zh. I. Alferov, V. M. Andreev, D. Z. Garbuzov, V. I. Korol'kov, D. N. Tret'yakov (1972)
Fundamental investigations of semiconductor heterostructures and creation of new devices on their basis.

V. V. Afrosimov, V. M. Dukel'skii, N. V. Fedorenko, O. B. Firsov (1972)
Investigations of elementary processes and inelastic scattering in atomic collisions.

V. M. Lobashev, V. A. Nazarenko (1974)
Experimental discovery and study of spatial parity non-conservation in nuclear EM transitions.

E. P. Mazets (1986)
Studies in the field of space physics.

Stalin Prize

Stalin Prize was established by the Soviet Government in 1939 and awarded annually till 1954 under personal supervision of the state leader J. V. Stalin for outstanding works in science, technology, literature, and arts. After the denunciation of the *cult of personality* in 1956 the Stalin Prize was abandoned. When establishing the novel State Prizes in 1966 (see below) the Soviet Government renamed the Stalin Prize as the State Prize. For the sake of historical reality, here the Prize name is retained.

N. Ya. Chernetsov, Yu. B. Kobzarev, P. A. Pogorelko (1940)
Development of the impulse radiolocation (radars).

A. I. Alikhanov, A. I. Alikhan'yan (1941)
Works on radioactivity.

A. P. Alexandrov, B. A. Gaev, I. V. Kurchatov, V. R. Regel', P. G. Stepanov, V. M. Tuchkevich (1942)
Development and realization of the countermine defense of ships.

A. F. Ioffe (1942)
Studies of semiconductors reported in the monograph *Semiconductors in Physics and Technology*.

N. N. Davidenkov (1943)
Outstanding many years' studies in science and technology.

E. F. Gross (1946)
Investigations of molecular scattering of light and microstructures of liquids and crystals.

G. N. Flerov (1946)
Discovery of spontaneous fission.

Ya. I. Frenkel (1947)
Development of a liquid state theory reported in the monograph *Kinetic Theory of Liquids*.

G. A. Grinberg (1949)
Investigations of mathematical physics reported in the monograph *Selected Problems of Mathematical Theory of Electrical and Magnetic Phenomena*.

A. P. Komar (1951)
Works on a classified topic[2].

B. T. Kolomiets (1951)
Development of thyristors.

L. I. Rusinov (1949)
Work on nuclear isomerism.

G. A. Smolenskii (1952)
Investigations of physical and chemical properties of ferroelectrics and ferrites.

K. V. Donskoi, B. A. Gaev, N. I. Ionov, B. P. Konstantinov, N. Yu. Lagunov, G. Ya. Ryskin, A. P. Votinov (1953)
Works on industrial production of light isotopes.

M. I. Kornfel'd (1953)
Works on nuclear reactor moderator.

A. N. Zaidel' (1954)
Works on a spectral analysis of isotopes.

State Prize

State Prize was established in 1966 and was annually awarded by the State Prize Committee under the Government for the most important achievements in science, technology, literature, and arts.

D. G. Alkhazov, D. S. Andreev, K. I. Erokhina, Yu. P. Gangrskii, A. P. Grinberg, I. Kh. Lemberg, K. A. Ter-Martirosian (1968)
Investigations of the Coulomb excitation of nuclei.

M. P. Petrov (1971)
Generation and study of high-temperature plasma in a tokamak.

V. K. Subashiev (1971)
Development of silicon solar cells.

A. P. Bedin, G. I. Mishin (1974)
Creation of experimental basis, development and application of methods of aeroballistic studies.

V. L. Gurevich (1974)
Establishment of fundamentals of acoustoelectronics.

Yu. M. Burdukov, D. N. Nasledov (1975)
Manufacture of gallium arsenide single crystals and investigation of their properties.

V. A. Isupov (1975)
Crystallography and crystallochemistry of ferroelectrics.

A. A. Kaplyanskii, V. N. Medvedev (1975)
Creation of original methods for investigation of complex impurity centers and defects in crystals.

M. I. D'yakonov, R. I. Dzhioev, A. I. Ekimov, V. G. Fleisher, V. I. Perel', V. I. Safarov, B. P. Zakharchenya (1976)
Discovery and investigation of original phenomena related to optical orientation of electron and nuclear spins in semiconductors.

[2]The compilers had no possibility to find out either the precise title of the work or its topic.

V. N. Shrednik (1978)
Fundamental studies on burst electron emission and autoelectronic processes responsible for it, development of a novel class of X-ray devices on their basis.

V. P. Shilo (1980)
Investigations of chalcogenide semiconductors.

V. V. Afrosimov, E. L. Berezovskii, Yu. S. Gordeev, A. B. Izvozchikov, A. I. Kislyakov, M. P. Petrov, A. N. Zinov'ev (1981)
Corpuscular diagnostics of high-temperature plasma.

A. R. Regel' (1981)
Investigations in the field of chemical thermodynamics.

B. L. Gel'mont, V. I. Ivanov-Omskii, V. K. Ogorodnikov, K. P. Smekalova (1982)
Prediction, discovery, and investigation of gapless semiconductors and exciton phases.

G. V. Ostrovskaya (1982)
Dynamics of current layers and solar activity.

V. E. Golant (1983)
Studies of high-power girotrons and their application in fusion research.

B. T. Kolomiets (1984)
Photoresistors CdS.

Zh. I. Alferov, A. T. Gorelenok, S. G. Konnikov (1984)
Development of isoperiodic heterostructures based on quaternary solid solutions of $A_{III}B_{V}$ compounds.

V. V. Lemanov (1984)
Works on acoustooptics.

A. V. Khomenko, M. P. Petrov, S. I. Stepanov (1985)
Fundamental investigations of photorefractive and liquid crystals for data processing optical systems.

G. T. Razdobarin (1986)
Creation of laser diagnostic methods and investigation of high-temperature plasma in physical experiment.

Yu. A. Astrov, G. B. Gorlin, L. G. Paritskii, I. D. Yaroshetskii (1986)
Development and investigations of optical information recording methods (semiconductor photography).

V. M. Andreev (1986)
Development and promotion of noncoherent optoelectronics to practice.

Yu. F. Biryulin, Yu. V. Shmartsev (1987)
Physicochemical fundamentals of isovalent doping of semiconductors.

Yu. L. Ivanov (1987)
Inverted distributions of hot charge carriers and generation of stimulated radiation in semiconductors in the millimeter, submillimeter, and far infrared regions.

D. Z. Garbuzov (1987)
Heterojunction-based light emitting diodes.

I. V. Grekhov, A. V. Gorbatyuk, S. V. Korotkov, A. F. Kardo-Sysoev, L. S. Kostina, S. V. Shenderei, N. S. Yakovchuk (1987)
Development of original principles of high-power switching by semiconductor devices.

V. N. Ageev, N. I. Ionov, M. A. Mittsev, U. Kh. Rasulev, V. I. Paleev, A. Ya. Tontegode, E. Ya. Zandberg (1988)
Studies of thermal desorption of neutral and charged particles from solid surfaces.

P. D. Altukhov, G. E. Pikus, A. A. Rogachev (1988)
Many-exciton complexes in semiconductors.

Yu. V. Koval'chuk, Yu. V. Pogorel'skii (1988)
Discovery of the phenomenon of pulsed oriented crystallization of solids.

A. G. Gurevich (1988)
Development of scientific fundamentals of spinwave RF electronics.

Lenin's *Komsomol* Prize

This Prize was established by the Communist League of Youth (*Komsomol* is Russian abbreviation) under Government support to award young scientists, engineers, and persons of arts and literature for the most outstanding works.

V. G. Ivanov, P. I. Kovalev, V. P. Meleshko, A. N. Mikhalev, Yu. V. Shelud'ko (1975)
Investigation of physical processes in flight of bodies in the atmospheres of the Earth and planets.

I. N. Arsent'ev, P. S. Kop'ev, V. A. Mishurnyi, V. D. Rumyantsev (1976)
Preparation and investigation of widegap $A_{III}B_V$ solid solutions and creation of effective injection emitters of visible radiation.

N. T. Bagraev, L. S. Vlasenko (1978)
Creation and investigation of original semiconductor materials based on doped germanium and silicon.

V. R. Larionov, B. V. Pushnyi (1978)
Development of heterostructures in the system Al-Ga-As for semiconductor devices of high efficiency.

E. Z. Gusakov (1981)
Physical effects in parametric plasma processes.

O. M. Fedorova, T. S. Tabarov, S. I. Troshkov, Yu. M. Zadiranov (1984)
Creation and investigation of semiconductor heterophotoconverters of concentrated radiation of high efficiency and autonomous modules for solar power stations on their basis.

S. A. Nikishin (1985)
Original solutions and applications of electroliquid epitaxy.

A. N. Raev (1986)
Automation of electron-beam epitaxy.

A. V. Andrianov, N. S. Averkiev, A. A. Bakun, Yu. B. Lyanda-Geller, M. N. Tkachuk (1988)
Circular photogalvanic effect in crystals.

Nominal Awards of the USSR (Russian) Academy

The Awards were established in honor of lifeworks of outstanding Russian scientists in various fields of knowledge, being conferred to researchers for the best works in these fields.

Lebedev Gold Medal

P. N. Lebedev (1866–1912) was a prominent Russian physicist of the early 1900s. This Medal is awarded for outstanding work in physics.

B. P. Zakharchenya (1996)
New effects of electron ordering and optical orientation of electrons by polarized light in semiconductors and quantum-sized semiconductor structures.

Kurnakov Prize

Outstanding Russian chemist, N. S. Kurnakov (1860–1941) is a classic of physicochemical analysis and founder of a school in chemistry and metallurgy. The Prize is awarded for outstanding works on non-organic and general chemistry and physicochemical analysis and its applications.

N. A. Gorynova (1958)
Works on physical chemistry of semiconductors and creation of a new class of semiconductor compounds.

Landau Prize

Scientist of world-wide reputation, L. D. Landau (1908–1968) was a most talented theorist. This Prize is awarded for works in theoretical physics (on nuclei and elementary particles).

V. N. Gribov (1971)
Works on the theory of complex momenta.

A. L. Efros, B. I. Shklovskii (1986)
Coulomb interaction and transport in disordered system with localized states.

Ioffe Prize

Prominent physicist and outstanding organizer of science, A. F. Ioffe (1880–1960) was a founder and the first director of the Institute. This Prize is awarded for works in various fields of physics.

A. R. Regel' (1972)
Works on electronic conduction of liquids.

L. S. Stil'bans (1975)
Studies of thermoelectric conversion and materials and creation of the world's first thermoelectric semiconductor refrigerators.

V. P. Zhuze (1984)
Works on semiconductor properties of a number of materials and compounds including a vast group of rare-earth semiconductor compounds.

G. E. Pikus (1987)
Works on the symmetry and new electron phenomena in semiconductors.

D. N. Mirlin (1990)
Spectroscopy of hot luminescence in semiconductors.

M. I. D'yakonov, I. A. Merkulov, V. I. Perel (1993)
Works on the theory of ordering of pulses of photoexcited electrons in semiconductor crystals and heterostructures.

Zh. I. Alferov, V. M. Andreev, V. D. Rumyantsev (1996)
Photoelectric converters of solar radiation based on heterostructures.

Konstantinov Prize

Outstanding scientist and organizer of research in physics and technology, B. P. Konstantinov (1910–1969) was the third director of the Institute. This Prize is awarded for works in physics, device fabrication, and automatization of research.

B. A. Mamyrin, V. M. Tuchkevich (1982)
Investigation and development of mass-spectrometric control of converter processing in ferrous and nonferrous metallurgy.

M. M. Larionov, K. M. Novik, V. V. Rozhdestvenskii (1989)
High-temperature plasma diagnostics using the SHF wave radiation and scattering.

Belopol'sky Prize

A. A. Belopol'sky (1854–1934) was a prominent Russian astronomer representing the famous Russian School of astronomy. This Prize is awarded for works in astrophysics.

Yu. N. Gnedin, A. Z. Dolginov, N. A. Silant'ev (1987)
Investigations of propagation and polarization of radiation in the space.

D. A. Varshalovich, S. A. Levshakov (1990)
Investigations of molecular clouds with large red shift.

Artsimovich Prize

One of the leading physicists of the Soviet atomic project, L. A. Artsimovich (1909–1973) headed thermonuclear fusion investigations. This Prize is awarded for works in the experimental physics.

V. E. Golant, S. V. Lebedev (1998)
Investigation of enhanced regimes of plasma confinement in a tokamak.

Members of National Academies of Sciences

The present Russian Academy of Sciences was founded as the St Petersburg Academy of Sciences by Peter the Great in 1725, established as the State supreme scientific institution, and was succeeded by the USSR Academy of Sciences (with its Soviet Republics' counterparts) in 1923 and the Russian Academy of Sciences in 1992. Now the Academy of Sciences is the highest scientific society and principal body coordinating research in natural and social sciences, technology, and production in Russia. Composed of outstanding Russian scholars, the Academy comprises elected members of three types—full members (academicians), corresponding members, and foreign members. Since its foundation the Academy has been under the auspices of the state and its membership has been officially believed the highest esteem of national scientists[1].

Below are listed the members of the USSR, Russian, and former Soviet Republic's Academies of Sciences, who have ever belonged to the Institute's staff. The field for every person includes a surname (followed by the first name and patronymic), the years of Institute's membership, academic ranks, the most prestigious scientific awards (Nobel Prize, national and academic prizes) and the highest USSR state decoration (Hero of Socialist Labor) with the years of election and/or awarding. (For brevity, the words Sciences are omitted in the titles of the academies.) The last field briefly describes person's main interests in physics and, in some cases, the most pronounced scientific efforts and merits.

Afrosimov, Vadim Vasil'evich (born in 1930)
Institute staff: 1953–present.
USSR Academy: Corresponding member, 1987.
Lenin Prize, 1972. State Prize, 1981.
Experimental atomic physics, high-temperature plasma diagnostics.

Ageev, Nikolai Vladimirovich (1903–1982)
Institute staff: 1926–1938.
USSR Academy: Corresponding member, 1946. Full member, 1968.
Physics of steels and alloy hardening, X-ray analysis of alloys.

Alexandrov, Anatolii Petrovich (1903–1994)
Institute staff: 1930–1946.
USSR Academy: Corresponding member, 1943. Full member, 1953.
President of the USSR Academy, 1975–1986.
Stalin Prizes, 1942, 1949, 1951, and 1953. Lenin Prize, 1959.
Kurchatov Gold Medal, 1968. Lomonosov Gold Medal, 1978. Vavilov Gold Medal, 1978. Ioffe Prize, 1980.
Hero of Socialist Labor, 1954, 1960, and 1973.
Physics of dielectrics and polymers, ship degaussing, nuclear and thermonuclear weapons, nuclear energetics.

[1] In 1992, all members of the USSR Academy became automatically members of the Russian Academy.

Alferov, Zhores Ivanovich (born in 1930)
Institute staff: 1953–present.
Director of the Institute: 1987–present.
USSR Academy: Corresponding member, 1972. Full member, 1979.
Vice-President of the Russian Academy, 1991–present.
Lenin Prize, 1972. State Prize, 1984.
Physics of semiconductors, heterojunctions, advanced semiconductor technology, first transistor in the USSR, heterolasers.

Alikhan'yan, Artem Isaakovich (1908–1978)
Institute staff: 1930–1945.
USSR Academy: Corresponding member, 1946.
Armenian Academy: Full member, 1943.
Stalin Prizes, 1941 and 1948. Lenin Prize, 1970.
Discovery of electron–positron pair in internal conversion of nuclei, cosmic rays, high-energy particles, accelerators.

Alikhanov, Abram Isaakovich (1904–1970)
Institute staff: 1927–1945.
USSR Academy: Corresponding member, 1939. Full member, 1943.
Armenian Academy: Full member, 1943.
Stalin Prizes, 1941, 1948, and 1953.
Hero of Socialist Labor, 1954.
Discovery of electron–positron pair in internal conversion of nuclei, beta-spectroscopy, cosmic rays, first Soviet heavy-water reactor, accelerators.

Andreev, Nikolai Nikolaevich (1880-1970)
Institute staff: 1926–1938.
USSR Academy: Corresponding member, 1933. Full member, 1953.
Hero of Socialist Labor, 1970.
Acoustics and applications, piezoelectricity, nonlinear acoustics.

Arkharov, Vladimir Ivanovich (born in 1907)
Institute staff: 1928–1932.
Ukrainian Academy: Full member, 1965.
Physics of metals, gas-phase corrosion, metal protection, alloys.

Aronov, Arkadii Girshevich (1939–1994)
Institute staff: 1991–1994.
USSR Academy: Corresponding member, 1990.
Theory of condensed matter, coherent phenomena in disordered conductors.

Artsimovich, Lev Andreevich (1909–1973)
Institute staff: 1933–1946.
USSR Academy: Corresponding member, 1946. Full member, 1953.
Stalin Prize, 1953. Lenin Prize, 1958. State Prize, 1971.
Hero of Socialist Labor, 1969.
High-energy electrons, electron–positron annihilation, isotope separation (EM method), controlled fusion (tokamaks).

Chernyshev, Alexander Alexandrovich (1882–1940)
Institute staff: 1918–1931.
USSR Academy: Corresponding member, 1929. Full member, 1932.
Lenin Prize, 1930.
Inventor of heated cathode, TV, Chernyshev's dischargers, ion inverter (plasma current control), high-frequency carrier telephony.

Davidenkov, Nikolai Nikolaevich (1879–1962)
Institute staff: 1925–1962.
Ukrainian Academy: Full member, 1939.
Stalin Prize, 1943.
Physics of metals and strength of solids, theory of mashining tools.

Denisyuk, Yurii Nikolaevich (born in 1927)
Institute staff: 1988–present.
USSR Academy: Corresponding member, 1970. Full member, 1992.
Lenin Prize, 1970. State Prize, 1982.
New method of holography and its applications.

Devyatkov, Nikolai Dmitrievich (born in 1907)
Institute staff: 1927–1930.
USSR Academy: Corresponding member, 1953. Full member, 1968.
Stalin Prize, 1949. Lenin Prize, 1965.
Hero of Socialist Labor, 1969.
Gas discharge, ultrahigh-frequency devices, modulated radioemission.

Dzhelepov, Boris Sergeevich (1910–1998)
Institute staff: 1931–1942.
USSR Academy: Corresponding member, 1953.
Nuclear spectroscopy, proton radioactivity, discovery of a number of isotopes.

Dzhelepov, Venedikt Petrovich (born in 1913)
Institute staff: 1939–1945.
USSR Academy: Corresponding member, 1966.
Stalin Prizes, 1951 and 1953.
Kurchatov Gold Medal, 1986.
Neutron physics, accelerators, elementary particles, high-energy physics.

El'yashevich, Mikhail Alexandrovich (1908–1996)
Institute staff: 1931–1932.
Belarus Academy: Full member, 1956.
Stalin Prizes, 1949 and 1950. Lenin Prize, 1966.
Optics and spectroscopy, optical monitoring of nuclear tests.

Flerov, Georgii Nikolaevich (1913–1990)
Institute staff: 1938–1941 and 1942–1943.
USSR Academy: Corresponding member, 1953. Full member, 1968.
Stalin Prizes, 1946 and 1949. Lenin Prize, 1967. State Prize, 1975.
Mendeleev Gold Medal, 1987.

Hero of Socialist Labor, 1949.
Spontaneous fission, methods of nuclear physics for geology, synthesis of superheavy elements, acceleration of heavy ions.

Fock, Vladimir Alexandrovich (1898–1974)
Institute staff: 1924–1936.
USSR Academy: Corresponding member, 1929. Full member, 1939.
Stalin Prize, 1946. Lenin Prize, 1960.
Mendeleev Prize, 1936.
Hero of Socialist Labor, 1968.
Quantum mechanics, Hartree–Fock method, relativistic quantum theory, general relativity, electrodynamics, mathematical physics.

Frank, Gleb Mikhailovich (1904–1976)
Institute staff: 1929–1933.
USSR Academy: Corresponding member, 1960. Full member, 1966.
Biology, medicine, biophysics, dosimetry, and radiology.

Frenkel, Yakov Il'ich (1894–1952)
Institute staff: 1921–1952.
USSR Academy: Corresponding member, 1929.
Stalin Prize, 1947.
Quantum theory of solids and liquids, theory of metals, Frenkel defects, theory of excitons, nuclear theory.

Gamov, Georgii Antonovich (1904–1968)
Institute staff: 1925–1933.
USSR Academy: Corresponding member, 1932.
Tunnel effect and alpha–decay, star energy sources, hot Universe model, genetic code.

Garbuzov, Dmitrii Zalmanovich (born in 1940)
Institute staff: 1964–present.
USSR Academy: Corresponding member, 1991.
Lenin Prize, 1972. State Prize, 1987.
Radiative and nonradiative recombination in bulk and QW III–V compounds, light emitting diodes and diode lasers, heterojunction solar and thermophotovoltaic cells.

Golant, Victor Evgen'evich (born in 1928)
Institute staff: 1961–present.
Director of Institute's Division: 1989–present.
USSR Academy: Corresponding member, 1984. Full member, 1990.
State Prize, 1983.
Artsimovich Prize, 1998.
Plasma physics and nuclear fusion. Diagnostics, heating, and confinement of tokamak plasma.

Grekhov, Igor' Vsevolodovich (born in 1934)
Institute staff: 1962–present.
USSR Academy: Corresponding member, 1991.
Lenin Prize, 1966. State Prize, 1987.
Power semiconductor electronics, superconductivity, surface physics.

Gribov, Vladimir Naumovich (1930–1997)
Institute staff: 1954–1971.
USSR Academy: Corresponding member, 1972.
Landau Prize, 1971.
Quantum field theory, high-energy particles, hadron physics, elementary particles, quantum chromodynamics.

Grinberg, Grigorii Abramovich (1900–1991)
Institute staff: 1918–1930 and 1941–1991.
USSR Academy: Corresponding member, 1946.
Stalin Prize, 1949.
Mathematical physics, EM wave diffraction, electron optics.

Gross, Evgenii Fedorovich (1897–1972)
Institute staff: 1944–1972.
USSR Academy: Corresponding member, 1946.
Stalin Prize, 1946. Lenin Prize, 1966.
Optics and spectroscopy (superfine structure of atomic spectra), discovery of Wannier–Mott exciton in semiconductors, and excitons in crystals.

Ioffe, Abram Fyodorovich (1880–1960)
Director of the Institute: 1918–1950.
USSR Academy: Corresponding member, 1918. Full member, 1920.
Vice-President of the USSR Academy, 1926–1929 and 1942–1945.
Stalin Prize, 1942. Lenin Prize, 1961.
Hero of Socialist Labor, 1955.
Mechanical and electrical properties of dielectrics (Ioffe effect related to mechanical strength, conduction laws in ion crystals), semiconductors (theory of rectification by metal–semiconductor contact, thermoelectric generators, Ioffe–Regel rule for the short-range order of a compound, a number of new semiconductor compounds), measurement of the electron charge, experimental confirmation of the quantum theory of photoeffect, classical monographies on semiconductors, manuals of general physics.

Kapitsa, Pyotr Leonidovich (1894–1984)
Institute staff: 1919–1921.
USSR Academy: Corresponding member, 1929. Full member, 1939.
Nobel Prize in Physics, 1978.
Stalin Prizes, 1941 and 1943.
Lomonosov Gold Medal, 1959.
Hero of Socialist Labor, 1945 and 1974.
Method for measuring the magnetic moment of the proton, superhigh magnetic fields, low-temperature physics, turbodetander technique, high-temperature plasma.

Kaplyanskii, Alexander Alexandrovich (born in 1930)
Institute staff: 1962–present.
USSR Academy: Corresponding member, 1987.
Lenin Prize, 1966. State Prize, 1975.
Optical spectroscopy of semiconductors and insulators, physics of nonequilibrium phonons, structural phase transitions in solids.

Kel'man, Veniamin Moiseevich (born in 1915)
Institute staff: 1945–1962.
Kazakh Academy: Full member, 1962.
Electron optics, nuclear physics (beta- and gamma-spectrometry).

Khariton, Yulii Borisovich (1904–1996)
Institute staff: 1921–1931.
USSR Academy: Corresponding member, 1946. Full member, 1953.
Stalin Prizes, 1949, 1951, and 1954. Lenin Prize, 1956.
Kurchatov Gold Medal, 1974. Lomonosov Gold Medal, 1982.
Hero of Socialist Labor, 1949, 1951, and 1954.
Chemical kinetics, nuclear chain reactions, explosives and detonation, nuclear and thermonuclear weapons.

Kikoin, Isaak Konstantinovich (1908–1984)
Institute staff: 1927–1937.
USSR Academy: Corresponding member, 1943. Full member, 1953.
Stalin Prizes, 1942, 1949, 1951, and 1953. Lenin Prize, 1959. State Prizes, 1967 and 1980.
Kurchatov Gold Medal, 1971. Lebedev Gold Medal, 1979.
Hero of Socialist Labor, 1951 and 1978.
Electrical and magnetic properties of metals, photomagnetic effect, anomalous Hall effect, isotope separation, nuclear and thermonuclear weapons, manuals of physics.

Kirpichev, Mikhail Viktorovich (1879–1955)
Institute staff: 1928–1930.
USSR Academy: Corresponding member, 1929. Full member, 1939.
Stalin Prize, 1941.
Boilers, heat engines, similarity theory, heat process modeling.

Kobeko, Pavel Pavlovich (1897–1954)
Institute staff: 1924–1952.
Substitute Director of the Institute in War years: 1941–1945.
USSR Academy: Corresponding member, 1943.
Ferroelectrics, physics of amorphous state (polymerization kinetics), polymers.

Kobzarev, Yurii Borisovich (1905–1991)
Institute staff: 1932–1933 and 1935–1943.
USSR Academy: Corresponding member, 1953. Full member, 1970.
Stalin Prize, 1940.
Popov Gold Medal, 1980.
Statistical radioelectronics, radars, theory of oscillations.

Komar, Anton Panteleimonovich (1904–1985)
Institute staff: 1930–1936 and 1950–1972.
Director of the Institute: 1950–1957.
Ukrainian Academy: Full member, 1948.
Stalin Prize, 1951.
Nuclear physics (photonuclear reactions), accelerators (first Soviet betatron, 1946), electron and ion microscopy.

Kondrat'ev, Victor Nikolaevich (1902–1979)
Institute staff: 1923–1931.
USSR Academy: Corresponding member, 1943. Full member, 1953.
Stalin Prize, 1946.
Elementary processes in chemistry, chain reactions, molecular structures.

Konstantinov, Boris Pavlovich (1910–1969)
Institute staff: 1926–1930 and 1940–1969.
Director of the Institute: 1957–1967.
USSR Academy: Corresponding member, 1953. Full member, 1960. Vice-President of the USSR Academy: 1967–1969.
Stalin Prize, 1953. Lenin Prize, 1958.
Hero of Socialist Labor, 1954.
Acoustics, light and heavy isotope separation, controlled fusion, astrophysics, holography.

Korsunskii, Moisei Izrailevich (1903–1976)
Institute staff: 1925–1930.
Kazakh Academy: Full member, 1962.
Physics of metals, photoconductivity, nuclear physics (multivalent ions).

Kostenko, Mikhail Polievktovich (1889–1976)
Institute staff: 1927–1930.
USSR Academy: Corresponding member, 1939. Full member, 1953.
Stalin Prizes, 1949 and 1951. Lenin Prize, 1958.
Hero of Socialist Labor, 1969.
Electric motors, converters, alternate-current generators.

Koton, Mikhail Mikhailovich (born in 1908)
Institute staff: 1946–1952.
USSR Academy: Corresponding member, 1960.
Stalin Prize, 1953.
Organic and organometallic chemistry, polymers, organic dielectrics.

Kravets, Torichan Pavlovich (1876–1955)
Institute staff: 1928–1930.
USSR Academy: Corresponding member, 1943.
Stalin Prize, 1946.
Photographic processes (photoemulsions), geophysics, history of physics.

Krutkov, Yurii Alexandrovich (1890–1952)
Institute staff: 1919–1936.
USSR Academy: Corresponding member, 1933.
Stalin Prize.
Theoretical physics, quantum theory, statistical mechanics, theory of elasticity.

Kurchatov, Igor' Vasil'evich (1903–1960)
Institute staff: 1925–1943.
USSR Academy: Full member, 1943.
Stalin Prizes, 1942, 1949, 1951, and 1954. Lenin Prize, 1957.
Joliot-Curie Gold Medal, 1959.

Hero of Socialist Labor, 1949, 1951, and 1954.
Physics of dielectrics, Rauchelle salts (ferroelectrics), photocells, discovery of nuclear isomerism, high-voltage accelerators and cyclotrons, ship degaussing, neutron physics, cyclotron, first Soviet nuclear reactor (1946), nuclear and thermonuclear weapons, nuclear energetics, controlled fusion.

Kurdyumov, Georgii Vyacheslavovich (1902–1996)
Institute staff: 1924–1933.
USSR Academy: Corresponding member, 1946. Full member, 1953.
Ukrainian Academy: Full member, 1939.
Stalin Prize, 1949.
Chernov Gold Medal, 1979.
Hero of Socialist Labor, 1969.
Physics of metals and steels, alloy hardening, Kurdyumov's effect in martensites.

Landau, Lev Davidovich (1908–1968)
Institute staff: 1927–1932.
USSR Academy: Full member, 1946.
Nobel Prize in Physics, 1962.
Stalin Prizes, 1941, 1949, and 1953. Lenin Prize, 1962.
Hero of Socialist Labor, 1954.
Various fields of theoretical physics (a number of effects and laws named after him), quantum mechanics, superfluids, theory of phase transition, superconductivity. Well-known course of theoretical physics.

Landsberg, Grigorii Samuilovich (1890–1957)
Institute staff: 1926–1935.
USSR Academy: Corresponding member, 1932. Full member, 1946.
Stalin Prize, 1941.
Molecular scattering, discovery of Raman scattering, selective scattering, molecular spectroscopy and spectral analysis.

Lashkarev, Vadim Evgen'evich (1903–1974)
Institute staff: 1929–1935.
Ukrainian Academy: Full member, 1945.
X-ray optics, semiconductor physics (discovery of p-n junction) and technology (photocells), photoelectric phenomena.

Latyshev, Georgii Dmitrievich (1907–1973)
Institute staff: 1943–1952.
Ukrainian Academy: Corresponding member, 1948.
Kazakh Academy: Full member, 1958.
Nuclear physics (accelerators, radiation interactions, nuclear spectroscopy).

Lazarev, Boris Georgievich (born in 1906)
Institute staff: 1928–1932.
Ukrainian Academy: Full member, 1951.
Stalin Prize, 1951.
Low-temperature physics (nuclear paramagnetism, superfluid He, superconductivity) and technology (superhigh magnetic fields, high pressures, He isotopes separation).

Ledentsov, Nikolai Nikolaevich (born in 1959)
Institute staff: 1982–present.
Russian Academy: Corresponding member, 1997.
Quantum-size heterostructures. MBE growth.

Leipunskii, Alexander Il'ich (1903–1972)
Institute staff: 1926–1929.
Ukrainian Academy: Full member, 1934.
Lenin Prize, 1960.
Hero of Socialist Labor, 1963.
Nuclear physics (high-voltage accelerators, fast-neutron reactors), nuclear energetics.

Lipatov, Lev Nikolaevich (born in 1940)
Institute staff: 1963–1971.
Russian Academy: Corresponding member, 1997.
Quantum field theory, quantum electrodynamics.

Lobashev, Vladimir Mikhailovich (born in 1934)
Institute staff: 1957–1972.
USSR Academy: Corresponding member, 1970.
Lenin Prize, 1974.
Nuclear physics, weak interactions.

Lukirskii, Petr Ivanovich (1894–1954)
Institute staff: 1918–1954.
USSR Academy: Corresponding member, 1929. Full member, 1932.
X-Rays, secondary electron emission, photoeffect, method of spherical capacitor, effect of neutrons on materials, nuclear reactions with photoemulsions.

Mamyrin, Boris Alexandrovich (born in 1919)
Institute staff: 1970–present.
Russian Academy: Corresponding member, 1994.
Konstantinov Prize, 1982.
Mass spectrometry, fundamental constants, helium isotopes in nature, physics and technology of tritium.

Mazets, Evgenii Pavlovich (born in 1929)
Institute staff: 1954–present.
Russian Academy: Corresponding member, 1990.
Lenin Prize, 1986.
Astrophysics, gamma-ray astronomy, physics of interplanetary space.

Migdal, Arkadii Beinusovich (Benediktovich) (1911–1991)
Institute staff: 1947–1949.
USSR Academy: Corresponding member, 1953. Full member, 1966.
Neutron–atom shell interactions, nuclear physics and its applications (nuclear photoabsorption, nuclear reactions, gigantic dipole resonance), quantum field theory, controlled fusion, neutron stars.

Mikheev, Mikhail Nikolaevich (1905–1989)
Institute staff: 1928–1932.
USSR Academy: Corresponding member, 1979.
Magnetic properties of materials, nondestructive test techniques, steels.

Nazarenko, Vladimir Andreevich (born in 1934)
Institute staff: 1958–1971.
Russian Academy: Corresponding member, 1997.
Lenin Prize, 1986.
Neutron physics.

Nemenov, Leonid Mikhailovich (1905–1980)
Institute staff: 1925–1944.
Kazakh Academy: Full member, 1962.
Stalin Prize, 1953.
Physics of dielectrics, first Soviet precision mass-spectrometer (1937), nuclear physics (cyclotrons, pionium).

Obreimov, Ivan Vasil'evich (1894–1981)
Institute staff: 1924–1930.
USSR Academy: Corresponding member, 1933. Full member, 1958.
Stalin Prize, 1946.
Vavilov Gold Medal, 1959.
Physics of crystals, molecular spectroscopy, low-temperature physics, kinetics of dislocation development.

Papaleksi, Nikolai Dmitrievich (1880–1947)
Institute staff: 1926–1935.
USSR Academy: Corresponding member, 1931. Full member, 1939.
Stalin Prize, 1942.
Mendeleev Prize, 1936.
Radiophysics, amplifiers, wave propagation, theory of oscillations, parametric generators, radiointerferometry.

Perel', Vladimir Idelevich (born in 1928)
Institute staff: 1958–present.
USSR Academy: Corresponding member, 1981.
State Prize, 1976.
Ioffe Prize, 1993.
Physics of solids, quantum optics, helicons, semiconductors (recombination waves).

Pomeranchuk, Isaak Yakovlevich (1913–1966)
Institute staff: 1939–1940.
USSR Academy: Corresponding member, 1953. Full member, 1964.
Stalin Prizes, 1950 and 1953.
Quantum electrodynamics, low-temperature conductivity of metals, nuclear physics, particle theory (pomerons).

Prikhot'ko, Antonina Fedorovna (born in 1906)
Institute staff: 1928–1930.
Ukrainian Academy: Full member, 1964.

Lenin Prize, 1966.
Hero of Socialist Labor, 1976.
Spectroscopy (molecular crystals), excitons, low-temperature physics.

Rasulev, Utkur Khasanovich (born in 1939)
Institute staff: 1965–1981.
Uzbek Academy: Corresponding member, 1984.
State Prize, 1988.
Surface physics and its applications.

Roginskii, Simon Zalmanovich (1900–1970)
Institute staff: 1923–1932.
USSR Academy: Corresponding member, 1939.
Stalin Prize, 1941.
Theory of catalysis, theory of active surfaces, colloid chemistry, kinetics of chemical reactions, explosives, radioactive indicators.

Rozhanskii, Dmitrii Appolinar'evich (1882–1936)
Institute staff: 1923–1929 and 1935–1936.
USSR Academy: Corresponding member, 1933.
Popov Prize, 1911.
Electric discharge, radiophysics, radars, wave propagation.

Semenov, Nikolai Nikolaevich (1896–1986)
Institute staff: 1920–1931.
USSR Academy: Corresponding member, 1929. Full member, 1932.
Vice-President of the USSR Academy of Sciences, 1963–1971.
Nobel Prize in Chemistry, 1956.
Stalin Prizes, 1941 and 1949. Lenin Prize, 1976.
Mendeleev Prize, 1936. Lomonosov Gold Medal, 1970.
Hero of Socialist Labor, 1966 and 1976.
Theory of heat breakdown of dielectrics, molecular beams, theory of combustion and detonation, chain reactions, chemical kinetics.

Shal'nikov, Alexander Iosifovich (1905–1986)
Institute staff: 1924–1935.
USSR Academy: Corresponding member, 1946. Full member, 1979.
Stalin Prizes, 1948, 1949, and 1953.
Lebedev Gold Medal, 1972.
Low-temperature physics (liquid He properties), superconductivity.

Shchelkin, Kirill Ivanovich (1911–1968)
Institute staff member in 1932.
USSR Academy: Corresponding member, 1953.
Stalin Prizes, 1949, 1951, and 1954. Lenin Prize, 1958. State Prize.
Hero of Socialist Labor, 1949, 1951, and 1953.
Combustion and detonation (detonation spin), nuclear and thermonuclear weapons.

Shchukin, Alexander Nikolaevich (1900–1990)
Institute staff: 1927–1929.
USSR Academy: Corresponding member, 1946. Full member, 1953.

Lenin Prize. Stalin Prize.
Hero of Socialist Labor, 1961 and 1975.
Wave propagation, short waves, amplifiers.

Shubnikov, Alexei Vasil'evich (1887–1970)
Institute staff: 1930–1932.
USSR Academy: Corresponding member, 1933. Full member, 1953.
Stalin Prizes, 1947 and 1950.
Hero of Socialist Labor, 1967.
Theory of symmetry, crystallography and its applications, Shubnikov's groups.

Sinel'nikov, Konstantin Dmitrievich (1901–1966)
Institute staff: 1924–1930.
Ukrainian Academy: Full member, 1948.
Nuclear physics (high-voltage accelerators), vacuum metallurgy of Be, Zr, and U, plasma physics, controlled fusion.

Skobel'tsyn, Dmitrii Vladimirovich (1892–1990)
Institute staff: 1924–1941.
USSR Academy: Corresponding member, 1939. Full member, 1946.
Stalin Prize, 1951. Lenin Prize, 1982.
Mendeleev Prize, 1936. Vavilov Gold Medal, 1952.
Hero of Socialist Labor, 1969.
First application of Wilson camera in cosmic ray studies, discovery of cosmic showers.

Smolenskii, Georgii Anatol'evich (1910–1986)
Institute staff: 1972–1986.
USSR Academy: Corresponding member, 1970.
Stalin Prize, 1952.
Electrical properties of dielectrics (discovery of a new class of ferroelectric crystals), physics of ferrites.

Spivak, Petr Efimovich (born in 1911)
Institute staff: 1936–1945.
USSR Academy: Corresponding member, 1964.
Stalin Prize, 1953.
Kurchatov Gold Medal, 1962.
Nuclear electronics, nuclear fission, neutron decay, neutrinos.

Starodubtsev, Sergei Vasil'evich (1914–1967)
Institute staff: 1946–1956.
Uzbek Academy: Full member, 1956.
Electronics, nuclear physics (X-ray resonance scattering), effect of radiation on solids, dosimetry, neutron physics (neutron generators).

Stepanov, Alexander Vasil'evich (1908–1972)
Institute staff: 1924–1929 and 1930–1972.
USSR Academy: Corresponding member, 1968.
Physics of solids, discovery of charged dislocations, Stepanov method of crystal growth from melts.

Strelkov, Petr Georgievich (1899–1968)
Institute staff: 1923–1936.
USSR Academy: Corresponding member, 1960.
Stalin Prize, 1943.
Bacteriological filters, thermophysics, low-temperature physics.

Styrikovich, Mikhail Adol'fovich (born in 1902)
Institute staff: 1927–1928.
USSR Academy: Corresponding member, 1946. Full member, 1964.
Heat engines, heat exchange, thermal and aerodynamic calculations of boilers.

Sumbaev, Oleg Igorevich (born in 1930)
Institute staff: 1957–1971.
USSR Academy: Corresponding member, 1979.
Konstantinov Prize, 1985.
Atomic and nuclear physics, gamma-ray diffraction.

Suris, Robert Arnol'dovich (born in 1936)
Institute staff: 1988–present.
Director of the Institute's Division, 1993–present.
Russian Academy: Corresponding member, 1997.
Solid state and semiconductor theory, solid state devices, technology of opto- and microelectronics. Statistical mechanics, phase transitions.

Talmud, David L'vovich (1900–1973)
Institute staff: 1930–1932.
USSR Academy: Corresponding member, 1934.
Stalin Prize, 1943.
Theory of surface, colloid chemistry, theory of macromolecules.

Tamm, Igor' Evgen'evich (1895–1971)
Institute staff: 1942–1946.
USSR Academy: Corresponding member, 1939. Full member, 1953.
Nobel Prize in Physics, 1958.
Stalin Prizes, 1946 and 1953.
Lomonosov Gold Medal, 1967.
Hero of Socialist Labor, 1953.
Photoelectric effect, classical and quantum electrodynamics, theory of electron radiation, nuclear physics, theory of elementary particles, thermonuclear weapons, course of theoretical physics.

Tuchkevich, Vladimir Maximovich (1904–1997)
Institute staff: 1935–1997.
Director of the Institute: 1967–1987.
USSR Academy: Corresponding member, 1968. Full member, 1970.
Stalin Prize, 1942. Lenin Prize, 1966.
Konstantinov Prize, 1982.
Hero of Socialist Labor, 1984.
Physics and technology of semiconductors (Si and Ge diodes and transistors, high-power semiconductor devices, new semiconductor materials), ship degaussing, isotope separation.

Val'ter, Alexander Filippovich (1898–1941)
Institute staff: 1927–1930.
USSR Academy: Corresponding member, 1933.
Dielectrics, electrical breakdown.

Val'ter, Anton Karlovich (1905–1965)
Institute staff: 1924–1928.
Ukrainian Academy: Full member, 1951.
Nuclear physics (electrostatic accelerators, elementary particles).

Varshalovich, Dmitrii Alexandrovich (born in 1934)
Institute staff: 1957–present.
Russian Academy: Corresponding member, 1994.
Belopolsky Prize, 1990.
Interstellar medium, stellar formation regions, quasars, cosmology.

Vernov, Sergei Nikolaevich (1910–1982)
Institute staff: 1934–1935.
USSR Academy: Corresponding member, 1953. Full member, 1968.
Lenin Prize, 1960. Stalin Prize, 1949.
Cosmic rays, outer radioactive belts, interplanetary space physics.

Vologdin, Valentin Petrovich (1881–1953)
Institute staff: 1926–1935.
USSR Academy: Corresponding member, 1939.
Stalin Prizes, 1943 and 1952.
Popov Gold Medal, 1948.
Electric generators, Hg rectifier, electric isolation, steel hardening, high-frequency currents.

Vonsovskii, Sergei Vasil'evich (1910–1998)
Institute staff: 1931–1932.
USSR Academy: Corresponding member, 1953. Full member, 1966.
State Prizes, 1975 and 1982.
Vavilov Gold Medal, 1982.
Hero of Socialist Labor, 1969.
Ferro- and paramagnetics, magnetic phenomena, metals and alloys and their superconductivity.

Vorob'ev, Alexei Alexeevich (born in 1931)
Institute staff: 1955–1971.
Russian Academy: Corresponding member, 1992.
State Prize, 1983.
Nuclear physics, elementary particles.

Vul'f, Yurii (Georgii) Viktorovich (1863–1925)
Institute staff: 1918–1925.
USSR Academy: Corresponding member, 1924.
Crystallography (Vul'f law of crystal growth), X-ray spectroscopy (Vul'f–Bragg law).

Vul, Bentsion Moiseevich (1903–1985)
Institute staff: 1930–1932.
USSR Academy: Corresponding member, 1939. Full member, 1972.

Stalin Prize, 1946. Lenin Prize, 1964.
Hero of Socialist Labor, 1969.
Dielectrics and semiconductors, electric discharge, studies of barium titanate properties, Si solar cells, semiconductor lasers.

Zakharchenya, Boris Petrovich (born in 1928)
Institute staff: 1952–present.
Director of the Institute's Division: 1989–present.
USSR Academy: Corresponding member, 1976. Full member, 1992.
Lenin Prize, 1966. State Prize, 1976.
Lebedev Gold Medal, 1996.
Optics of semiconductors, excitons, optoelectronics, superconductivity.

Zel'dovich, Yakov Borisovich (1915–1987)
Institute staff: 1931
USSR Academy: Corresponding member, 1946. Full member, 1958.
Stalin Prizes, 1943, 1949, 1951, and 1953. Lenin Prize, 1957.
Kurchatov Gold Medal, 1977.
Hero of Socialist Labor, 1949, 1953 and 1957.
Theory of adsorption, theory of combustion, shock waves, detonation, chemical kinetics, nuclear chain reactions and critical masses, ballistics, nuclear and thermonuclear weapons, elementary particles, astrophysics and cosmology.

Zhurkov, Serafim Nikolaevich (1905–1997)
Institute staff: 1930–1997.
USSR Academy: Corresponding member, 1958. Full member, 1968.
Hero of Socialist Labor, 1975.
Polymers, kinetic theory of strength of solids.

Part VI

Honorary Membership

Charter of Honorary Membership[1] at the Ioffe Institute

The title of Honorary Member of the Ioffe Institute shall be given to Russian and foreign scientists who have made distinguished contributions to the research areas represented at the Ioffe Institute. Honorary Members of the Ioffe Institute shall be elected by the Institute's Scientific Council at the annual celebration sessions in September. The scientists who have been awarded the Title shall be presented with a diploma and a medal.

Status

Honorary Member is welcome to promote research activities at the Institute and to develop contacts between the Institute and other research centers of the international scientific community. Honorary Member has all the rights of a member of the Institute's Scientific Council except the right for decisive vote. Honorary Member may participate in the research conducted at any of the Institute's laboratories, being provided with accommodation for a period of one month per year and offered a financial compensation as large as a month's salary of an Institute's principal researcher.

Nomination

The candidates for the Honorary Membership should not be on the Institute's staff at the moment of nomination.

Nominations for the Honorary Membership can be made only by

- members of the Ioffe Institute's Scientific Council, including the elected Honorary Members;
- members of the Russian Academy of Sciences working at the Institute.

A special commission shall be appointed by the Institute's Scientific Council for consideration of the candidates. The commission shall circulate the information concerning the candidates among all the Council members at the session prior to the annual celebration session in September.

Election and Awarding

Honorary Members shall be elected by ballot at the September celebration session of the Scientific Council. The elected Honorary Members shall be informed officially about their election. The Honorary Members receive the Charter of Honorary Membership, the list of the elected Honorary Members and an invitation to visit the Ioffe Institute in September next year to be presented with a diploma and a medal. The information about the election of new Honorary Members shall be submitted for publication to the *Poisk* newspaper and the Ioffe Institute's periodicals.

[1] Established in 1992 with the first honorary appointments of five Russian and five foreign nominees.

Honorary Members of the Ioffe Institute

Abragam, Anatole (born in 1914)
Collége de France, ***France***.
Year of election: 1992.
Discoverer of nuclear ferromagnetism and dynamical polarization of nuclei. Honorable contributions in physics of spin interactions in solids. Valuable services in the advancement of scientific knowledge.

Alexandrov, Anatoly Petrovich (1903–1994)
Russian Research Center "Kurchatov Institute," Moscow, ***Russia***.
Year of election: 1992.
Important works on solid state physics, ship degaussing, atomic physics, and nuclear energetics. President of the USSR Academy of Sciences (1975–1986).

Andreev, Alexander Fedorovich (born in 1939)
Kapitsa Institute of Physical Problems, Moscow, ***Russia***.
Year of election: 1996.
Outstanding physicist who greatly contributed to the theory of condensed matter phenomena at low temperature. Pioneering works on superfluidity and superconductivity, quantum crystals, and low-temperature phase transitions.

Basov, Nikolai Gennadiyevich (born in 1922)
Lebedev Physical Institute, Moscow, ***Russia***.
Year of election: 1994.
Nobel Prize in Physics for fundamental research that led to development the maser and laser, 1964.
Outstanding works on quantum radiophysics, non-linear optics, inertial fusion.

Bimberg, Dieter (born in 1942)
Institut für Festkörperphysik, Technische Universität, Berlin, ***Germany***.
Year of election: 1995.
Eminent achievements in the area of heterostructures, quantum-well structures, and heterolasers. Recognized contributor to the advancement of the Institute's works.

Esaki, Leo (born in 1925)
University of Tsukuba, ***Japan***.
Year of election: 1992.
Nobel Prize in Physics for experimental discovery of tunneling phenomena in solids, 1973.
One of the first to study tunnel effects in solids (Esaki diodes). Pioneering works in the field of semiconductor superlattices.

Fortov, Vladimir Evgenievich (born in 1946)
High Energy Density Research Center, Moscow, ***Russia***.
Year of election: 1997.
Advanced researches in plasma physics, energetics, and shock waves. Knowledgeable services in the state scientific bodies.

Grimmeiss, Hermann (born in 1930)
Department of Solid State Physics, University of Lund, ***Sweden***.

Year of election: 1997.
Well-known specialist in semiconductor heterostructures, quantum wells, and superlattices. Research into optical and electronic properties of novel semiconductors.

Holonyak Jr, Nick (born in 1928)
University of Illinois at Urbana-Champaign, ***USA***.
Year of election: 1992.
Distinguished works on semiconductor optoelectronics and inventions of semiconductor lasers and light diodes.

Keldysh, Leonid Veniaminovich (born in 1931)
Lebedev Physical Institute, Moscow, ***Russia***.
Year of election: 1992.
Outstanding theorist in the field of electronic phenomena in semiconductors, phonon wind, electron–hole conduction.

Khariton, Yuly Borisovich (1904–1996)
Center for Nuclear Research, Arzamas-Sarov, ***Russia***.
Year of election: 1992.
Head of Soviet projects on nuclear weapons. Pioneering works on chemical kinetics, chain reactions, and nuclear phenomena.

Klitzing, Klaus von (born in 1943)
Max-Planck-Institut für Festkörperforschung, Stuttgart, ***Germany***.
Year of election: 1992.
Nobel Prize in Physics for the discovery of the quantized Hall effect, 1985.
Discoverer of the quantum Hall effect. Important investigations in low-dimensional structures.

Kurdyumov, Georgii Vyacheslavovich (1902–1996)
Russian Academy of Sciences, Moscow, ***Russia***.
Year of election: 1993.
Prominent scientist in the field of metal physics (Kurdyumov effect) and metal and alloy processing.

Landwehr, Gottfried (born in 1929)
Physikalisches Institut der Universität Würzburg, ***Germany***.
Year of election: 1993.
Recognized for his research on magnetic phenomena in semiconductors and conspicuous service rendered to the advancement of knowledge as a Dean of the Faculty of Physics at the Würzburg University.

Mott, Sir Nevil F. (1905–1996)
Cavendish Laboratory, Cambridge University, England, ***United Kingdom***.
Year of election: 1993.
Nobel Prize in Physics for his researches on the magnetic and electrical properties of noncrystalline solids, 1977.
Classic of atomic physics. Outstanding works on metals, semiconductors, superconductivity, and other fields of modern physics.

Ossipyan, Yury Andreevich (born in 1931)
Institute of Solid State Physics, Chernogolovka, ***Russia***.
Year of election: 1993.
Important studies in the field of solid state physics (metals, crystals, dislocations) and high-temperature superconductivity. Great merits in science organization.

Peierls, Sir Rudolf Ernst (1907–1995)
Department of Physics, University of Oxford, ***United Kingdom***.
Year of election: 1993.
Outstanding theorist and author of classical investigations into the nature of crystals and metals, quantum theory of solids, and nuclear physics. Many years' relations with Russian physicists.

Pozhela, Yuras K. (born in 1925)
Semiconductor Physics Institute, Vilnius, ***Lithuania***.
Year of election: 1993.
Works on semiconductor physics, including plasma processes and effects of electric fields, partly, in collaboration with the Institute. Greatly merited contribution to the Lithuanian–Russian scientific relations.

Prokhorov, Alexander Mikhailovich (born in 1916)
Institute of General Physics, Moscow, ***Russia***.
Year of election: 1992.
Nobel Prize in Physics for fundamental research that led to development the maser and laser, 1964.
Leading researches in quantum electronics and non-linear optics.

Rebane, Karl K. (born in 1926)
Institute of Physics of Estonia, ***Estonia***.
Year of election: 1992.
Notable contributions to several fields of spectroscopy of condensed matter. Approvable efforts toward fruitful cooperation of Estonian and Russian physicists.

Scharmann, Arthur (born in 1928)
I. Physikalisches Institut der Justus-Liebig-Universität Giessen, ***Germany***.
Year of election: 1994.
Prominent researcher in the fields of optical properties and luminescence of solids, electronic processes in crystals and their applications. Approvable contributor to the German–Russian scientific collaboration, co-author of works performed at the Ioffe Institute.

Velikhov, Evgenii Pavlovich (born in 1935)
Russian Research Center "Kurchatov Institute," Moscow, ***Russia***.
Year of election: 1995.
Prominent specialist on plasma physics, fusion, energy conversion, quantum electronics. Valuable cooperation with the Ioffe Institute's scientists.

Vonsovskii, Sergei Vasilievich (1910–1998)
Institute of Physics of Metals, Ekaterinburg, ***Russia***.
Year of election: 1993.
Classic of ferro- and antiferromagnetism, theory of metals and semiconductors, superconductivity.

Wagner, Friedrich (born in 1943)
Max-Planck-Institut für Plasmaphysik, Garching bei München, ***Germany***.
Year of election: 1996.
Valuable work on fusion and high-temperature plasma. Head of the tokamak and stellarator experiments.

Отпечатано с готового оригинал-макета в типографии ПИЯФ РАН
188350, Гатчина Ленинградской обл., Орлова роща
Зак. 371, тир. 750, уч.-изд. л. 13,3; 21.08.1998 г.

Erratum

Part III **Scientific Reviews of Advanced Areas**

Page	Printed	Correct
65, reference [14]	**71** 2468 (1993);	**71** 2465 (1993);
65, reference [14]	**67** (1995).	**67** 1137 (1995).
65, reference [29]	*Sov. Phys. Solid State*	*Sov. Phys. JETP*
86, line 11 from bottom	80 kA cm^{-2}	80 A cm^{-2}
87, line 3 from top	90 kA cm^{-2}	90 A cm^{-2}
120, figure caption		(a) 1990, (b) 2000.
123, line 14 from bottom	1.6 kV	2.6 kV
125, line 15 from top	4.2×10^{-2}	2.3×10^{-2}
125, line 9 from bottom	of hundreds	
128, line 5 from top	exceeds	is equal to
128 line 16 from top	20 MW	80 MW
128, line 11 from bottom	SOS	DSRD
128, line 18 from bottom	6 GW	0.45 GW
135, line 22 from bottom	[15–18].	[16–18].
142, line 20 from top	[32, 33]	[41, 42]
142, line 22 from top	properties [32, 33]:	properties:
161, line 11 from top	deformation ($\dot{\varepsilon}$)	deformation (ε)
173, line 1 after figure	[12]	[11]
174, line 16 from top	[13–16]	[12–15]
174, line 24 from top	group.	group [16].
176, reference [11]	(1967)	(1997)
183, line 15 from bottom	α-particles. The clouds	α-particles [13]. The cloud
184, figure 5	êadius	radius
185, figure caption	1—stoichiometric composition, 2—by ion flux,	1—by ion flux, 2—stoichiometric composition,
189, reference [9]	M. I. Pergamet	M. I. Pergament
189, reference [11]	Workshop in Varenna (Italy)	p. 353, Plenum Press
191, table 1, last row	sore	size
193, line 12 from bottom	spectrum [8].	spectrum [8] (see Fig. 2).
207, line 27 from top	accelerated particles [48].	accelerating force [48].
214, line 8 after figure	could be detected	were detected
221, last line	[1, 2]	[1–3]
222, line 5 from bottom	[3]	[4]
231, reference [7]	and L. V. Khabarin	L. V. Khabarin, I. L. Kamenskii, and I. N. Tolstikhin

Part V **Personalia**

Page	Printed	Correct
265, new line after line 5 from top		Vice-President of the USSR Academy, 1989–1991.
265, new line after line 6 from top		Ioffe Prize, 1996.
273, line 8 from top	1986.	1974.